Tourism Development in Japan

This significant and timely volume focuses on the unique trajectory of tourism development in Japan, which has been characterized by an historical emphasis on promoting both domestic and international tourism to Japanese tourists, followed by the more recent policy of competing aggressively in the international incoming tourist market.

Initial chapters present an overview of past and present tourism, including policy and research perspectives. Thematic perspectives on tourism and specific contexts and places in which tourism occurs are then examined. Strains of Japanese tourism such as sport, surf, forest, mountain, urban, tea, pilgrimage and even whaling heritage tourism are among those analyzed. The book also explores tourism's role in confronting difficult pasts and presents, and the challenges facing the development of tourism in contemporary Japan. A short postscript outlines some of the challenges and possible future directions tourism in Japan may take in light of the COVID-19 crisis.

Written by a team of well-known editors and contributors, including academics from Japan, this volume will be of great interest to upper-students and researchers and academics in development studies, cultural studies, geography and tourism.

Richard Sharpley is Professor of Tourism and Development at the University of Central Lancashire, Preston, UK, and Distinguished Professor in the Center for Tourism Research at Wakayama University, Japan.

Kumi Kato is a professor at the Faculty of Tourism, Wakayama University, Japan. She currently chairs a committee for sustainable tourism at the Japan Tourism Agency.

Contemporary Geographies of Leisure, Tourism and Mobility
Series Editor: C. Michael Hall, *Professor at the Department of Management, College of Business and Economics, University of Canterbury, Christchurch, New Zealand*

The aim of this series is to explore and communicate the intersections and relationships between leisure, tourism and human mobility within the social sciences.

It will incorporate both traditional and new perspectives on leisure and tourism from contemporary geography, e.g., notions of identity, representation and culture, while also providing for perspectives from cognate areas such as anthropology, cultural studies, gastronomy and food studies, marketing, policy studies and political economy, regional and urban planning, and sociology, within the development of an integrated field of leisure and tourism studies.

Also, increasingly, tourism and leisure are regarded as steps in a continuum of human mobility. Inclusion of mobility in the series offers the prospect to examine the relationship between tourism and migration, the sojourner, educational travel, and second home and retirement travel phenomena.

The series comprises two strands:

Contemporary Geographies of Leisure, Tourism and Mobility aims to address the needs of students and academics, and the titles will be published in hardback and paperback. Titles include:

Authenticity in North America
Place, Tourism, Heritage, Culture and the Popular Imagination
Edited by Jane Lovell and Sam Hitchmough

Tourism Development, Governance and Sustainability in The Bahamas
Edited by Sophia Rolle, Jessica Minnis and Ian Bethell-Bennett

Cultural and Heritage Tourism in the Middle East and North Africa
Complexities, Management and Practices
Edited by C. Michael Hall and Siamak Seyfi

Tourism Development in Japan
Themes, Issues and Challenges
Edited by Richard Sharpley and Kumi Kato

For more information about this series, please visit: www.routledge.com/Contemporary-Geographies-of-Leisure-Tourism-and-Mobility/book-series/SE0522

Tourism Development in Japan

Themes, Issues and Challenges

Edited by
Richard Sharpley and Kumi Kato

Routledge
Taylor & Francis Group
LONDON AND NEW YORK

First published 2021
by Routledge
2 Park Square, Milton Park, Abingdon, Oxon OX14 4RN

and by Routledge
52 Vanderbilt Avenue, New York, NY 10017

Routledge is an imprint of the Taylor & Francis Group, an informa business

British Library Cataloguing-in-Publication Data
A catalogue record for this book is available from the British Library

Library of Congress Cataloging-in-Publication Data
Names: Sharpley, Richard, 1956- editor. | Kato, Kumi, editor.
Title: Tourism development in Japan: themes, issues and challenges/
edited by Richard Sharpley and Kumi Kato.
Description: Abingdon, Oxon; New York, NY: Routledge, 2021. |
Series: Contemporary geographies of leisure, tourism and mobility |
Includes bibliographical references and index.
Subjects: LCSH: Tourism–Japan.
Classification: LCC G155.J27 T67 2021 (print) |
LCC G155.J27 (ebook) | DDC 338.4/79152–dc23
LC record available at https://lccn.loc.gov/2020021053
LC ebook record available at https://lccn.loc.gov/2020021054

ISBN: 978-0-367-22147-8 (hbk)
ISBN: 978-0-367-56774-3 (pbk)
ISBN: 978-0-429-27351-3 (ebk)

Typeset in Times New Roman
byNewgenPublishingUK

Contents

List of figures vii
List of tables ix
Notes on contributors x

1 Introduction: tourism in Japan – from the past to the present 1
RICHARD SHARPLEY AND KUMI KATO

2 Tourism research on Japan – overview of major trends: Japanese and English-language materials 19
YUMIKO HORITA AND KUMI KATO

3 Urban development and tourism in Japanese cities 47
YUMIKO HORITA

4 Transition of forest tourism policies in Japanese national forest management 64
YUMI OURA

5 A systematic review of sport tourism research in Japan 82
EIJI ITO AND TOM HINCH

6 Mobilizing stoke: a genealogy of surf tourism development in Miyazaki, Japan 102
ADAM DOERING

7 Japan's mountain tourism at a crossroads: insights from the North Japan Alps 119
ABHIK CHAKRABORTY

8 International exchange in tea tourism: reconceptualizing Japanese green tourism for sustainable farming communities 140
AMNAJ KHAOKHRUEAMUANG

9 Pilgrimage tourism in regional communities: the case of Tanabe City and Kumano Kodo 160
RICARDO NICOLAS PROGANO AND KUMI KATO

10 Confronting difficult pasts: the case of 'kamikaze' tourism 179
RICHARD SHARPLEY AND KUMI KATO

11 Whaling heritage and tourism development – 'sliced, diced and boiled down' 200
SIMON WEARNE

12 Debating sustainability in tourism development: resilience, traditional knowledge and community: a post-disaster perspective 224
KUMI KATO

13 International tourists in Japan: their increasing numbers and vulnerability to natural hazards 239
HAYATO NAGAI, KAEDE SANO, BRENT W. RITCHIE AND TAKASHI YOSHINO

14 The expansion of peer-to-peer accommodation rentals in Japan: issues and challenges 257
TAKAHIRO IKEJI AND HAYATO NAGAI

Postscript 276
RICHARD SHARPLEY AND KUMI KATO

Index 280

Figures

4.1 Volume of wood production of national forests (1945–2017) 68
4.2 Number of ski resorts and its visitors in Nagano prefecture (1980–2018) 71
4.3 Number of workers in NF management (1970–2018) 77
6.1 Seagaia Ocean Dome first week of demolition. 1 September 2016 108
6.2 Surf City Miyazaki's yoga studio and Roxy standup paddle boards for rent. August 2017 110
7.1 Location of the North Japan Alps 122
7.2 The Kamikochi valley with Azusa River and Hotaka Range in autumn 124
7.3 The Hakuba ridgeline with the three main peaks of Mt. Shirouma-Yari (left), Mt. Shakushi (centre) and Mt. Shirouma (right) 125
8.1 A conceptual framework of green tourism based on the four elements of tea cultural commodification in tea tourism 145
8.2 The green tourism route map of Tawara targeting domestic tourists from the city 147
8.3 The reconceptualized framework of Japanese green tourism 155
10.1 Replica kamikaze plane, Chiran Peace Museum 187
10.2 Statue of a kamikaze pilot, Chiran Peace Museum 188
10.3 Mural in Chiran Peace Museum 189
10.4 Wreckage of a Japanese 'Zero', Chiran Peace Museum 189
10.5 Original runway, Uzurano airfield 193
10.6 Memorial to 63 kamikaze pilots, Uzurano airfield 193
10.7 Entrance to an air-raid shelter 194
10.8 Reconstructed gate house 195
11.1 Spine of Frank Hawley's *Miscellanea Japonica II, Volume 1, Whales and Whaling in Japan* 206
11.2 'The Harpooners Carouse' 206
11.3 Sekobune 207
11.4 The coming of the Black Ships (kurofune) 208

11.5 Whale Wars DVD covers. 'The second coming of The Black Ships' 209
11.6 Single sekobune, profile 217
11.7 Six panels of Kishū Kumano-ura Hogei-zu (Whaling in the Kumano Sea) 217
13.1 Number of international tourists visiting Japan, 2003–2017 241
13.2 Top ten source markets for Japanese inbound tourism, 2017 242
13.3 Number of deaths and missing persons caused by natural disasters in Japan organized by prefecture, 2013–2017 245
13.4 Promotion of the mobile safety app 'Safety tips' at JR Namba Station in Osaka 250
14.1 Growth in inbound tourism to Japan 263
14.2 Growth in number of international visitors staying in Kyoto Prefecture 267
14.3 Protest sign in a residential area of Kyoto 268
14.4 Protest sign in a residential area of Kyoto 269
14.5 Growth in hotel and simplified accommodation rooms in Kyoto 270

Tables

1.1 Japan: international arrivals and outbound tourism 1965–2018 2
1.2 Domestic tourism in Japan 2012–2017 (millions) 3
2.1 Number of books on tourism in the NDL search database 22
2.2 Number of articles with words frequently used in articles of JITR and JAST 25
2.3 Database search result (English-based publications) 26
2.4 Journals with Japan related articles 27
2.5 Database search result: books and chapters in English 28
3.1 Top 10 prefectures attracting large numbers of tourists and day visitors and cities with the largest numbers within their prefectures 55
3.2 Kankō machizukuri cities, towns and villages 57
5.1 Results of the systematic review (sorted by year) 87
13.1 Number of fatalities and missing persons caused by natural disasters in Japan, 1993–2017 244
14.1 Comparison of Japanese Hotels and Inns Act and Private Housing Accommodation Law 265

Contributors

Abhik Chakraborty currently works as Lecturer in the Faculty of Tourism, Wakayama University, Japan. His research interests encompass complex systems, landscape level analysis of tourism impacts, vulnerability and integrity of natural resources in protected areas, ecotourism and conservation, mobility and change in mountain regions, and World Natural Heritage sites. He is a member of the World Commission on Protected Areas of the IUCN and a reviewer of UNESCO World Natural Heritage site nominations in that capacity. He is also a keen traveler and spends most of his summer in the ridgeline of North Japan Alps and mountains in the Hokkaido Island in the north.

Adam Doering is an Associate Professor of tourism studies at Wakayama University, Japan. He currently serves as the Sustainability Research Unit Sub-leader for the Center for Tourism Research and is a co-founding member of the Critical Tourism Studies Asia Pacific (CTS-AP) research group. Adam has published broadly on a variety of topics including sustainable tourism, lifestyle sports in East Asia, Japanese surf culture and tourism destinations, sport tourism, continental philosophy for tourism studies, and Destination Management Organizations (DMO) in Japan. His current research examines the role of surfing and surf tourism in Fukushima Prefecture's post-disaster resurgence.

Tom Hinch was, until his recent retirement, a Professor in the Faculty of Kinesiology, Sport, and Recreation at the University of Alberta, Canada. His continuing research interests focus on sport based travel and sustainable tourism development more generally. He studies the way that people who travel for sport connect to the places that they visit. His research has shown that characteristics such as embodiment, emotional engagement and celebration of community make sport a strong tourist attraction. Such characteristics tend to foster meaningful connections between visitors and host destinations and can be used as 'place makers'.

Yumiko Horita is a professor at Wakayama University, Japan. She holds her PhD in global environment science from Kobe University. Her research interests include tourism as urban regeneration, social inclusion, building consensus among community and urban policy. Her ongoing research focuses on the relation between local residents' everyday life and tourism space/development and on the impacts of the phenomenon of peer-to-peer accommodation on local communities.

Takahiro Ikeji is a researcher at the Japan Travel Bureau Foundation, a think tank in Japan. He provides research and advice to both national and local governments on issues concerning tourism and community development. His research interests include destination marketing, destination finance, tourism policy and overtourism. Currently, he is working on the research focusing on hotel taxation policies and employment at Japanese-style accommodation (*Ryokan*), and short-term rentals.

Eiji Ito is an Associate Professor in the Faculty of Tourism and acting director of the Center for Tourism Research, Wakayama University, Japan. He was awarded his BSc and MSc by Juntendo University, Japan, and gained his PhD at the University of Alberta, Canada. His areas of research include leisure studies (the social psychology of leisure), sport tourism, and cultural psychology. He has received two Young Distinguished Scholar Awards from the Japan Society of Physical Education, Health and Sport Sciences in 2014 and the Japanese Society of Lifelong Sports in 2017.

Kumi Kato, PhD, is a professor at the Faculty of Tourism, Wakayama University, Japan. She currently chairs a committee for sustainable tourism at the Japan Tourism Agency. Her research area includes sustainability, community resilience and ethics related to tourism, and she is currently engaged in projects with communities in Iitate and Soma, Fukushima to support recovery from the 2011 tsunami and in Kumano region, promoting pilgrimage and slow tourism. Recent journal publications include 'Gender and sustainability: exploring ways of knowing: An ecohumanities perspective' (2019); 'Debating sustainability in tourism development – Resilience, traditional knowledge and community: A post-disaster perspective' (2017); 'Slow (walking) tourism and community: Kumano pilgrimage trail' (2017, with N. Progano); and 'Australia's whaling discourse: Global norm, green consciousness and identity' (2015: John Barrett Award for Best Paper).

Amnaj Khaokhrueamuang is Associate Professor in Tourism at the School of Management and Information, University of Shizuoka, Japan. His research interests focus on rural tourism-related issues associated with community development, agricultural extension, culture and heritage, and sustainability. He is currently working on developing projects and academic collaboration networks with Thailand's universities and organizations to promote the strong relationships between Japan and Thailand.

Hayato Nagai is a Lecturer in the Faculty of Tourism at Wakayama University, Japan, and a member of the Center for Tourism Research at the same university. He joined Wakayama University in 2016 after the completion of his PhD at the University of Queensland in Australia. His PhD project focused on Asian working holiday-makers staying in Australia and explored their perceptions of travel related risks. His current research interests include tourism risk management, destination marketing and management and youth tourism market especially in the Asia-Pacific region. He has published his work in international tourism journals such as *Annals of Tourism Research*, *Journal of Vacation Marketing* and *Journal of Hospitality and Tourism Management*. He currently serves on the editorial committee of *Kankogaku Hyoron* (Tourism Studies Review), the official journal of the Japan Society for Tourism Studies.

Yumi Oura, PhD (Nagoya), is a professor at the Faculty of Tourism, Wakayama University, Japan. She serves as a member of Forest committee at Wakayama Prefecture. Her research areas include forest tourism, rural tourism and local resources management. She is currently engaging in projects with communities in Aridagawa and Kamitonda, Wakayama, to support conservation activities of rice terraces, satoyama and family-run agriculture business. Recent journal publications include 'Transition of forest tourism policies in Japanese national forest management' (2018) and 'Policy development and direction of urban-rural interaction after th 1990's in Japan' (2012).

Ricardo Nicolas Progano is Lecturer at the Center for Tourism Research, Wakayama University, Japan. His research interests include religious tourism, regional tourism and cultural tourism. In particular, his research has focused on spiritual and pilgrimage tourism in Japan from multiple perspectives. He combines his backgrounds in Asian studies (University of El Salvador) and Tourism studies (Wakayama University) leveraging his language skills in Spanish, English and Japanese.

Brent W. Ritchie is Associate Dean (Research) in the Faculty of Business, Economics and Law at the University of Queensland, Australia. He is also a Distinguished Professor in the Center for Tourism Research at Wakayama University, Japan. Brent's research interests include tourism risk, crisis and recovery management from a consumer, organizational and destination perspective. Brent also has interests associated with resident support for tourism and major events. His work has been published in outlets such as *Tourism Management*, *Journal of Travel Research*, *Annals of Tourism Research* and the *Journal of Sustainable Tourism*. He is on the editorial board of eight journals including the *Journal of Travel Research* and *Annals of Tourism Research*.

Kaede Sano joined the Faculty of Tourism at Wakayama University, Japan, in April 2015, following two years as an Assistant Professor in the Faculty of Commerce at Doshisha University. She received her PhD in 2013 from Kobe University in Japan. Dr. Sano's research interests include tourism marketing, tourism risk management, social media marketing and social customer relationship management (social CRM). Her research focuses on understanding how tourists search for travel-related information through social media and how tourists reduce perceived risks by communicating with each other in a digital environment of user-generated content. Dr. Sano and her colleagues are currently using Global Positioning System (GPS) data to track international tourists' spatial and temporal behaviors in Osaka.

Richard Sharpley is Professor of Tourism and Development at the University of Central Lancashire, Preston, UK, and Distinguished Professor in the Center for Tourism Research at Wakayama University, Japan. He is co-editor of the journal *Tourism Planning & Development* and a member of the editorial boards of a number of other tourism journals. His principal research interests lie within the fields of tourism and development and the sociology of tourism and he has published widely in both. His books include *Tourism and Development: Concepts and Issues, 2nd Edition* (2014, with David Telfer), *Tourism and Development in the Developing World, 2nd Edition* (2016, with David Telfer), *Tourism, Tourists and Society, 5th Edition* (2018), *Tourism, Development and Environment: Beyond Sustainability* (2009), *Mass Tourism in a Small World* (2017, with David Harrison) and, most recently, *A Research Agenda for Tourism and Development* (2019, also with David Harrison).

Simon Wearne was born in Ballarat, Victoria, Australia, in 1956. Educated in Victoria and Western Australia and graduating from Curtin University in 1977 (design/cinematography), he was an award winning (Emmy nominated 2008) cinematographer for over 40 years, specializing in current affairs and documentaries on social justice, environment, politics, arts and sport. With a strong background in community inclusive activism focused on environmental protection, his current PhD research focuses on Japanese traditional whaling culture including the first industrial whaling developed in Wakayama in 1606. The research involves the conservation of cultural and natural values and an initiative for UNESCO industrial heritage to recognize Japanese traditional whaling. Simon has teaching and research collaborations with universities and other education institutions in Japan as well as with tourism industry consultants locally, nationally and internationally.

Takashi Yoshino has been a professor at Wakayama University, Japan, since 2013. He received his Bachelor of Engineering in Electronics Engineering and his Masters of Engineering in Electric Engineering from Kagoshima University, Kagoshima, Japan, in 1992 and 1994 respectively. He received his PhD in Information Science from Tohoku University, Sendai, Japan, in 2001. Since joining Wakayama University, has developed various types of communication support system and information sharing systems including: a tour guide support system; a tourist information extraction and visualization system; multilingual communication support systems; disaster prevention information sharing systems; and a medical information sharing system. More generally, his research interests include computer supported collaborative work.

1 Introduction

Tourism in Japan – from the past to the present

Richard Sharpley and Kumi Kato

Introduction

In 2018, Japan received almost 31.2 million international tourist arrivals (JNTO, 2019a).[1] Moreover, at the time of writing, a total of 10.98 million international arrivals had been recorded during the first four months of 2019, representing 4.4 percent growth on the same period in the previous year (JNTO, 2019b). On the one hand, such figures might appear unremarkable. Indeed, as the world's third largest economy with highly developed international transport links and a wealth of cultural and national attractions, Japan's tourism sector might be considered to be performing relatively unfavourably compared with other countries. For example, in 2017, the UK attracted 37.6 million international arrivals and Germany 37.4 million, whilst Thailand received 35.4 international arrivals that year (UNWTO, 2018). Moreover, worldwide international tourist arrivals increased by an estimated 6 percent in 2018, with growth of between 3 and 4 percent forecast for 2019 (UNWTO 2019).

On the other hand, when contextualized within both the recent history of Japan's tourism sector in particular and the growth of global international tourism more generally, these figures can only be described as dramatic. As can be seen from Table 1.1, following more than 40 years of average growth reflecting the wider expansion of the international tourism sector, the annual number of international tourist arrivals in Japan over the last decade has almost quadrupled. More specifically, between 2012 and 2017 in particular, an average annual growth rate of around 28 percent was achieved. This compares with an overall increase in worldwide tourist arrivals of approximately 50 percent over the last decade and, between 2005 and 2017, an average annual growth rate of 4.2 percent (UNWTO, 2018). In short, in recent years not only has Japan significantly outperformed the global international tourism sector as a whole, but also has done so despite experiencing a more than average 18.7 percent decline in arrivals in 2009 following the global economic crisis and, most notably, a 27.8 percent decline in 2011 as a result of the east coast earthquake and tsunami, causing the

Table 1.1 Japan: international arrivals and outbound tourism 1965–2018

Year	*International arrivals*	*Growth on previous year (%)*	*International tourism receipts (US$mn)*	*Outbound tourists*	*Growth on previous year (%)*
1965	366,649	3.9	-	158,827	23.6
1970	854,419	40.4	-	663,467	34.6
1975	811,672	6.2	-	2,466,326	5.6
1980	1,316,632	18.3	-	3,909,333	–3.2
1985	2,327,,047	10.3	-	4,948,366	6.2
1990	3,235,860	14.1	-	10,997,431	13.8
1995	3,345,274	–3.5	4,894	15,298,125	12.7
2000	4,757,146	7.2	5,970	17,818,590	8.9
2005	6,727,926	9.6	15,554	17,403,565	3.4
2006	7,334,007	9.0	11,490	17,534,565	0.8
2007	8,346,969	13.8	12,422	17,294,935	–1.4
2008	8,350,835	0.0	13,781	15,9897,250	–7.6
2009	6,789,658	–18.7	12,537	15,445,684	–3.4
2010	8,611,175	26.8	15,356	16,637,224	7.7
2011	6,218,752	–27.8	12,533	16,994,200	2.1
2012	8,358,105	34.4	16,197	18,490,638	8.8
2013	10,363,904	24.0	16,865	17,472,748	–5.5
2014	13,413,467	29.4	20,790	16,903,351	–3.3
2015	19,737,409	47.1	27,285	16,213,766	–4.1
2016	24,039,700	21.8	33,428	17,116,420	5.6
2017	28,691,073	19.3	36,979	17,889,292	4.5
2018	31,191,856	8.7	-	18,954,026	6.0

Source: Adapted from JNTO (2019a) and Knoema (2019).

nuclear disaster at the Fukushima Daiichi Nuclear Power Station. In addition, in 2015, the number of international arrivals in Japan was, for the first time since 1970, greater than the number of outbound tourists, an important milestone given the country's widely recognized policy during the 1980s of promoting outbound tourism (Soshiroda, 2005).

It should be noted that the figures for both inbound and outbound tourism remain dwarfed by the scale and value of domestic tourism in Japan (see Table 1.2). Latest figures indicate that, in 2017, total spending on domestic tourism amounted to JPY 21,113 trillion (around US$195 billion), giving continuing credence to Nelson Graburn's observation some 35 years ago that 'The Japanese are inveterate tourists, not only throughout much of the world but particularly within Japan itself … the country is, perhaps, the best organized in the world for mass internal travel' (Graburn, 1983: 2).

To return to the issue of the remarkable and rapid increase in international tourism to Japan, this has not of course occurred by accident. As discussed in more detail later in this chapter, since the late 1990s, the country has been seeking to develop what Funck and Cooper (2015: 46) refer to as a national 'tourism culture'. Initially, this was driven by the belief that encouraging

Table 1.2 Domestic tourism in Japan 2012–2017 (millions)

	2012	*2013*	*2014*	*2015*	*2016*	*2017*
Total trips	612.75	630.95	595.22	604.72	641.08	647.51
Overnight stays	315.55	320.42	297.34	312.99	325.66	323.33
Day trips	297.20	310.53	297.88	291.73	315.42	324.18

Source: Adapted from OECD (2018) and JTA (2018).

greater numbers of international visitors would help Japan to 'fulfil its long-cherished role of fostering long-standing friendship and trust among nations' (Funck & Cooper, 2015: 46); the 1997 'Welcome Plan 21' sought to increase international arrivals to 8 million by 2007, a target which, as can be seen from Table 1.1, was achieved. More recent policies, however, focusing on transforming Japan into a 'tourism nation' (MLIT, 2012: 2), have succumbed to the more typical goal of increasing tourism as a means of supporting economic growth and development (Sharpley & Telfer, 2015), although they also continue to highlight the objectives of enhancing community well-being and international understanding (MLIT, 2012). Specifically, the 2016 'Tourism Vision to Support the Future of Japan' (MILT, 2016) firmly places the objective of tourism policy on achieving national economic growth and regional revitalization through continuing growth in international arrivals. Hence, the 2007 'Tourism Nation Promotion Basic Plan' established a target of 10 million international arrivals by 2010, although this was to be frustrated by the 2009 global economic crisis. The subsequent new 'Tourism Nation Promotion Basic Plan', approved in 2012, proposed targets of 18 million arrivals by 2016 and 25 million by 2020 (MLIT, 2012), both of which have since proved to be highly conservative. Hence, the 2020 target has been revised ambitiously upwards to 40 million, rising to 60 million by 2030 (MLIT, 2016; Murai, 2016).

Since 2013, the increase in tourism to Japan has been facilitated by both the implementation of policy measures and a number of external factors. For example, airline deregulation and the relaxation of visa requirements for visitors from ASEAN nations and, in particular, China, have boosted regional tourist numbers, whilst the depreciation of the Yen and the rapid growth in China's outbound tourism have also played a significant role (Andonian et al., 2016). Consequently, not only do tourists from East Asia, specifically South Korea, China, Taiwan and Hong Kong, together currently comprise almost three quarters of Japan's international tourism market, but China alone accounts for more than 8.3 million, or almost 27 percent, of all international tourist arrivals. As such, questions may be raised with regard to this arguably excessive dependence on regional markets as well as the viability (or indeed desirability) of hosting 60 million tourists in a country that, in popular destinations such as Kyoto, is already experiencing so-called 'overtourism'.

Irrespective of such debates, however, the important point is that development of tourism in Japan, as well as the academic study of it, is clearly at a significant juncture. Although having welcomed international tourists in steadily growing but nevertheless, in global terms, relatively limited numbers during the latter half of the 20th century, Japanese tourism policy was primarily focused on promoting domestic and outbound tourism. Moreover, and perhaps not coincidentally, academic attention was (and to a great extent, remains) primarily concerned with understanding the motivations, behaviour and impacts of Japanese tourists travelling either domestically or overseas (e.g., Guichard-Anguis & Moon, 2009), with many studies building on the foundation provided by Graburn (1983) in his seminal anthropological study of Japanese domestic tourism. In contrast, with some notable exceptions (e.g., Funck, 2012), relatively few studies address inbound tourism.

As detailed above, however, the last decade has witnessed not only a fundamental policy shift in Japan towards competing aggressively in international tourist markets, supported in no small measure by the country hosting a number of major international sports events including the Rugby World Cup in 2019 and the (subsequently postponed to 2021) Summer Olympics in 2020, but also an unprecedented growth in international arrivals. Consequently, the country faces a number of opportunities and challenges with regards to the development and consequences of tourism, opportunities and challenges that have not gone unrecognized. Specifically, recent years have witnessed increasing emphasis being placed on the academic study of tourism in Japan through both the provision of related tertiary education and a growing body of research (see Chapter 2 in this volume for a review of extant tourism research in Japan).

The purpose of this book, then, is to consider contemporary tourism in Japan at this pivotal point in its development. More specifically, it seeks to explore issues relevant to the role of tourism as established in the 2012 Tourism Nation Promotion Basic Plan and the subsequent 2016 Tourism Vision referred to above, not least its potential contribution to local community development and the 'enhancement of mutual international understanding' (MLIT, 2012: 4), as well as a number of challenges that must be addressed. In so doing, it does not intend to provide a comprehensive analysis of the future development of tourism in Japan, a task well beyond the scope of a volume such as this. Rather, it offers contributions that, organized into thematic sections, adopt a novel perspective on Japan's future as a 'tourism nation'. And perhaps the most appropriate way to introduce that future is to contextualize it in a brief review of the historical development of tourism and tourism-related policy in Japan up to the present day.

Tourism in Japan: a historical perspective

The history of tourism in Japan is, as Funck and Cooper (2015: 10) observe, 'dealt with comprehensively in the Western and English-language literature', with oft-cited works by commentators such as Ishimori (1989), Soshiroda (2005) and Vaporis (1995) drawing on equally extensive Japanese language sources on the subject. A useful brief summary of tourism's historical development is provided by March (n.d.) and in a recent chapter by Crow-Delaney (2019) whilst, in contrast, Funck and Cooper (2015) devote two detailed chapters of their book to the emergence of, first, a travel culture and, second, policy and planning for tourism in Japan. For the purposes of this chapter, therefore, the following sections summarize briefly the key themes to emerge from unsurprisingly similar stories in the literature that trace the evolution of tourism in Japan. The historical development of tourism as a sociocultural phenomenon and of tourism policy are addressed together within the context of distinctive periods in the history of tourism although, inevitably, it is in more modern times that specific policies have not only been in evidence but also have had more influence on the trajectory of tourism in the country.

The roots of Japanese tourism: the early years

As Funck and Cooper (2015: 3) note, the roots of contemporary tourism in Japan are widely considered to lie in the Meiji era from 1868 onwards, in some ways reflecting the origins of modern mass tourism elsewhere during the latter half of the 19th century. However, as Ishimori (1989: 181) explains, 'travel for pleasure dates back to ancient times in Japan's history, when only the ruling class could afford it'. For example, there is evidence of travel to hot springs as early as the 8th century whilst, by the 12th century, former emperors were undertaking religious-based travel to Kumano, as were the samurai (the military nobility). Nevertheless, by the 14th century, some wealthier farmers were also able to visit shrines and temples such as those at Ise, Kumano and Kōyasan – both now popular contemporary destinations for domestic and international tourists (see Chapter 9) – although generally, the costs and challenges of travel restricted wider participation in this nascent tourism.

It was during the Edo or Tokugawa era (1603–1868), the first of arguably four distinctive periods in its development in Japan, that tourism not only became more accessible to the population at large but also began to emerge 'as a form of recreation among the masses – a national pastime' (Vaporis, 1995: 26). A number of factors were influential in this transformation, not least the increasing affluence amongst farmers and peasants; the Edo era, distinctive for national peace and stability after more than a century of civil war, ushered in a long period of economic growth and prosperity. Equally, although official constraints were placed on it, primarily to limit the time that peasants would be away from their fields, travel was allowed (and regulated through the issue of permits) for the purposes of pilgrimage or visiting spas for health reasons (Vaporis, 1995).

Most significant during this period, however, was the development of transport infrastructure, in particular the expansion of the road network in response to the government's requirement that all feudal lords should reside in Edo (now Tokyo) for at least one year in every two. Consequently, not only were a number of major routes established, known collectively as the Five Highways or *Gokaidō*, the most important of which was the *Tōkaidō* linking Kyoto and Edo, but also a total of more than 200 post stations, providing accommodation and, subsequently, meals and other personal services for male guests, were developed along the routes. As Ishimori (1989: 184) suggests, this was fundamental to the emergence of 'travel for pleasure', as was similarly the provision of lodging and other facilities at then major destinations, such as the Ise Shrine in present day Ise City, Mie Prefecture. According to Ishimori (1989: 185), in 1724 there were 615 lodgings available at the Ise Outer Shrine whilst reflecting, for example, the development of British and European spas in the 18th century (Towner, 1996), towns also grew around major shrines and temples. In addition to accommodation, other facilities and entertainment were available in these towns including theatres, restaurants and souvenir shops – the tradition amongst contemporary Japanese tourists, described by Graburn (1983: 44–46), of buying gifts or *miyage* for family and friends at home dates back to this period – whilst there were also numerous brothels (Ishimori, 1989).

These developments support Watkins' (2008: 98) contention that 'Within the context of Japan's culture, history and religious beliefs, there exists no clear line between tourism and pilgrimage; the two developed in tandem'. In other words, as in many other countries, early tourism on a relatively large scale in Japan was manifested in religiously motivated travel. This was encouraged in part by the system of permits referred to above and considered in more detail by Vaporis (1995) but perhaps more so by, on the one hand, the efforts of shrines and temples to attract visitors by dispatching priests, or *sendatsu*, around the country to guide groups of pilgrims to their temple/shrine, in effect acting as precursors of today's travel agents (Funck & Cooper, 2015: 14) and, on the other hand, local co-operative associations, or *kō*. Linked to particular shrines, *kō* were established around the country with the purpose of both promoting worship at those shrines and raising funds to provide financial support for members, typically the head of the family, to travel there. Nevertheless, as pilgrimage became increasingly popular during the Edo era, 'sightseeing and pleasure seeking was at least as important, if not more important than the religious goal' (Fromanek, 1998, cited in Watkins, 2008: 95) or, putting it another way, spiritual experiences became secondary to more earthly pleasures.

One such pleasure was visiting hot springs/spas, or *onsen*. As discussed later in this chapter, visiting hot springs is fundamental to contemporary domestic tourism in Japan; according to Kamata (2016), there are more than 3,000 hot spring destinations around the country, varying from small, Japanese-style inns (*ryokan*) with attached *onsen* to major spa resorts, such as Beppu on the southern island of Kyushu. Not surprisingly, much academic attention is devoted to hot springs as a contemporary wellness tourism resource in Japan (e.g., Erfurt-Cooper, 2014; Rátz, 2013), but visiting hot springs as a manifestation of health tourism in Japan was, as noted above, one of the earliest forms of tourism in Japan. For example, Arima, located near Mount Rokkō north east of Kobe city, is a popular hot spring destination that, according to Guichard-Anguis (2002), was patronized by emperors during the 8th century. As narrated by Funck and Cooper (2015: 17–20), tourism to hot springs became more popular during the Edo era not only amongst the elite in society but also amongst the wider population, not coincidentally reflecting the growth in pilgrimage and other religiously motivated travel. Indeed, it can be seen very much as a by-product of pilgrimage tourism although interest in the health benefits of hot springs was enhanced by the publication of books and guides that listed and compared the qualities of *onsen* around the country. Equally, Ishimori (1989) attributes the more general development of tourism in Japan at this time in part to the increasing availability of guidebooks and other publications, including, for example, the *Miyako Meisho Zue*, or 'Illustrated Book of Famous Places in the Capital' (Ishimori, 1989: 189), published in 1780. Along with the infrastructural developments described above and the evolving culture of travel in the country, these undoubtedly contributed to the foundation for the further development of tourism in Japan during the second half of the 19th and into the 20th century.

The expansion of domestic and incoming tourism: 1868–1940

The second distinctive period in the evolution of tourism in Japan commenced at the beginning of the Meiji era (1868–1912) with both the relaxation of the restrictions placed on domestic travel during the preceding Edo era and the opening up of country to both inbound and outbound travel. As a consequence, domestic travel became much easier with travel permits no longer required and the removal of checkpoints along major routes, whilst the foundations were established for the development of international tourism to and from Japan that was to continue through to the outbreak of the Pacific War.

Key to the initial growth in tourism during this period was the development of transport infrastructure which, as the potential significance of international tourism as a source of foreign exchange came to be recognized, became the focus of increasing private and state investment. The first railway in the country was completed in 1872, running from Yokohama northwards up to Shinagawa and subsequently extended to Shinbashi, both now districts of present-day Tokyo. As a result, Yokohma became a popular destination, with almost 1.5 million passengers using the Shinbashi–Yokohama line in 1874 (Funck & Cooper, 2015: 32). Other routes, such as the line between Kobe, Osaka and Kyoto, were also constructed whilst private railway companies went on to develop routes primarily to already popular tourist destinations and attractions, further stimulating growth in domestic tourism. By the early 20th century, these railway companies were also investing in other developments, such as hotels, department stores and recreational facilities, hence playing a major role in the expansion of the tourism sector.

Domestic land travel was also simplified by the construction of bridges which had, for security purposes, been banned during the Edo era, meaning it was no longer necessary to rely on ferries to cross rivers. At the same time, opportunities for inbound tourism were enhanced by developments in sea transport; by 1870, for example, four of the world's major steamship companies had offices in Yokohama and tourist travel to Japan had become a realistic proposition for Westerners (at least for those with sufficient financial and temporal resources). Then, in the early 1920s, three small companies established domestic air transport services; in 1928, these were combined into Japan's first national flag carrier, Japan Air Transport Corporation, which commenced commercial operations the following year (Century of Flight, n.d.). Haneda airport was subsequently opened in Tokyo in 1931. Also during the 1930s, large ocean liners offering affordable passage were enabling increasing numbers of American tourists to cross the Pacific to visit Japan.

In addition to improvements in transport infrastructure, the accommodation sector also began to expand during this second period in Japan's tourism history, perhaps the most notable development being the introduction of Western-style hotels. In contrast to traditional Japanese *ryokan* with their communal baths and other facilities, these new hotels offered private rooms and European cuisine. According to Funck and Cooper (2015: 34), the first such hotel opened in Yokohama in 1863, followed by one in Tokyo's foreign-settlement area in 1869, both amongst the small number of locations in which foreigners were then permitted to reside in Japan under the Harris Treaty[2]. These included the ports of Hakodate, Niigata, Yokohama, Kobe and Nagasaki (Soshiroda, 2005: 1106) where, unsurprisingly, Western-style hotels were also built in the latter half of the 19th century. Such hotels were also developed in the places where foreigners spent their summer holidays, such as Nikkō and Hakone, the latter being the location of the Fujiya Hotel which became well-known amongst overseas and Japanese tourists alike. Interestingly, Funck and Cooper (2015) describe how these early foreign tourists in Japan also had a major influence on outdoor recreation in Japan by 'importing' their own tourism and leisure activities. For example, the first golf course was developed in 1903 in the Rokkō mountains near Kobe, the first step towards Japan becoming the world's second largest market after the US for the sport, with 2,450 golf courses and up to 10 million golfers (Hays, 2009) – though its popularity is now on the wane. Similarly, Japanese mountains, long visited on pilgrimages, became 're-invented' as places of adventure and to experience nature.

Some of the first official tourism policies also emerged during this period. Until the late 1800s, no specific attempts had been made to promote Japan to overseas tourist markets although, as Soshiroda (2005: 1112) notes, foreign visitors had been invited to an 'Expo' in Kyoto in 1872. Then, in 1893 the 'Welcome Society' was established, ostensibly to promote and support the development of inbound tourism through, for example, encouraging the upgrading of accommodation facilities and the provision of travel information. However, it is generally regarded to have been indicative of 'an era during which international tourism policies became a part of diplomatic policies, as the government aimed to use them to raise the international status of Japan' (Soshiroda, 2005: 1102) in general, and to challenge the perceived inequality of the Harris treaty in particular. The subsequent founding of the Japan Tourist Bureau (JTB) by the Ministry of Railways in 1912 was, then, the first explicit move towards promoting international tourism to Japan as well as providing services to visitors. The JTB implemented overseas promotional campaigns, establishing offices in many countries as well as providing information services for international visitors at Japan's major railway stations (Soshiroda, 2005). It also engaged in commercial activities, setting it on the path to becoming today's JTB Corporation, the largest travel agency in Japan.

By the early 1930s, increasing emphasis was placed on enhancing Japan's foreign exchange earnings through tourism. The responsibility for promoting international tourism was taken over by the Board of Tourist Industry, created by the Ministry of Railways in 1930 (and subsequently disbanded in 1942). In addition to establishing offices in London, New York and Paris, the Board implemented a number of specific strategies to increase inbound tourism, including encouraging sports teams to visit and organizing what are now more generally referred to as familiarization visits amongst key decision makers. The key target market was America, supported by the introduction of tourist class accommodation on Pacific liners referred to above (see March, 2007) and, by 1936, international tourist spending 'represented Japan's fourth largest source of foreign exchange revenue' (Funck & Cooper, 2015: 35). Travel from the US to Japan continued to be promoted up to 1940 but the country had already forfeited the opportunity to host the 1940 Olympics for military reasons (it is interesting that, 80 years later, major international sporting events are again contributing to inbound tourism – see also Chapter 5) and in 1941, war broke out in the Pacific following the Japanese attack on Pearl Harbor.

The post-war years to the 1990s

In the years immediately following the end of hostilities in the Pacific, limited travel was permitted into and out of Japan; the first (American) leisure tourists arrived in Yokohama in December 1947 (March, 2007) and, from the following year, week-long conducted tours were introduced for small groups of, again, American visitors. Over the next decade or so, however, both inbound and, to a lesser extent, outbound tourism steadily grew whilst Tokyo's hosting of the Summer Olympics in 1964 signified Japan's re-establishment in the international community.

A number of factors supported this growth in tourism during this period, not least the development of air transport. Two US-based airlines commenced operations to Japan in 1947 and Japan Airlines was found in 1951, becoming officially recognized as the country's national carrier in 1953. Initially, it flew on domestic routes but in February 1954, it carried 18 passengers on its first international flight to San Francisco, with other international routes subsequently being established. Domestic air services also increased during the 1950s and, by 1958, there were 80 scheduled services linking 21major cities (March, 2007). In addition, not only were a number of travel agencies established, primarily to meet the needs of increasing numbers of international visitors, but also a variety of policies were implemented to support inbound tourism and facilitate its contribution to economic recovery. For example, a number of cities were designated for development as international tourist destinations whist policies were also introduced to stimulate hotel construction (Soshiroda, 2005: 1109). However, it was the period during the build-up to the Summer Olympics in 1964 that arguably witnessed the most significant developments in transport, communications and accommodation whilst the 1963 Basic Tourism Law provided comprehensive policies for the further development of tourism in Japan as well as recognizing the role of tourism in enhancing the country's international relations. In April 1964, all Japanese citizens could travel feely abroad.

The Japan National Tourist Organization was also established in 1964 with the aim of promoting and supporting inbound tourism. However, during the 1970s and into the 1980s, the focus was primarily on supporting and developing domestic and outbound tourism; indeed, in terms of inbound tourism policies, the period between 1970 and 1987 is referred to by Soshiroda (2005: 1110) as 'one of stagnation, during which the focus of policies in general shifted to national tourism and natural conservation', as well as encouraging local and regional economic development through appropriate laws and policies. This was epitomized, perhaps, by the Resort Law of 1987 which supported significant investment in local tourism projects, such as the Seagaia Resort discussed in Chapter 6 in this book (see also Funck, 1999). Not only were policies designed to increase and shape the use of the population's leisure time but also rising prosperity allowed people to take advantage of these policies. As a consequence, domestic tourism boomed during the 1970s and into the 1980s. Steady growth was witnessed in outbound tourism although, by the mid-1980s, the number of Japanese tourists travelling overseas annually was only about 5 million, representing roughly 4 percent of the population. Not only did this compare unfavourably with the level of outbound tourism from other major economies (e.g., almost 40 percent of the UK population and 34 percent of West Germans travelled overseas), but also Japan was enjoying a significant balance of payments surplus with the rest of the world.

In 1987, therefore, the '10 million' programme was launched, the aim being to double the number of outbound Japanese tourists by 1991. In addition to supporting the economies of destination countries, this was seen as a means of enhancing mutual understanding between nations, a persistent theme in Japan's tourism policies. Facilitated by a strengthening of the Yen, the target of this unique policy was reached a year early and, as can be seen from Table 1.1 above, outbound tourism from Japan continued to grow rapidly up until the end of the 20th century. The level of inbound tourism, however, remained relatively weak and so, suffering from the economic effects of the so-called 'lost decade' in Japan's economic development from 1990, as well as the belief that international relations in general, and understanding of the Japan and the Japanese in particular, could be strengthened by encouraging more people to visit the country, from 1997 the focus of tourism policy turned explicitly to increasing inbound tourism.

From 1997 to the present: growing inbound tourism

As pointed out earlier in this chapter, the most recent period in the development of tourism in Japan, from 1997 onwards, is most notable for the policy of seeking a rapid and sustained increase in international tourist arrivals. As such, this period signifies not only a significant reversal in preceding tourism policies but also the deliberate opening up of the country to the internal community in general and the potential economic benefits of tourism in particular. In part, this has been driven by the long-held desire to enhance international relations (Funck & Cooper, 2015: 46) but it primarily reflects the need to stimulate economic growth and the inter-related challenges of an ageing and declining population, urbanization and rural decline (Crowe-Delaney, 2019).

This new direction in tourism policy commenced with the 'Welcome Plan 21' in 1997 although, following the announcement of the goal of increasing international arrivals to 10 million by 2010 and the establishment of the 'Tourism Nation' initiative by then prime minister Junichiro Koizumi in 2003, a series of policies targeting an increase international arrivals were implemented, including both the launch of the *Visit Japan Campaign* and the appointment of a minister with responsibility for the tourism nation in the same year (JTA, 2010). The Campaign has supported a variety of initiatives, including the production and promotion of videos based initially upon the *Yokoso Japan* (Welcome Japan) theme and, from 2010, under the new *Japan. Endless Discoveries* logo. In 2018, a new multilingual campaign video *Enjoy My Japan* was launched, specifically targeting long-haul visitors from Europe, North America and Australia in order to address the dominance of regional markets (see JNTO, 2019c). Arguably, however, the most significant development in this period was the establishment of the Japan Tourism Agency (*Kankō-chō*, or JTA) on 1 October 2008 within the Ministry of Land, Infrastructure, Transport and Tourism. At the time of writing (2019), the JTA employs almost 220 staff and co-ordinates activities focusing the development of the 'tourism nation', such as enhancing tourism infrastructure, training and promotion.

In terms of policy, the 'Tourism Nation Promotion Basic Plan' of 2012, which established a target of 25 million international arrivals by 2020 (in 2017, more than 28 million arrivals were recorded) was superseded by the

'Tourism Vision to Support the Future of Japan' which increased the targets to an ambitious 40 million arrivals by 2020 and 60 million by 2030, as well as significantly increasing the number of guest nights outside the major metropolitan areas. Overall, the Vision states that:

> Objective: world-class tourist industry! Japan is ready to welcome international visitors throughout the country. The government wishes to foster dynamic multicultural exchange so that Japan can truly open to the world, rapidly develop new services and innovations in the tourism sector and thereby create a positive cycle which enhances regional economies and industry.
>
> (MLIT, 2016: 3)

Thus, the development of mutual understanding and effective international relations remains an implicit objective, although the three 'basic visions' or pillars of the policy emphasize the focus its economic goals: (i) maximizing the value of tourism resources in order to make tourism the foundation of regional revitalization; (ii) developing tourism as the nation's core industry through promoting innovation in the tourism industry and increasing global competitiveness; and (iii) developing a stress-free travel environment for all tourists to promote satisfying experiences (from MLIT, 2016: 3). Underpinning the achievement of these objectives are ten specific strategies including: the utilization of cultural and natural heritage, including national parks; the revitalization and promotion of traditional hot spring spas and rural towns; improving soft-infrastructure, such as the online environment and smart entry at the immigration; and, reforming the system of work and holidays in Japan to encourage greater participation in domestic tourism for a more vibrant society. Given these ambitious targets, it is acknowledged that more effective of management of tourism within the country is required. Hence, since 2015, an increasing number of destination management organizations (DMOs) have been established. There are currently 136 DMOs registered with the JTA in the three categories of Wider Region (*Kōiki,* 10), Inter-regional (Chiiki-renkei, 69) and Regional (*Chiiki,* 57), promoting and managing regional tourism.

The inevitable outcome of these policies has been rapid growth in international arrivals; as noted earlier in this chapter, in 2015, the number of inbound tourists exceeded that of outbound tourists for the first time in 50 years although, as is typical in many countries, the domestic tourism market remains much larger than the international market. Nevertheless, remarkable growth has, to date, been achieved, the 10 million mark being exceed in 2013, the 20 million mark in 2015 and the 30 million mark in in 2018 (see Table 1.1 above). Further growth is expected, encouraged by a series of international mega events being hosted in Japan – the World Rugby Cup (2019), the Olympic Games (2021), the World Masters Games (2021) and the Osaka and Kansai Expo (2025). It must, however, be questioned whether the

2030 target of 60 million international arrivals is realistically achievable or, indeed, desirable.

Indeed, as the consequence of rapid growth in and concentration of tourism, some of the more well-known destinations, such as Kyoto, have begun to suffer the symptoms of so-called overtourism: congestion, concentration and resident complaints (see Chapter 14 for more detail). In addition, regardless of whether visitors are international or domestic, the influence of social media, symbolized by words such as *Bae-ru* or *Bae* (looks good on Instagram) is becoming increasingly evident. Visiting sites where one can take spectacular, impressive or unique photographs and then upload them on Instagram or other social media platforms has become a major motivation for visitors, who arrive in destinations simply via word-of-mouth (or *photograph-on-eyes*) is as evident in Japan as in any other destination. In a sense, Japan is experiencing a new kind of mass tourism, where visitors gather to experience the places but at random times and in an uncontrolled manner, and also to experience places not developed as tourism destinations including residential areas and local amenities.

As a consequence, recent years have witnessed increasing concern for and activities related to the sustainability of tourism. For example, during 2017, the UN designated International Year of Sustainable Tourism for Development, activities included participation in the ASEAN+3 Tourism Leaders Summit and the ASEAN-Japan Centre organized an accessible tourism workshop. The Japanese version of the UNWTO promotional video on responsible tourism, *Travel, Enjoy, Respect* was also launched with 'Hello Kitty' as the promotional ambassador. At the same time, greater attention was accorded to the Sustainable Development Goals (SDGs), with the national government selecting 30 cities as 'SDG Future Cities'. More specifically, in June 2018, the JTA established the Sustainable Tourism Promotion Office and conducted a national survey into the state of tourism, focusing on specific elements including transport, accommodation and infrastructure. The results of this survey, including responses from 138 local governments that manage major tourism destinations and interviews with key domestic and overseas municipalities, was published in June 2019 (JTA, 2019). Subsequently, it was decided that the JTA should develop a set of internationally recognized sustainable tourism indicators as a national guideline and facilitate its local implementation. A set of indicators-based Global Sustainable Tourism Council destination criteria was developed during 2019, with a plan for it to be trailed in several destinations in Japan during 2020.

In summary, then, during this most recent period, tourism has been adopted as a strategic economic and social stimulus to address the challenges of a decreasing and ageing national population (the Japanese population is forecast to decline by 30 percent over the next 50 years, with 40 percent of the population being aged over 65 by 2060). The rapid increase in international arrivals is undoubtedly bringing much needed economic benefits, as well as stimulating an international outlook (catering for diversity) and a deeper understanding of tourism and its social benefits. At the same time, however, there is still disparity in the benefits received from tourism, particularly in regional areas where adequately trained human resources remain insufficient, although *chiho-sōsei* (regional innovation and revitalization) is, as explained above, encouraged in recent tourism policies. Nevertheless, the need undoubtedly exists for Japan to shift its policy and strategies from simply quantitatively increasing tourism arrivals to more of a focus on qualitative growth, utilizing the country's abundant natural and cultural resources such as spiritual sites, spas, traditional lifestyles and cuisines, cultural heritage, arts, crafts and skills deeply rooted in regional environment and traditions which hold natural sustainability as their core. Such community-based and people-focused tourism is likely to emphasize Japan's quality much more clearly and strongly.

Structure of the book

It is within this context of more recent developments that this book offers a number of contemporary perspectives on tourism in Japan. Comprising a total of 14 chapters, including this introduction and brief historical overview of tourism policy and practice, the book addresses a number of themes, the first of which is a 'past to the present' focus on both the academic study of tourism in Japan and the evolution of policy in two specific tourism contexts.

In Chapter 2, Kumi Kato and Yumiko Horita provide a review of research articles and books on tourism and Japan published in both Japanese and English, the purpose being to trace the development of and identify the main themes and trends in the research from both a Japanese and international perspective. In so doing, they point to potential future directions in research into tourism in Japan, as well as justifying the content and focus of this book. The following two chapters then explore policies for and the development of tourism in two distinctive contexts, namely urban areas and forests. In Chapter 3, Yumiko Horita explores how urban tourism spaces in Japan have been developed alongside more general urban development, examining significant aspects and implications of the urban tourism development process and focusing in particular on how local communities in cities have become more engaged in the development and use of urban spaces for tourism and leisure. A similar theme is evident in Chapter 4 in which Yumi Oura analyses critically the historical development of the national forest management and policy in Japan and considers its relationship with tourism policy. She notes in particular that, since the 1990s, transformations in national forest administration policy have brought about the introduction of collaborative management in all aspects of forest management within the new concept of 'forests for people', concluding that such initiatives, including public participation in management, should be implemented more widely to support the further successful development of forest tourism.

The following five chapters explore different forms of tourism in Japan from a variety of perspectives. With the country hosting a number of international sports events from 2019 onwards, Chapter 5 contributed by Eiji Ito and Tom Hinch, explores sport tourism research in Japan. Based upon a systematic review of sport tourism research articles published in Japanese journals, it identifies trends in research methods and thematic patterns, going on to discuss avenues for future research, in particular for sustainable sport tourism research in Japan. Notably, it suggests that the integration of interdisciplinary knowledge in tourism, sport and other related studies is the key in helping to encourage sustainable sport tourism development after the hosting of major events. In Chapter 6, Adam Doering then focuses on surf tourism. Specifically, he offers a genealogy of domestic surf tourism development in contemporary Japan by tracing the historical production of one of Japan's most prominent surf tourism destinations, Miyazaki Prefecture. Examining how surf tourism came to be mobilized as a tool for revitalizing the local economy, he concludes with a reflection on how surf tourism development may (dis)connect local surfing communities. This is followed by Abhik Chakraborty's assessment in Chapter 7 of the emergence of and future challenges facing mountain tourism. Noting the transformation of mountains from sacred places to the focus of tourism development, he draws on two case studies to explore how mountain landscapes are under pressure from both visitor numbers and climate change, but finds hope in indications of a heightened ecological consciousness and sense of place can be detected among local tourism stakeholders. In Chapter 8, Amnaj Khaokhrueamuang turns to rural tourism and, specifically, tea tourism. Observing that in recent years the production of tea in Japan has been in decline, he laments that limited attention has been paid to the potential of revitalizing this traditional industry through tea tourism. However, through a case study of an exchange programme between Thai and Japanese tea farmers, he develops a new model of tea tourism that conceptualizes its potential to the sustainability of tea growing communities. Pilgrimage tourism is the focus of Chapter 9, in which Ricardo Nicolas Progano and Kumi Kato explore this form of tourism in the specific context of the World Heritage nominated pilgrimage trail, the Kumano-kodo in Wakayama. Making reference to the evolving meaning of spirituality and tourism both globally and in Japan, this chapter explores the significance of today's spiritualities in destination management from the perspective of local communities.

Chapter 10 is the first of two chapters that consider how Japan confronts difficult pasts. Here, Richard Sharpley's chapter explores the manner in which the controversial Kamikaze strategy is commemorated in contemporary Japan. Framed within a review of literature that reveals competing approaches to (re)presenting and commemorating Japan's war history, he compares two heritage sites, Chiran Peace Museum and the Uzurano airfield near Kasai, to identify the extent to which the commemoration of Kamikaze pilots contributes to international understanding and reconciliation. Then, in Chapter 11, Simon Wearne provides the political and historical context for an ongoing project that seeks to present, through the development of a cultural tourism attraction, a more nuanced and informed understanding of Japan's whaling heritage. In so doing, the project eventually seeks to contribute towards breaking the current impasse in contemporary international whaling discussions whilst promoting and maintaining the traditional sustainable whaling culture of Taiji in Wakayama Prefecture.

The final three chapters of the book collectively address different challenges facing the future of tourism in Japan. With the country's susceptibility to natural disasters, in Chapter 12 Kumi Kato considers how resilience-based tourism may contribute to post-disaster community sustainability. With social sustainability-oriented tourism concepts and resilience planning as a framework, she proposes that tourism development in the early recovery phase is vital in helping communities maintain their connection with their places, which is argued to be the core of resilience. Such resilience is closely related to traditional ecological knowledge found in personal stories, monuments, folktales, literature and arts. This, she suggests, lays a foundation for a sustainable recovery and reconstruction, in which tourism can play a vital role. Chapter 13 is similarly concerned with the relationship between tourism and natural disasters in Japan, though from the perspective of the vulnerability of international tourists. Specifically, Hayato Nagai, Kaede Sano, Brent Ritchie and Takashi Yoshino reveal the country's high risk of exposure to a range of natural hazards, such as earthquakes, tsunamis, typhoons and floods, before going on to highlight and discuss the challenges and issues that national and local government and also the tourism industry face in relation to reducing the vulnerability of international tourists to these hazards. Finally, in Chapter 14, Takahiro Ikeji and Hayato Nagai address the challenges surrounding the development of peer-to-peer (P2P) accommodation in Japan. After reviewing the rapid expansion of the supply of P2P accommodation both globally and in Japan, the chapter considers relevant policies in Japan before going on to highlight potential solutions to the problems of P2P accommodation through a case study of Kyoto.

The book is then brought to a close with the addition of a short postscript that acknowledges the onset of the coronavirus pandemic and the subsequent crisis for tourism as the writing of the book was being completed. Whilst the pandemic will have both short and long-term implications for tourism in Japan, not least necessitating a review of the country's growth-oriented tourism policies outlined in this chapter as well as of its significant dependence on particular markets, it may also represent an opportunity to develop a more sustainable approach to tourism which, as many of the chapters in this book explicitly suggest, adopts a more community-focused approach.

Notes

1 The Japanese National Tourism Organisation provides extensive, detailed and up-to-date statistical data on tourism in Japan. See https://statistics.jnto.go.jp

2 The Harris Treaty of 1858, negotiated by Townsend Harris, the first US Consul in Japan, was an agreement that secured commercial and diplomatic rights for the US in Japan. It was widely considered to disadvantage Japan, hence subsequent attempts to re-establish the country's national status.

References

Andonian, A., Kuwabara, T., Yamakawa, N. and Ishida, R. (2016) *The Future of Japan's Tourism: Path for Sustainable Growth Towards 2020.* Tokyo: McKinsey & Company.

Century of Flight (n.d.) Early Japanese civil aviation. Available at: www.century-of-flight.net/new%20site/commercial/Japanese%20civil%20aviation.htm (Accessed 11 June 2019).

Crowe-Delaney, L. (2019) Japanese tourism in the late 20th and early 21st centuries. In T. O'Rourke and M. Koščak (Eds), *Ethical and Responsible Tourism: Managing Sustainability on Local Tourism Destinations*. London: Earthscan/Routledge, pp. 163–182.

Erfurt-Cooper, P. (2014) Wellness tourism: a perspective from Japan. In C. Voigt and C. Pforr (Eds), *Wellness Tourism: A Destination Perspective*. London: Routledge, pp. 235–254.

Fromanek, S. (1998) Pilgrimage in the Edo period. In S. Linhart and S. Früstück (Eds), *The Culture of Japan as Seen Through Its Leisure*. Albany: State University of New York Press, pp. 165–194.

Funck, C. (1999) When the bubble burst: Planning and reality in Japan's resort industry. *Current Issues in Tourism*, 2(4), 333–353.

Funck, C. (2012) The innovative potential of inbound tourism in Japan for destination development: a case study of Hida Takayama. *Contemporary Japan*, 24(2), 121–147.

Funck, C and Cooper, M. (2015) *Japanese Tourism: Spaces, Places and Structures.* New York/Oxford: Berghahn Books.

Graburn, N. (1983) *To Pray, Pray and Play: The Cultural Structure of Japanese Domestic Tourism.* Aix en Provence: Centre des Hautes Études Touristiques.

Guichard-Anguis, S. (2002) From curing and playing to leisure: two Japanese hot springs – Arima and Kinosaki Onsen. In J. Hendry and M. Raveri (Eds), *Japan at Play: The Ludic and the Logic of Power*. London: Routledge, pp. 245–258.

Guichard-Anguis, S. and Moon, O. (Eds) (2009) *Japanese Tourism and Travel Culture*. London: Routledge.

Hays, J. (2009) Recreational golf in Japan. Availa88ble at: http://factsanddetails.com/japan/cat21/sub143/item784.html (Accessed 11 June 2019).

Ishimori, S. (1989) Popularization and commercialization of tourism in early modern Japan. *Senri Ethnological Studies*, 26, 179–194.

JNTO (2019a) *Japan Tourism Statistics: Trends in Visitor Arrivals to Japan.* Japan National Tourism Organisation. Available at: https://statistics.jnto.go.jp/en/graph/#graphgeninbound--travelers--transition (Accessed 31 May 2019).

JNTO (2019b) *Japan Tourism Statistics; Latest Information.* Japan National Tourism Organisation. Available at: https://statistics.jnto.go.jp/en/graph/#graph--latest--figures (Accessed 31 May 2019).

JNTO (2019c) *Enjoy My Japan*. Japan National Tourism Organisation. Available at: www.enjoymyjapan.jp/en/?utm_country=gb (Accessed 22 December 2019).

JTA (2010) Paving the Way to a Tourism Nation. Japan Tourism Agency. Available at: www.mlit.go.jp/kankocho/en/about/keiki.html (Accessed 22 December 2019).

JTA (2018) Travel and Tourism Consumption Trend Survey. Japan Tourism Agency. Available at: www.mlit.go.jp/kankocho/en/kouhou/page01_000307.html (Accessed 3 June 2019).

JTA (2019) Towards an Advanced Sustainable Tourism Nation. Japan Tourism Agency Available at: www.mlit.go.jp/common/001293012.pdf (Accessed 4 December 2019).

Kamata, H. (2016) A segmentation analysis of Japanese spa tourists. *Journal of Tourism & Services*, 8, 20–37.

Knoema (2019) Japan International Tourism Receipts. Knoema Corporation. Available at: https://knoema.com/atlas/Japan/Tourism-receipts (Accessed 31 May 2019).

March, R. (2007) How Japan solicited the West: the first hundred years of modern Japanese tourism. In I. MacDonnell, S. Grabowski and R. March (Eds), *Tourism: Past Achievements, Future Challenges. Proceedings of CAUTHE 2007*. Sydney: University of Technology, pp. 843–852.

March, R. (n.d.) The Historical Development of Japanese Tourism. Available at: www.inboundtourism.com.au/pdf/japanese-tourism-history.pdf (Accessed 28 May 2019).

MLIT (2012) *Tourism Nation Promotion Basic Plan*. Ministry of Land, Infrastructure and Transport. Available at: www.mlit.go.jp/common/000234920.pdf

MLIT (2016) Meeting of the Council for a Tourism Vision to Support the Future of Japan. Ministry of Land, Infrastructure and Transport. Available at: www.mlit.go.jp/common/001172615.pdf (Accessed 22 December 2019).

Murai, S. (2016) Japan doubles overseas tourist target for 2020. *Japan Times*. Available at: www.japantimes.co.jp/news/2016/03/30/national/japan-doubles-overseas-tourist-target-2020/#.XPFfHy2ZNBx (Accessed 31 May 2019).

OECD (2018) Japan: Domestic, Inbound and Outbound Tourism. Available at: www.oecd-ilibrary.org/docserver/tour-2018-table56-en.pdf?expires=1559567594&id=id&accname=guest&checksum=4FDC482F48988DD6781F3F06CB054C2B (Accessed 3 June 2019).

Rátz, T. (2013) Hot springs in Japanese domestic and international tourism. In M. Smith and L. Puczkó (Eds), *Health and Wellness Tourism*. London: Routledge, pp. 345–349.

Sharpley, R and Telfer, D. (2015) *Tourism and Development: Concepts and Issues*, 2nd Edition. Bristol: Channel View Publications.

Soshiroda, A. (2005) Inbound tourism policies in Japan from 1859 to 2003. *Annals of Tourism Research*, 32(4), 1100–1120.

Towner, J. (1996) *An Historical Geography of Recreation and Tourism in the Western World 1540–1940*. Chichester: John Wiley & Sons.

UNWTO (2018) *Tourism Highlights 2018 Edition*. Madrid: UN World Tourism Organisation. Available at: http://doi.org/10.18111/9789284419876

UNWTO (2019) International tourist arrivals reach 1.4 billion two years ahead of forecasts. *Press Release PR 19003*: UN World Tourism Organisation. Available at: www.2.unwto.org/press-release/2019-01-21/international-tourist-arrivals-reach-14-billion-two-years-ahead-forecasts (Accessed 31 May 2019).

Vaporis, C. (1995) The early modern origins of Japanese tourism. *Senri Ethnological Studies*, 38, 25–38.

Watkins, L. (2008) Japanese travel culture: an investigation of the links between early Japanese pilgrimage and modern Japanese travel behaviour. *New Zealand Journal of Asian Studies*, 10(2), 93–110.

2 Tourism research on Japan – overview of major trends

Japanese and English-language materials[1]

Yumiko Horita and Kumi Kato

Introduction

This chapter reviews research publications related to tourism and Japan based on a literature survey of materials published in both Japanese and English. In both cases, the reviews are based upon selected academic books and refereed journal articles that were identified in a major online search using keywords *tourism* (or *kankō* in Japanese) and *Japan*. The Japanese publications identified include 2,463 books and 727 government publications published between 1940 and 2016, and 471 articles published in two journals, JAST (*Japan Academic Society of Tourism,* 1961–2015) and JITR (*Japan Institute of Tourism Research*, 1987–2015). The search of the English language materials revealed 21 books and 7 book chapters published between 1983 and 2013, and 224 articles in 97 journal titles published between 1975 and 2016. Industry reports, magazine articles, book reviews, conference proceedings or unpublished theses are excluded from the search.[2]

The purpose of this chapter is twofold. First, it seeks to identify the principal research trends in both Japan and globally, examining the two groups of materials (Japanese and English language) separately as well as comparatively, and referring to some of the major events in its chronological development of Japan's tourism. The second purpose is to examine how some of the various themes in this book, including regions, community and places in Japan, are portrayed (or not) in the studies that are collectively analysed in this review. Hence, the overall aim this chapter, in addition to highlighting the breadth of research into tourism in Japan (the majority of which has been published in Japanese), is to contextualize the book as a whole within that research. In so doing, it seeks to situate the specific themes and issues discussed in the book within the broader development of inherent themes in research into tourism in Japan.

For convenience, more contemporary trends in the research are analysed on a decade-by-decade basis, as are the major topics emerging in the collection. It is important to note, however, that the methods of analysis employed in the two groups (Japanese and English literatures) are not exactly the same, whilst the authors also acknowledge that this review is by no means exhaustive, not least because the search not only was limited to certain keywords, but also excluded works published in other languages, in other forms of publication and in other disciplines. Nor is it the intention of this review to quantify the results, to categorize them in a rigid manner or to provide a conclusive analysis. Nevertheless, it is hoped that this 'snapshot' will provide a unique perspective on tourism research in focusing on Japan and point to future directions in this field of research. The following section commences with a review of trends and themes in tourism-related research in Japanese language publications before the chapter, then goes on to consider the relevant English language literature.

Japanese language research publications

Early stages of tourism research published in Japanese

According to Shiota (1974a), research on the social phenomenon of tourism in general commenced with the compilation of tourism statistics in European countries at the end of the 19th century. As the number of tourists from the US to European countries increased between the end of the 19th and the beginning of the 20th centuries, tourism statistics and research on the economic effect of tourism also began to be developed. In Japan in particular, the Board of the Tourist Industry, established within the Ministry of Railways in 1930, and the Japan Tourist Bureau, an extra-governmental body established in 1912, started to undertake research into tourism in foreign countries as well as gathering statistics on foreign visitors entering Japan. At the same time, they supported the translation of early European academic books on tourism, such as those by Mariotti (1927), Glücksmann (1935) and Bormann (1930).

As detailed previously in Chapter 1, following the end of the Second World War, the development of tourism in Japan quickly restarted with the promotion of international tourism in particular becoming a key pillar of economic recovery measures focused on earning foreign currency for post-war reconstruction. The number of published works related to the business of tourism and the development of tourism projects increased up to the 1960s, coinciding with the government's tourism policy. This was the period during which the economic benefits of tourism increased significantly and, consequently, major research interests in tourism were directed towards seeking means of maximizing the benefits of tourism and examining its impacts on society. Shiota (1974b) describes this research trend as follows:

> tourism research [was] always linked with the practical consciousness. This is the evidence that tourism research had the task of seeking means of increasing gains from tourism activities because tourism, in particular international tourism, was regarded as [a] phenomenon [with] economic

> gain[s]. The necessity of tourism research was clearly defined by the time, therefore, and tourism research became less speculative and more empirical.
>
> (Shiota, 1974b: 296, authors' translation)

Indeed, a number of articles and books published during the 1960s and 1970s focused on the practical aspects of tourism businesses and projects and how to promote them. Specifically, the themes of tourist behaviour, tourism resources, tourism businesses and projects and tourism policy are predominant in the literature of that time, with many works focused on identifying and explaining the social and economic aspects of tourism phenomena. Subsequently, a number of textbooks on tourism appeared during the 1970s and 1980s, drawing on the outcomes of this earlier research (e.g., Koike & Ashiba, 1988; Maeda, 1973; Suzuki, 1974).

Nevertheless, some commentators pointed out the significance of and the need for a more theoretical approach to the study of tourism during this empirical research-oriented period (Maeda, 1987; Shiota, 1974b). Their work advocated the potential of theoretical tourism research, the arguments presented being categorized by Shiota (1974b) in two ways. On the one hand, tourism research was considered to be a new, separate and defined sphere of academic endeavour with the potential to develop and become established; on the other hand, tourism was regarded as an applied sphere of existing individual science. Subsequently, however, few articles that build on these arguments can be found. One reason for this, as Maeda (1987) pointed out, was that there existed no basic consensus amongst researchers with regards to capturing or identifying complex tourism phenomena, to developing the concepts to explain them, as well as to the research methods for exploring them. Moreover, academia at that time made no effort to develop such a consensus, explaining in part the lack of academic works considering tourism from a more conceptual perspective. Indeed, it was only from the beginning of the 1990s that Japan witnessed an unprecedented expansion in tourism research, both empirical and conceptual. The next section, however, provides an overview of the trends in tourism research from the 1980s, preceding that expansion.

Research trends from the 1980s

A survey of books on tourism written in Japanese was conducted through a search of the database of the National Diet Library (NDL). The number of books and government publications were counted according to the classification of NDL. The search keyword used was *kankō* and the search coverage, the NDL catalogue[3] (Table 2.1).

The search results included academic books as well as those targeting a more general readership, along with government publications. As can be seen from Table 2.1, the total number of books has gradually increased since

Table 2.1 Number of books on tourism in the NDL search database

	Total number of books	*Number of government publications*
–1950	17	1
1951–1970	135	12
1971–1980	287	83
1981–1990	456	164
1991–2000	667	186
2001–2010	901	281
Total	2463	727

1981. Until 1970, the majority of the books had been published by extra-governmental bodies, such as the Japan Travel Bureau Foundation (JTB Foundation) and the Japan Tourism Association. In addition to these books, the number of government publications focusing on tourism development increased from the early 1970s. Reflecting the expansion of the tourism sector, government and extra-governmental bodies played a role at this early stage by conducting necessary basic research on tourism phenomena.

The quantitative growth in academic tourism research commenced in the 1990s, although some pioneer research had been undertaken prior to that. Two academic research associations were formally established in the 1990s in addition to the Japan Academic Society of Tourism (JAST) which was the first to be established in 1960. The Japan Foundation for International Tourism (JAFIT) was established in 1993, and the Japan Institute of Tourism Research (JITR) was launched in the following year. The latter was based on the activities of the Federation of Researchers for Tourism in Japan from 1986 onwards.

The next section examines the principal characteristics of tourism research by period from the 1980s. The analysis focuses on the books published not by organizations but by individual authors.

(i) 1981–1990. During this period, the proportion of the total number of books comprising government publications was relatively high (36 percent) whilst, as already mentioned, the majority of books were published by extra-governmental organizations with only a few academic books to be found. A distinctive feature of the research during this decade was that a number of books explicitly addressing resort development appeared towards the end of this period. In fact, a total of 15 books containing the word 'resort' in either the title or subtitle were published in 1989 and 1990 alone. Many of these tended to promote and explore the implications of resort development and investment (e.g., Maruta, 1989; Nabeta, 1989; Tanabe, 1990) and, thus, were clearly were responding to the rising expectations of the booming property market under the Comprehensive Resort Areas Development Act enacted in 1987 (see Chapter 1).

(ii) 1991–2000. Since the 1990s, the number of books written by academic researchers gradually increased, reflecting the fact that, on one hand, many Japanese academic researchers had finally embarked on the study of tourism and, on the other hand, tourism studies had begun to be recognized as a legitimate field of academic significance. One important factor driving this expansion of tourism research was the enactment of the Comprehensive Resort Areas Development Act in 1987 referred to above and discussed elsewhere in this book (Chapters 1 and 6) although, once this 'notorious' piece of legislation was enacted, criticisms of resort developments rapidly increased, particularly for their reckless development and consequential negative impacts on the natural environment and local communities (e.g., Asato, 1991; Dohi, 1991; Futaba & Naruse, 1992; Ohno, Nakayama, & Sasaki, 1991; Sato, 1990). This marked a turning point away from the previous dominant support for promoting tourism development and business/projects.

Two key features in this period can be identified. First, within the increasing number of books, tourism was explored from a variety of perspectives from several academic fields and disciplines. For example, studies employed approaches from the service and hospitality field (Fukunaga & Suzuki, 1996; Hattori, 1994), marketing (Hase, 1996; Yamagami, 1997), anthropology (Hashimoto, 1999; Ishimori, 1996; Yamashita, 1996), sociology (Ishihara, Yasufuku & Yoshikane, 2000; Kitagawa 1998; Naitou, 1996; Yamauchi, 1997) and the regional/local community (Mizoo, 1994; Nishioka, 1996; Sasaki, 1998). Second, an increasing number of works appeared that examined the impacts of foreign tourism in the context of particular fields of study (Asamizu, 1999; Imai and Yamauchi, 1999; Kuji 1993; Matsui, 1993; Naitou, 1998; Yamanaka, 1992).

(iii) 2001–2010. During this period, the Japanese government developed a number of formal tourism policies, discussed in more detail in Chapter 1. Previously, the government's tourism initiatives carried neither the validity nor the influence compared to those for manufacturing and construction works. However, the Visit Japan campaign began in 2003; the Tourism Nation Promotion Basic Law was then enacted in 2007 and the Japan Tourism Agency was established in 2008 as an extra-ministerial bureau of the Ministry of Land, Infrastructure Transport and Tourism. With this governmental promotion of and support for tourism, tourism research also flourished, with the increase in both tourism and tourism research reflecting the fact that a service- and knowledge-based economy was replacing the dominant secondary manufacturing economy in post-war Japan. At the same time, tourism's significance was also becoming socially and economically recognized. In particular, with advances in information technology and transportation, public consciousness with regards to travel and tourism behaviour began to change, with a consequential substantial impact on the Japanese economy and modern society. This growth of tourism research represents the current boom in tourism in Japan, which can be described in two ways.

First, the books in this period tend to employ the word 'tourism (*tsūrizumu*)' in their title or subtitle. The Japanese word '*kankō*', which means tourism, had previously often been used, but the number of book titles including the word '*tsūrizumu*' has increased since the latter half of 2000s; the search identified 35 books with *tsūrizumu* in their title or subtitle. With the increase in usage of the word tourism/*tsūrizumu*, new words that combine a word and tourism, such as food tourism, event tourism or health tourism, have also emerged. Notably, this spread in the use of the word '*tsūrizumu*' represents the expansion and diversification of tourist behaviour based on broad individual interests into what is collectively referred to as 'new tourism'; in contrast, the more traditional '*kankō*' more specifically connotes sightseeing and the 'pleasure jaunt'. Some academics, therefore, may use *tsūrizumu* consciously or unconsciously to imply current trends in tourism in Japan. Moreover, several book titles included the word 'new' (Hase, 2003; Suda, 2006; Yoshida, 2010). Not only the diversification of the objectives and direction of the tourist gaze, but also the exploration of alternative ways of developing tourism based on a reflection on the past processes could be regarded as such new movements.

Second, a large number of books dealing with culture, intercommunication between guests and hosts, and regional and local-based development in relation to tourism are adding to the literature (Hashimoto & Sato, 2003; Iguchi, 2002, 2005; Kitagawa, 2003, 2004; Mizoo, 2007; Nishimura, 2009; Yahagi, 2004; Yasumura, 2006). With the decline of key regional industries, falling populations and a loss of vitality, regional cities, towns and villages have faced the challenges of recovering from a devastating economic and social situation. As in other parts of the developed world, tourism has become a significant tool for such regional/local revitalization, regarded as a new type of *machizukuri* movement (*machi* –community, town or region; - *tsukuri/zukuri*-making). Reflecting this trend, the contents and theme of tourism research and books tends to focus on intercommunication between tourists and local people and the work experience of local industry. The review also found that some books examine tourist behaviours and regional- and local-based tourism management.

Trends in articles in tourism journals

In order to obtain a picture of research trends in Japanese language journal articles, a survey of articles was conducted. In total, 490 articles in two journals were found, 471 of which were analysed. The two journals are the *Journal of Tourism Research* and *The Tourism Studies*, which are published by JAST and JITR and were established in 1960 and 1994, respectively.

Table 2.2 shows the results of the survey, summarizing the words frequently used in article titles or subtitles. The themes and approaches that the articles adopt are varied and mixed; hence, it was difficult to define them into categories, but keywords were considered as the useful indicators of the trends. As Table 2.2 shows, the words 'region/local' are most frequently used (47),

Table 2.2 Number of articles with words frequently used in articles of JITR and JAST*

	1961–1970	*1971–1980*	*1981–1990*	*1991–2000*	*2001–2010*	*2011–2015*	*Total*
JITR (1986~2015)	-	-	6	38	50	43	137
JAST (1961~2015)	57	40	67	71	84	15	334
Total number	57	40	73	109	134	58	471
Hotel, ryokan, accommodation, guesthouse	9	3	4	5	14	3	38
Oversea travel	0	2	0	0	4	2	8
Aviation and airplane	2	3	4	3	3	0	15
Service	0	1	3	2	3	0	9
Tourism industry	2	1	2	3	1	1	10
Tourism business	5	3	3	4	1	1	17
Region/Local	1	3	4	13	18	8	47
Destination	5	0	7	6	10	7	35
Development	5	3	5	7	10	2	32
Tourism	1	0	0	6	10	7	24
Culture	0	0	0	3	8	3	14
Behaviour	1	1	0	5	8	4	19
Policy	3	0	3	3	8	0	17
Tourism resource	2	0	2	0	4	1	9
Image	0	0	0	2	4	2	8

Note: The articles published by the former organization of the JITR (established in 1986) are included.

* The *Journal of Tourism Research* has 350 articles published between Volume 1 in 1961 and Volume 56, Number 1 in 2015. This excludes articles written in foreign language, or those equivalent to essays and research notes. As for *The Tourism Studies*, the 140 articles published between Volume 1 in 1987 and Volume 27, Number 1 in 2015 are targeted. Some articles may be counted twice or more if the article title has plural words listed in the table.

followed by words related to accommodation, such as hotel and *ryokan* (38). The word 'destination' is also often used (35). Similarly, the word 'development' appears frequently (32). A similar trend is also found in book titles with the exception of the words related to accommodation. A content analysis of articles in *Tourism Studies* published by JITR reveals similar findings to this study, indicating that theme and field were weighted towards those related to tourism development and business (Choi, 2009). Indeed, a number of articles focus specifically on how to promote and manage tourism as a tool for regional/local development and revitalization. As pointed out previously, this is one of the features reflected in a typical Japanese socioeconomic situation driven by development. Differences in stances adopted towards region/local and destination and development can be found depending on the year of publication and characteristics of the journal; however, more recent articles tend to approach development from a regional- or local-oriented perspective (Nakako, 2010; Tabira & Asoh, 2007; Yamazaki & Soshiroda, 2010). This is evidence of the popularity of the *machizukuri* concept in tourism development in Japan.

One specific feature of tourism research published in journals is the number of articles related to accommodation, such as hotels, guesthouses and other lodgings. Books addressing this theme were not found in the NDL search, although earlier articles in the 1960s considered accommodation from the perspective of hotel management and, in the 1980s, from an economics approach (Hikima, 1964; Inagaki, 1983). Since the 2000s, the sheer volume of publications and their themes has diversified, to include management strategy, marketing and the labour market related to accommodation industry (e.g., J. Kim, 2005; L. Kim, 2004).

English language research publications

Review of English-based publications – books and articles

'English-based' is defined here as materials written in English for international journals for an international audience. A major database search was carried out with *Japan* and *Tourism* as keywords, locating 22 books and 10 book chapters, and 226 articles in 97 journals. Journals were limited to those specified as refereed or peer-reviewed and screened for suitability in content. The books selected for this review were published between 1983 and 2016, and articles between 1983 and 2016 with one exception published in 1975. The majority of work was published during the 2000s and, in particular, after 2010 (see Table 2.3).

Within this review, the earliest book published is *To Pray, Pay and Play. The Cultural Structure of Japanese Domestic Tourism* (Graburn, 1983a) and the most recent is *Japanese Tourism: Spaces, Places and Structure* (Funck & Cooper, 2013). Books and chapters are identified as those providing an overview of 'Japan's tourism' or 'tourism', and others focusing on specific types or aspects of tourism as described in the next section. In the journals, the earliest article identified is 'International tourism and intercultural communication: The case of Japanese tourists in Singapore' in the *Southeast Asian Journal of Social Science* (Riaz, 1975). More than 85 percent of the articles were written during the 2000s and the number written after 2010 was more than double that written in the previous decade. This is partially due to the fact that four special issues were published: two in *Tourism Review International*

Table 2.3 Database search result (English-based publications)

	1970s	*1980s*	*1990s*	*2000s*	*2010~*	*Total*
Books (Book chapters)	–	1	3(2)	9(2)	9	22(10)
Journal articles	1	9	20	68	126	224

Table 2.4 Journals with Japan related articles

Journal of Travel & Tourism Marketing (Special issues in 2000 and 2016)	36
Tourism Review International (Special issues in 2006 and 2007)	14
Annals of Tourism Research	13
Tourism Management	8
Tourism Economics	7
Journal of Travel Research	5
Journal of Sustainable Tourism	5
Asia Pacific Journal of Tourism Research	5
Journal of Tourism & Hospitality Research	4
Tourism Geographies	4

(2006/2007) and two in *Journal of Travel & Tourism Marketing* (2000/2016). Although 97 journals were identified, 45 percent of articles were published in the 10 journals listed in Table 2.4, with the remaining titles each offering just three or less.

Among the journal article titles, 27 contain the word *tourism*, 10 *hospitality*, 5 *geography,* 4 *leisure*, 2 *travel* and 1 *recreation.* A total of 14 articles were found in Japan-specific journals, such as *Journal of Japanese Studies*, *Japan Forum* and *Japan Review*, on topics such as popular culture, heritage, religion or history. The countries with the largest number of first author affiliation[4] were Japan (64) and US (51), followed by Australia (11), Canada (8), Taiwan (7), Singapore (5) and France (5). In recent years, there has been a clear 'Asian wave' in tourism research as reported in Leung et al. (2015). In the following section, the content of the publications is reviewed for the publication year and content, pointing out some of the characteristics.

Books and articles providing overviews on Japan's tourism development

Books, although limited in number, may be classified according to the key topics identified in Table 2.5. In two of the more recent publications on tourism and Japan, *Tourism in Japan: An Ethno-semiotic Analysis* (Berger, 2010) and *Japanese Tourism and Travel Culture* (Guichard-Anguis & Moon, 2009), the focus is on the cultural aspects of travelling in Japan. Berger (2010) presents Japan's tourism, culture and everyday life through an 'ethnosemiotic' analysis, making reference to a wide range of iconic social phenomena, from pop culture to traditional gardens. In Guichard-Anguis and Moon's (2009) edited collection, introduced at the outset as '*tabi-no-bunka*' or culture of travel, tourism is considered to be essentially synonymous with travelling; chapters cover a variety of forms of domestic and international tourism approached from anthropological and ethnographic perspectives. Alternatively, Leheny (2000, 2003) writes on Japan's leisure policy as the *rules of play*, which define how people spend their leisure time from perspectives based on the experience of working in the Japanese government's travel policy formulation, including

Table 2.5 Database search result: books and chapters in English

	1980s	*1990s*	*2000s*	*2010~*	*Total*
Books (Book chapters)	1	3(2)	9	9(5)	21(7)
- Overview: tourism in Japan			1	2	3
- Domestic tourism/leisure/social changes	1	1(1)	1		3(1)
- Japanese tourists overseas/outbound		2	2	1(1)	5(1)
- Inbound/outbound marketing		(1)			(1)
- Heritage tourism/history			3	2	5
- Popular culture			2	1(1)	3(1)
- Sustainability/conservation				3(2)	3(2)
- Disaster and reconstruction				(1)	(1)

a segment on post-war Japan. Publications on leisure activities refer to Japan's work culture, often portrayed stereotypically as 'workaholics living in rabbit hutches' (Fuess, 2012: 250; Graburn, 1983a; Linhart & Frühstück, 1998). A broader overview of tourism, not limited to travel or leisure, is provided in *Japanese Tourism: Spaces, Places and Structure* by the two Japan-based authors (Funck & Cooper, 2013) based on historical, geographical and socio-economic analyses. By providing a global context, the analysis provides a scope for future development, referring to changing social situations and tourism trends (Funck & Cooper, 2013).

There has also been a recent increase in tourism-related publications – that is, books which do not include the word 'tourism' in the title but which are nevertheless clearly relevant to recent tourism development in Japan or, perhaps more appropriate to say, relevant to focus of this book. For example, *Sustainability in Contemporary Rural Japan. Challenges and Opportunities* (Assmann, 2016) deals with the environmental and social sustainability issues Japan faces, including population decline and ageing, youth settlement in rural areas and post-disaster recovery. In this book, tourism is situated as a tool for the revitalization and development of regional areas. Other publications relating to sustainability issues include *Parkscapes: Green Spaces in Modern Japan* (Havens, 2011), which presents a political, economic and social history of green spaces in Japan over two centuries, ranging from urban green space to national parks, also including government policies. In contrast, *Hearing Monkeys to Paradise: How Macaque Troops are Managed for Tourism in Japan* (Knight, 2011) presents interesting ethnographic insights into human–wildlife relationship through the management of wild monkeys, which are one of Japan's popular tourism attractions.

Knight (1994, 1995, 1996) also documents changes in rural life through his examination of forest resources (Knight 2000) and, more recently, popular wild monkeys (Knight, 2010, 2011). In particular, Knight paid early attention to rural areas in Japan from a social anthropological perspective. One of the keywords in Japanese tourism, namely *machizukuri* (community/town/regional development), is used by Knight in his study of rural areas in Wakayama (Knight, 1994). Similarly, the term *muraokoshi* (awakening the village) was employed by Arlt (2006) in his discussion of sustainable tourism in Japan. Works on regional branding (e.g., Jones, Nagata, Nakajima & Masuyama, 2009) also relate this 'revitalization' movement. Attention to rural areas also relates to Japan's ageing and declining population, as illustrated by many such as Harada (1999). In addition, domestic policies on nature-based tourism and sustainability are documented by Jones (2012a, 2012b, 2016).

Characteristics identified by period

Works published in 1980s and 1990s

Many of the earlier works published in the 1980s and 1990s focus on inter-cultural, communication and linguistic aspects of tourism encounters, including the aforementioned 'International tourism and intercultural communication: the case of Japanese tourists in Singapore' (Riaz, 1975), and 'The language of Japanese tourism' (Moeran, 1983), the latter presenting analysis of promotional language used in travel brochures for both domestic and international tourism. An anthropological interpretation of Japanese domestic tourism is provided by Graburn (1983a), whose discussion on the relationship between tourism and pilgrimage appeared earlier (Graburn, 1983b). Inter- and cross-cultural communication analysis reflects trends in communication and organizational culture theory by scholars such as Hofstede (1980, 1991) and Hall (1976, 1983), especially pointing to the collectivist nature of Japanese society.

Research during the 1990s is notable for its focus on 'international engagement, mostly Japanese tourists overseas' (e.g., Bucklay, Mirza, & Witt, 1989; Chon, Inagaki, & Ohashi, 2000; Morris, 1990) with Hawaii being one of the popular destinations (Lewis, 1996). In cross-cultural behaviour analyses, the specific nature of Japanese tourists is highlighted (e.g., Reisinger & Turner, 1999; Reisinger & Waryszak, 1994). The behaviour and economic benefits of Japanese tourists in specific destinations is discussed in the context of destinations such as Australia, Alaska, Canada, Hawaii, Scotland, Taiwan, Palau and the US (e.g., Agrusa, 2000; Andersen, Prentice, & Watanabe, 2000; Childs, 1991; Lewis, 1996; Marriott, 1991; Milner, Collins, Tachibana, & Hiser, 2000; Reynolds, 2000; Tsai & Wang, 1998; Yamaguchi, Emenheiser,& Yamashita, 2000). Cross-cultural communication and its gap (e.g., Ziff-Levin, 1990) also lead to various studies on 'how to deal with Japanese customers effectively' (e.g., Nishiyama, 1996). Specific types of tourism, such as Asian sex tourism (Leheny, 1995), are related to Japan's work patterns and organizational culture, as noted in reviews of leisure development in Japan (Cole, 1992; Linhart & Frühstück, 1998).

The impacts of international and domestic tourism on Japanese society are recognized, especially in rural areas (e.g., Knight, 1995, 1996), on traditional art (*kyogen* – Kominz, 1988) and social structure (Moon, 1998). In addition, major policies for tourism, such as implementation of the *Comprehensive Regional Resort Preparation Plan, 1987* ('Resort Act') (Rimmer, 1992) and its failure in the time of the 'bubble burst' and the failure of 'Resort Law', are noted (Funck, 1999) as a factor that assigned tourism a specific role of economic revitalization especially in rural areas.

Works published in the 2000s and from 2010 onwards

As noted previously, over 86 percent (180) of the journal articles reviewed here were published during the 2000s, many appearing in just four special issues. Various aspects of Japan's tourism are analysed in these articles, but this review identifies some major themes, as classified below.

- i International engagement, including outbound/inbound tourism, bilateral relations and comparative analysis of tourist behaviour and tourism operation
- ii Japan's social change and tourism, particularly related to diversity and ageing
- iii Heritage tourism, including world heritage and war memorial sites (domestic/international) and dark tourism
- iv Popular culture, including digital contents, technology and food and culinary tourism
- v Sustainability, including green, rural and community-based tourism, as an approach to regional development/revitalization and environmental impacts and management
- vi Issues associated with the 3.11 East Japan earthquake and tsunami

International engagement

A total of 54 articles refer to Japanese travellers going overseas, their behaviour, consumption patterns or activity choices. Within this review, countries featured as a destination or compared with Japan are: China (8 articles), South Korea (6), Taiwan (4), Hong Kong (3), US (3), Hawaii (3), Singapore (2), Canada (2), Guam (2) and one each for Australia, Alaska, Germany, Scotland, UK, Romania and Palau. Among the bilateral engagement studies, those with Korea and China have specific features, including K-pop motivated trips (Kim, Mayasari & Oh, 2013), health and medical tourism (Kim, Lee & Ko, 2016), post-colonial bilateral flows between the two countries (Kim & Prideaux, 2012) or souvenir purchasing behaviour (Park, 2000). Visits by Japanese travellers to historic sites are also examined, including former battle sites in China (Ruoff, 2014), Japan's colonial expansion sites such as Palau (Yamashita, 2000), or Guam and Chichijima (Schumann, 2010).

Tourism to island resorts such as Hawaii and Guam, promoted extensively during the 1980s and still maintaining an image of a 'tropical paradise', is studied through its evolving discourse by Yamaguchi and Yoshihara (2004). Similarly, Guam as a popular wedding venue (Schumann, 2012) and Hawaii as a popular marathon destination (Agrusa & Lema, 2009) are studied for marketing strategy, local economic contribution and destination choice (Martin, 2010). Outbound tourism, or going overseas, has a clear association with 'freedom seeking' and educational meaning as in the study carried out by Middeleer (2016). Bui and Wilkins (2016) study the mobility of young Japanese, providing implications for curricula development as well as future mobility, whilst Japanese retirees seeking lifestyle migration in Malaysia are explored by Ono (2015). Other studies examine factors influencing destination choices, including risk factors such as foot-and-mouth disease (Yamada, 2014).

Soshiroda (2005) provides a comprehensive analysis of Japan's inbound tourism policies from 1853 to 2003 (Soshiroda, 2005), structured around five temporal periods and ten further phases, marked by significant policy introductions such as the establishment of the Welcome Society (1893), the Japan Travel Bureau (1912) and the International Tourism Office (1930), the introduction of Basic Tourism Law (1963), and the Olympic Games (1964) and Expo (1970) policies to improve domestic infrastructure, the 'Discover Japan' campaign promoting domestic tourism, and more recent policies such as Ten Million Plan (1987), Welcome Plan 21 (1996) and Visit Japan Campaign (2003). Arlt (2006) defines domestic tourism in Japan through policies such as 'Discover Japan' and points to sustainability-oriented tourism in rural and agriculture-based tourism.

The government's inbound tourism campaign is evaluated on its success and failure (Handler & Ito, 2008; Uzama, 2012), as are broad marketing strategies in Japan targeting international travellers (Uzama, 2009; Zhang, 2004; Zhang & McCornac, 2014). A rapid increase of inbound tourism, especially a large influx of Asian tourists, is noted (Larke & Hyun 2016), and a large number of articles address Korean tourists visiting Japan, mostly on the elements considered for selecting tourism products (Lee & Hyun, 2016; Ryu, Hyung & Shim, 2015) or visa requirements (Lee, Song & Bendle, 2010).

Foreign visitors' insights in Japan are revealed through storytelling and discourse amongst tourists visiting Tokyo (Martin & Woodside, 2011), whilst tourism impacts on local communities are studied at various destinations, including Takayama (Araya, Naoi & Iijima, 2016), Takayama (Funck, 2012) and Hiroshima (Wu, Funck & Hayashi, 2014). These studies present objective analyses of impacts on society, while personal stories (Guichard-Anguis, 2007, 2011) also exist, along with the anthropological works provided in 1990s by Knight (1995, 1994).

Social change and tourism

Specific kinds of social change noted in this period include an ageing and declining population, diversity, and tourism impacts on traditional life and minority groups. Depopulation, particularly in rural areas, previously noted in Palmer's (1988) study of the planned relocation of a community, and the consequential decline of rural life, traditional festivals and other intangible heritage is observed while in some cases it has been found to stimulate the revival of regional cultures, such as heritage food (Love, 2007), performing arts and festivals (Foster, 2013; Hashimoto, 1998). Research has also found that rural areas are becoming a source of nostalgia, social memory and national identity (Love, 2007). Substantial work has been devoted to age and tourism/leisure, specifically aged travellers' behaviour, travel choices and motivation (Harada, 1999; Hongsranagon, 2006; Ryu, Hyung & Shim, 2015; You & O'Leary, 2000), the impact on insurance (Henderson, 2007) and future forecasting (Mak, Carlile & Dai, 2005). Increased attention is given to social diversity within Japanese society, including ethnic tourism in the Ainu community (Chang, Su & Chang, 2011; Hiwasaki, 2000), the Brazilian community (Maruyama, Woosnan, & Boley, 2016; Woosnan, Maryuama & Boley, 2016) and gender and tourism (Nakata & Momsen, 2010). As part of diversity, Henderson (2016) examined industry responses to Muslim travellers.

Heritage tourism/history

Two kinds of heritage sites are found in the research in this review, one relating to pilgrimage routes, the other associated with war and memorial sites, as well as a broader discussion on heritage, its authenticity and economic benefits (McMorran, 2008; Yasuda, 2010). Pilgrimage, as earlier discussed in Graburn (1983a, 1983b, 1993), laid the foundation for modern tourism and has now become an important tourism activity internationally (Guichard-Anguis & Moon, 2009; Matsui, 2014). Two of the most well-known pilgrimage routes in Japan are studied: the Shikoku pilgrimage (Kubota, 1988), and the Kii mountain range (Guichard-Anguis, 2011: Jimura, 2016). Shikoku has been nominated as Japan Heritage (2015), and the Kii mountain pilgrimage has been World Heritage since 2004. Both are attracting significant numbers of international tourists.

Visits to sites associated with death and tragedy is contentious, provoking ethical debates among visitors, as extensively debated in the field of dark tourism (Stone, 2013, 2012; Stone & Sharpley, 2009). Such debates are found in Broderick (2010) and Selmi, Tur and Dornier (2012) referring to the case of Hiroshima. Contention between education and tourism at such sites are presented in the contexts of Minamata, a former mercury poisoning site (McMorran, 2015), and at Hiroshima and Nagasaki (Yoshida, Bui & Lee, 2016). Hiroshima and Nagasaki as tourism destinations for domestic travellers were analysed through Japanese guidebooks (Siegenthaler, 2002). Another war-related site, Okinawa, was studied by Suzuki, who defines Okinawa as a peace education site primarily for school tours with the local guides as interpreters (Suzuki, 2016, 2012). The term 'activist bubble' is employed to imply the contentious nature of Okinawa as a 'tourist site', and the role of tourism in promoting sense of justice and peace. More recently, a site nominated in the World Industrial Heritage List, Gunkan-jima, has been studied as a contentious tourist site associated with the history of claimed forced labour (Hashimoto & Telfer, 2016). Although not specifically related to tourism, an interesting study of Australian photography of the post-war military occupation has also been published (Gerster, 2015). Interestingly, the journal *Intersection: Gender, History and Culture in the Asian Context* published a Japan-special issue on 'Arts and Media Responses to the traumatic effects of war on Japan' which included an article on Hiroshima as a tourism site (Broderick, 2010).

Popular culture

Research into popular culture as a tourism resource includes publications focusing on so-called 'contents' ('digital media contents', Seaton & Yamamura, 2015) tourism, the use of technology and food/culinary tourism. Contents tourism is most popularly discussed in the context of locations associated with particular Anime, drama or movies (Denison, 2010; Okamoto, 2015; Seaton, 2015; Yamamura, 2015). Terms such as *Rekijo* (young women interested in history) (Sugawa-Shimada, 2015) are added to already commonly used Japanese terms such as *Anime* and *Otaku*. Contents-motivated international trips are also observed in UK (Iwashita, 2006, 2008). The use of particular Apps, camera phones and blogs is also studied (Foster, 2009; Sasaki & Nishii, 2012; Song & Kim, 2016). Food as popular culture also attracts international and cross-cultural attention. Kim writes extensively on various types of noodle culture in Japan (e.g., Kim, 2016; Kim & Iwashita, 2016), including food competition as a means of promotion (Kim, 2015). Food-culture, specifically *ramen*, as one of the now universal foods along with sushi, is discussed through a review of five books (White, 2015) and dining out and gourmet-motivated domestic travel as a popular leisure activity in Japan is also discussed (Tussyadiah, 2006).

Sustainability/conservation

Although the concept of sustainability is widely applicable, covering environmental, social and economic spheres, two broad themes are identified in this review: first, green, rural and community-based tourism and approaches to regional development, and second, tourism's environmental impacts and management. Under the green or rural tourism heading, emphasis is given to community initiatives and empowerment based on social capital, and multi-stakeholder partnerships that can be claimed as a basis for sustainable tourism development. In particular, resident (community) initiatives and empowerment are the focus of much of the research, its significance being highlighted by Boley, Maruyama and Woosnam (2015: 112) for 'its far-reaching intrinsic and extrinsic implications to not only resident attitudes towards tourism, community well-being, and international development goals, but also to destination competitiveness through including residents in marketing initiatives aimed at highlighting the destination from a local perspective'. Various case studies, from Kyushu to Hokkaido (Hashimoto & Telfer, 2010; Miyakuni, 2006; Shikida, Yoda, Kino & Morishige, 2010), also explore approaches to conservation (Hiwasaki, 2006) and the revival of rural areas (Chen & Qui, 2013).

Issues relating to tourism's environmental and social impacts are also discussed. Gielen, Kurihara and Moriguchi (2002), for example, measured tourism's environmental impacts in Japan, reporting that tourism is responsible for 17 percent of national greenhouse gas emissions and 13 percent of the national primary energy use. Japan's conservation approach in the context of tourism is discussed by scholars such as Jones (2012a, 2012b), focusing on nature-based tourism policy and development and Knight (2000, 2010, 2011). An interesting insight is given through the example of conflicts between fishing rights and marine leisure (Funck, 2006), or volcanic threats in Japan (Jones, 2016). Within the journals listed here, no article deals specifically with the relationship between climate change and tourism. One article published in a Japanese journal (*Journal of the Meteorological Society of Japan*), however, examines the correlation between tourist numbers and climate change using a Tourism Climate Index (Kubokawa et al., 2014). Perhaps the most comprehensive overview of the Japan's environmental sustainability, albeit specific to rural and regional areas, can be found in Assmann's (2016) *Sustainability in Contemporary Rural Japan. Challenges and Opportunities*. Matsui's (2014) *Geography of Religion in Japan* provides a good background to the historic and cultural context.

Disaster and reconstruction – issues associated with the 3.11 East Japan earthquake and tsunami

The Great East Japan Earthquake and Tsunami that occurred on 11 March 2011 clearly had a major effect on Japanese society. Research identified in this review address related topics including perceptions of risk and safety (Chew & Jahari, 2014; Handler, 2016); the event's socioeconomic and environmental impacts (Bachev, 2014); decision-making on travel (Chen, Chen & Cheng, 2012; Wu & Walters, 2016); the role of tourism and leisure in post-disaster recovery (Kono & Shinew, 2015; Leelawat, Suppasri & Imamura, 2015; Muscat, Nakanishi, & Blackman 2015); and, the consequences of debris reaching other parts of the world (Ghaderi & Henderson, 2013). Henderson (2013) addresses the multi-dimensionality of large-scale crises such as the Great East Japan Earthquake and points to society's vulnerability as well as resilience, asserting the importance of planning and collaboration based on past experiences. Korstanje and Tarlow (2013), based on a study of one diaspora community in Buenos Aires, discuss nationalism and consumerism in relation to the disaster in Fukushima.

Discussion

This chapter has identified the major trends in research on Japan's tourism based on a review of publications written in Japanese and English. It reveals that tourism research on Japan has increased significantly in both volume and the range of topics addressed, in particular since the early 2000s. Although the search was limited to the keyword, *tourism*, the term itself contains a diverse range of meanings, some specifically applied to Japan's social phenomena such as rural decline and ageing, implementation of governmental policies as well as socioeconomic and political trends at the domestic, regional and global levels.

Despite this apparent limitation in the data used, some major trends emerge from the review, both within and comparatively across the two language groups of publications. Three points can be emphasized here, reflecting to the two aims specified at the outset of this chapter. First, the term *kankō* today has a wide range of meanings, but frequently refers to the specific role of tourism in region/local destination development as evidenced in the frequently used terminology in Japanese articles. Machizukuri is perhaps one of the most prominent concepts in the Japanese approach to tourism as community and regional development or revitalization, becoming more popular particularly since the rural decline became obvious in the 2000s. The term may be most appropriately translated as 'sustainable community development' with a focus on social capital, community unity and resilience based on regional knowledge, wisdom and sense of place.

The second is the recent rise of new tourism, marked with the use of term tsūrizumu, as in *contents tourism, green tourism* or *ecotourism*, being incorporated into the Japanese language. Although these concepts are global, they are still new as evident in this review, and the continuing localization process into Japan's social and cultural context will make interesting contributions to the field. With a series of international sport mega-events scheduled in Japan from 2019 and beyond, *sport tourism* in particular is expected to gain popularity as tourism activity and as a research topic, as reviewed by Hinch and Ito in this book (see Chapter 5). Related to this, the third point is that it is highly likely that a large volume and variety of work related to tourism in Japan exists in other disciplines, such as sociology, anthropology, geography, environmental studies and town planning as well as economics and business. Although such a review is beyond the scope of this chapter, this indicates diverse approaches and opportunities exist for tourism research in Japan to cross-fertilize and grow, incorporating knowledge from other disciplines and language sources.

Notes

1 A previous version of this chapter appears in *Tourism Planning & Development*, 15(1), 3–25.

2 Although the term *tsūrizumu* is used often interchangeably with *kankō* in Japanese, and the results included some publications employing this term, this review used only *kankō* as the search word.
3 NDL catalogue retrieved from http://iss.ndl.go.jp/ (9 August 2016).
4 Affiliation does not necessarily indicate the author's home country or nationality; in many cases, the affiliation was during a study or fellowship period.

References

Japanese publications

Asamizu, M. (1999) *Ōsutoraria no kankō to syokubunka* [*Tourism and Food Culture in Australia*]. Tokyo: Gakubunsha.

Asato, E. (1991) *Yureru seiki:rizōto kaihatsu to shima no kurashi* [Wobble siècle: Resort development and life in Okinawa islands]. Okinawa: Okinawa Times.

Bormann, A. (1930) *Die Lehre vom Fremdenverkehr* [The teaching of Tourism]. Berli n: Verkehrswissenschaftliche Lehrmittelgesellschaft bei d. Deutschen Reichsbahn.

Choi, K. (2009) A content analysis of subject areas and research methods used in Journal of JITR, The Tourism Studies, 21(1), 27–34. Retrieved from: http://ci.nii.ac.jp/naid/110009844513

Dohi, T. (1991) *Rizōto saisei to chiiki shinko* [*Regeneration of Resort and Regional Development]*. Kyoto: Gakugei Publishing.

Fukunaga, A. and Suzuki, Y. (Eds.) (1996) *Hosupitarithi sangyouron: kokyaku manzoku kanri no jidai wo koete* [*Introduction to the Hospitality Industry: Beyond the Management of Customer Satisfaction*]. Tokyo: Chūoukeizaisha.

Futaba, K. and Naruse, T. (Eds.) (1992). *Rizōto kara naihatsuteki chiikidukuri he: tango rizōto de towareteiru koto* [*From Resort to Spontaneous Regional Development: Disputed Points in Tango Resort*]. Kyoto: Kyoto jichitaimondai kenkyusho.

Glücksmann, A. (1935) *Allgemeine Fremdenverkehrskunde* [General Tourism]. Berlin: Stämpfli.

Hase, M. (1996) *Kankō mākethingu: riron to jissen* [*Tourism Marketing: Theory and Practice*]. Tokyo: Doubunkan.

Hase, M. (2003) *Atarashii kankō shinkou: hassou to senryaku* [*New Tourism Promotion: Ideas and Strategy*]. Tokyo: Dōbunkan.

Hashimoto, K. (1999) *Kankō jinruikaku no senryaku:bunka no urikata urarekata* [*Strategy of Tourism Anthropology: Selling and Being Sold Culture*]. Kyoto: Sekaishisosha.

Hashimoto, K. and Sato, Y. (Eds.) (2003) *Kankō kaihatsu to bunka: minami karano toikake* [*Tourism Development and Culture: Questions from the South*]. Tokyo: Kokonshoin.

Hattori, K. (1994) *Shin gainen toshiteno hosupitarithi manejimento* [*Hospitality Management as a New Concept*]. Tokyo: Maruzen.

Hikima, S. (1964) *Ryokan no keiei to sabisu no kaizen narabini sekkyaku jyugyoin no taigu kaizen nitsuite* [Service improvement of hotel and remuneration raising of the service-section employee especially those of the chief c1ass.]. *Journal of Tourism Research*, 3, 51–54.

Iguchi, M. (2002) *Kankō bunka no shinko to chiiki shakai* [*Promotion of Tourism Culture and Local Community*]. Tokyo: Minerva.

Iguchi, M. (2005) *Machizukuri/kankō to chiiki bunka no souzō* [*Community Development / Tourism and the Creation of Regional Culture*]. Tokyo: Gakubunsha.
Imai, M. and Yamauchi, Y. (1999) *Taishū kankō no seitaishi: nihon to igirisu* [*Ecological History of Mass Tourism*]. Hiroshima: Keisuisya.
Inagaki, T. (1983) *Hoteru toushi kettei nitaisuru hedonikku apurōchi no tekiyousei* [Applicability of hedonic approach to the investment decisions in hotel industry: A normative extension of the hedonic approach]. *Journal of Tourism Research*, 13, 2–9.
Ishihara, T., Yasufuku, E. and Yoshikane, H. (Eds) (2000) *Atarashii kankō to chiikishakai* [*New Tourism and the Local Community*]. Tokyo: Kokonshoin.
Ishimori, S. (1996) *Kankō no 20 seiki* [*Twentieth Century Tourism*]. Tokyo: Domesu.
Kim, L. (2004) Nihon no hoterugyo niokeru koyou no jyunannsei ni kansuru kenkyū [Research on flexibility of employment in hotel industry in Japan]. *The Tourism Studies*, 15(2), 9–18.
Kim, J. (2005) Keiei senryaku teki shiten kara mita syukuhakutokkagata hoteru no tokuchō; pojishoningu apurōchi to sigen bēsu apurōchi wo chūshin ni [A Study on the features of 'shukuhakutokkagata hotels' from the business strategy theoretical perspective: Based on positioning approach and resource based approach]. *The Tourism Studies*, 17(1), 9–18.
Kitagawa, M. (1998) *Kankō to Shakai: tūrizumu heno michi* [*Tourism and Society: A Pathway to Tourism*]. Hikone: Sunrise.
Kitagawa, M. (2003) *Kankō kōryū shinjidai* [*New Ear of Tourism Intercommunication*]. Hikone: Sunrise.
Kitagawa, M. (2004) *Kankō bunkaron* [*Tourism Culture*]. Tokyo: Minerva.
Koike, Y. and Ashiba, H. (Eds) (1988) *Kankō gaku gairon* [Introduction to Tourism]. Kyoto: Minerva.
Kuji, T. (1993) *Hawai rizōto senso: nikkei kigyo no rankaihatu to jyumin no teiko* [Resort War in Hawaii: Reckless Exploitation by Japanese-affiliated Companies and Residents' Resistance]. Tokyo: Marujyusha.
Maeda, I. (Ed.) (1973) *Kankō gairon* [*Introduction to Tourism*].Tokyo: Gakubunsha.
Maeda, I. (1987) A methodological consideration for tourism study: From view point of the behavioural sciences. *The Tourism Studies Quarterly*, 1(1/2), 6–13.
Mariotti, A. (1927) *Lezioni di economia turistica: Per gli studenti della facolta di scienze politiche della R. Universita di Roma* [Lessons of tourist economy: For students of the Faculty of Political Science, Rome University]. Rome: Edizioni Tiber.
Maruta, Y. (1989) *Rizōto kaihatsu keikaku ron: Chiiki keisei to rizōto kompurekkusu* [*Planning for Resort Development: Formation of Regional and Resort Complexes*]. Tokyo: Soft-science.
Matsui, Y. (1993) *Ajia no kankō kaihatsu to nihon* [Japan and Tourism development in Asia]. Tokyo: Shinkansha.
Mizoo, Y. (1994) *Kankō wo yomu: chiikishinko heno teigen* [*Understanding Tourism: Suggestions for Regional Development*]. Tokyo: Kokonshoin.
Mizoo, Y. (2007) *Kankō machizukuri genba karano hōkoku:niiharu mura/sado shi/ kotohira chō/Kawagoe shi* [Report on the spot of tourism and community development]. Tokyo: Hara Shobo.
Nabeta, K. (1989) *Rizōto shin jidai no toushi senryaku* [*Investment Strategy in the New Era of Resorts*]. Tokyo: Sannō Institute of Management Publication Department.
Naitou, Y. (1996) *Kankō to gendai* [*Tourism and the Modern Era*]. Tokyo: Kindaibungeisha.

Naitou, Y. (1998) *Kankō to ajia* [*Tourism and Asia*]. Tokyo: Gakubunsha.

Nakako, F. (2010) Chiiki kankō no hatten ni okeru bariafurī tūrizumu no yakuwari: sōsyaru kyapitaru no kōchiku to shimin katudō [The role of barrier-free tourism in the development of community based tourism: Establishment of social capital and civil activities]. *The Tourism Studies*, 21(2), 41–46.

Nishimura, Y. (Ed.) (2009) *Kankō machizukuri: machi jiman kara hajimaru chiiki manejimento* [*Tourism and Community Development: Local Management Stemmed from Town Pride*]. Kyoto: Gakugei Publishing.

Nishioka, H. (1996) *Kankō to chiikikaihatsu* [*Tourism and Regional Development*]. Tokyo: Naigai Publishing.

Ohno, T., Nakayama, K. and Sasaki. K. (1991) *Rizōto kaihatu wo tou* [*Question the Development of Resort*]. Tokyo: Shinihon Shuppansha

Sasaki, H. (1998) *Kankō to chiiki* [*Tourism and Region*]. Tokyo: Ninomiya shoten.

Sato, M. (1990) *Rizōto rettou* [*Resort Archipelago*]. Tokyo: Iwanami Shoten.

Shiota, M. (1974a) *Kankō no seiritu to tenkai* [Establishment and development of tourism study]. In T. Suzuki (Ed.), *Gendai kankō ron* [*Modern Tourism*]. Tokyo: Yūhikaku, Ch. 2, pp.15–29.

Shiota, M. (1974b) *Kankōron no seikaku to taikei.* [Features and systems of tourism theory] In T. Suzuki (Ed.), *Gendai kankō ron* [*Modern Tourism*]. Tokyo: Yūhikaku, Ch. 17, pp. 265–280.

Suda, H. (2006) *Atarashii kankō: sangyō kankō, kaido kankō and toshi kankō* [*New Tourism: Industry Tourism, Old Way Tourism and City Tourism*]. Tokyo: Kotsushimbunsha.

Suzuki, T. (Ed.) (1974) *Gendai kankō ron* [Modern Tourism]. Tokyo: Yūhikaku.

Tabira, A. and Asoh, K. (2007) *Gūrin tūrizumu niyoru chiikidukuri: Oita ken Ajimu cyo wo jirei toshite* [Green tourism and regional promotion: A case study of Ajimu town in Oita]. *Journal of Tourism Research*, 48, 9–20.

Tanabe, M. (1990) *Rizōto jigyou yoken: jitsugen heno sohuto burēku* surū [*Foreseeing Resort Projects: Soft-Breakthrough for Implementation]*. Tokyo: Shibata Shoten.

Yahagi, H. (2004) *Sangyō isan to machizukuri* [*Industrial Heritage and Community Development*]. Kyoto: Gakugei Publishing.

Yamagami, T. (1997) *Kokusai kankō mākethingu* [*International Tourism Marketing*]. Tokyo: Hakuto-shobo.

Yamanaka, H. (1992) *Imēji no rakuen:Kankō hawai no bunkashi* [*Image of Paradise: Cultural History of Tourism in Hawaii*]. Tokyo: Chikumashobo.

Yamashita, S. (1996) *Kankō jinruigaku* [*Tourism Anthropology*]. Tokyo: Shinyō-sha.

Yamauchi, Y. (Ed.) (1997) *Posuto modan no shōhi to kankō: kankō no jidai* [*Consumption and Tourism in the Post-Modern Era of Tourism*]. Hiroshima: Keisuisya.

Yamazaki, T. and Soshiroda, A. (2010) Monogatari tema no enyo niyoru kankō machizukuri: Matsuyama si wo jirei toshite [Tourism and city planning implemented by referring to the theme of narrative: A case study of Matsuyama city]. *The Tourism Studies*, 22(1), 37–43.

Yasumura, K. (2006) *Kankō machizukuri no rikigaku* [*Dynamics of Community Development*]. Tokyo: Gakubunsha.

Yoshida, H. (2010) *Atarashii kankō no jidai: kankō seisaku, onsen, nyū tūrizumu zensō [New Tourism Era: Tourism Policy, Hot Springs and the Mirage of New Tourism*]. Tokyo: Hara shobo.

English publications

Agrusa, R. (2000) Legalisation of gambling in Hawaii and its potential effects on Japanese intention to visit. A philosophical inquiry. *Journal of Travel & Tourism Marketing*, 9(1/2), 211–217.

Agrusa, J. and Lema, J. (2009) The Japanese influence on Hawaii's largest event, 2008 Honolulu Marathon. *Journal of Tourism Challenges and Trends*, 2(2), 21–28.

Andersen, V., Prentice, R. and Watanabe, K. (2000) Journey for experiences: Japanese independent travellers in Scotland. *Journal of Travel & Tourism Marketing*, 9(1/2), 129–151.

Araya , A., Naoi, T. and Iijima, S. (2016) The effects of Westerners' presence on Japanese evaluation of historic district: The case of Takayama City. *Journal of Travel & Tourism Marketing*, 33(5), 628–644.

Arlt, W. G. (2006) Thinking through tourism in Japan. *Tourism and Hospitality Planning & Development*, 3(3), 199–207.

Assmann, S. (Ed.) (2016) *Sustainability in Contemporary Rural Japan. Challenges and Opportunities*. New York: Routledge.

Bachev, H. (2014) Socio-economic and environmental impacts of March 2011 earthquake, tsunami and Fukushima nuclear accident in Japan. *Journal of Environmental Management and Tourism*, (10), 127–222.

Berger, A. (2010) *Tourism in Japan: An Ethno-semiotic Analysis*. New York: Channel View Publications.

Boley, B., Maruyama, N. and Woosnam, K. (2015) Measuring empowerment in an eastern context: Findings from Japan. *Tourism Management*, 50, 112–122.

Broderick, M. (2010) Topographies of trauma: Dark tourism, World Heritage and Hiroshima, *Intersections: Gender and Sexuality in Asia and the Pacific*, Issue 24. Retrieved from: http://intersections.anu.edu.au/issue24/broderick.htm

Bucklay, P., Mirza, H. and Witt, S. (1989) Japan's international tourism in the context of its international economic relations. *The Service Industries Journal*, 9(3), 357–383.

Bui, H. and Wilkins, H. (2016) The mobility of young Japanese: The travel-migration nexus. *Journal of Travel & Tourism Marketing*, 33(5), 581–596.

Chang, J., Su, W. and Chang, C. (2011) The creative destruction of Ainu community in Hokkaido, Japan. *Asia Pacific Journal of Tourism Research*, 16(5), 505–516.

Chen, B. and Qui, Z. (2013) Green tourism in Japan. Opportunities for a GIAHS Pilot Site. *Journal of Resources and Ecology*, 4(3), 285–292.

Chen, F., Chen, M. and Cheng, C. (2012) A study of the students' travel Japan intentions from Departments of Applied Japanese in Taiwan after 3/11 East Japan earthquake. *Journal of Information & Optimization Sciences*, 33(2/3), 363–384.

Chew, E. and Jahari, S. (2014) Destination image as a mediator between perceived risks and revisit intention: A case of post-disaster Japan. *Tourism Management*, 40, 382–393.

Childs, I. (1991) Interviewing Japanese tourists in Australia. *Japanese Studies*, 11(1), 65–70.

Chon, K. S., Inagaki, T. and Ohashi, T. (Eds) (2000) *Japanese Tourists: Socio-economic Marketing and Psychological Analysis.* New York: Routledge.

Cole, R. (1992) Work and leisure in Japan. *California Management Review*, 34(3), 52–63.

Denison, R. (2010) Anime tourism: Discursive construction and reception of the Studio Ghibli Museum. *Japan Forum*, 22(3–4), 545–563.

Foster, M. (2009) What time is this picture? Cameraphones, tourism and the digital gaze in Japan. *Social Identities*, 15(3), 351–372.

Foster, M. (2013) Inviting the uninvited guest: Ritual, festival, tourism and the Namahage. *Journal of American Folklore*, 126(501), 302–334.

Fuess, S. (2012) Leisure in Japan. 1988–2006. A revival? *Journal of Family and Economic Issues*, 33(2), 250–260.

Funck, C. (1999) When the bubble burst: Planning and reality in Japan's resort industry. *Current Issues in Tourism*, 2(4), 333–353.

Funck, C. (2006) Conflicts over space for marine leisure: A case study of recreational boating in Japan. *Current Issues in Tourism*, 9(4&5), 459–480.

Funck, C. (2012) The innovative potential of inbound tourism in Japan for destination development. A case of Hida Takayama. *Journal of German Institute of Japanese Studies*, 2(2), 121–147.

Funck, C. and Cooper, M. (Eds) (2013) *Japanese Tourism: Spaces, Places and Structure*. New York: Berghahn.

Gerster, R. (2015) Capturing Japan: Australian photography of the post-war military occupation. *History of Photography*, 39(3), 279–299.

Ghaderi, Z. and Henderson, J. (2013) Japanese tsunami debris and the threat to sustainable tourism in the Hawaiian Islands. *Tourism Management Perspectives*, 8, 98–105.

Gielen, D., Kurihara, R. and Moriguchi, Y. (2002) The environmental impacts of Japanese tourism and leisure. *Journal of Environmental Assessment Policy and Management*, 4(4), 397–424.

Graburn, N. (1983a) *To Pray, Pay and Play. The Cultural Structure of Japanese Domestic Tourism*. Aix en Provence: Centre des Hautes Études Touristiques.

Graburn, N. (1983b) Anthropology of tourism. *Annals of Tourism Research*, 10(1), 9–33.

Graburn, N. (1993) Tourism in cross-cultural perspective. *Annals of Tourism Research*, 20(2), 367–368.

Guichard-Anguis, S. (2007) Japanese Inns (Ryokan) in Atami (Japan) and the Shaping of Coming 'Traditions'. *Tourism Review International*, 11(1), 19–31.

Guichard-Anguis, S. (2011) Walking through the world heritage forest in Japan. The Kumano Pilgrimage. *Journal of Heritage Tourism*, 6(4), 285–295.

Guichard-Anguis, S. and Moon, O. (2009) *Japanese Tourism and Travel Culture*. New York: Routledge.

Hall, E. (1976) *Beyond Culture*. New York: Doubleday.

Hall, E. (1983) *The Dance of Live. The Other Dimension of Time.* New York: Doubleday.

Handler, I. (2016) The impact of the Fukushima disaster on Japan's travel image. An exploratory study of Taiwanese travellers. *Journal of Hospitality and Tourism Management*, 27, 12–17.

Handler, I. and Ito, N. (2008) Missed connection. Japan inbound tourism campaign. *Econtros Cientificos: Tourism & Management Studies*, 4, 50–60.

Harada, M. (1999) Ageing and leisure in Japan. *World Leisure & Recreation*, 41(3), 30–32.

Hashimoto, A. and Telfer, D. J. (2010). Developing sustainable partnerships in rural tourism: The case of Oita, Japan. *Journal of Policy Research in Tourism, Leisure and Events*, 2(2), 165–183.

Hashimoto, A. and Telfer, D. J. (2016) Transformation of Gunkanjima (Battleship Island): From a coalmine island to a modern industrial heritage tourism site in Japan. *Journal of Heritage Tourism,* 12(2), 107–124.

Hashimoto, H. (1998) Recreating and reimagining folk performing arts in contemporary Japan. *Journal of Folklore Research*, 35(1), 35–46.

Havens, T. (2011) *Parkscapes: Green spaces in Modern Japan*. Honolulu: University of Hawaii Press.

Henderson, J. (2007) Population ageing, tourism and travel insurance. *Tourism Recreation Research*, 32(3), 79–82.

Henderson, J. (2013) The Great East Japan Earthquake and tourism. A preliminary case study. *Tourism Recreation Research*, 38(1), 93–98.

Henderson, J. (2016) Muslin travellers. Tourism industry's response and the case of Japan. *Tourism Recreation Research*, 41(3), 339–347.

Hiwasaki, L. (2000) Ethnic tourism in Hokkaido and shaping of Ainu Identity. *Pacific Affairs,* 73(3), 393–412.

Hiwasaki, L. (2006) Community-based tourism: A pathway to sustainability for Japan's protected areas. *Society & Nature Resources*, 19(8), 675–692.

Hofstede, G. (1980) *Culture's Consequences: International Differences in Work-related Values.* Beverly Hills, CA: Sage.

Hofstede, G. (1991) *Cultures and Organizations: Software of the Mind.* London: McGraw-Hill.

Hongsranagon, P. (2006) Needs of, and readiness of respond to the needs of Japanese pensioners in long-stay tourism in Chiangmai Province, Thailand. *Tourism Economics*, 12(3), 475–485.

Iwashita, C. (2006) Media representation of UK as a destination for Japanese tourists: Social constructionism and tourism. *Tourism and Hospitality Research*, 4(4), 331–340.

Iwashita, C. (2008) Role of films and television dramas in international tourism: The case of Japanese tourists to the UK. *Journal of Travel & Tourism Marketing*, 24(2–3), 139–151.

Jimura, T. (2016) World heritage site management: A case study of sacred sites and pilgrimage routes in the Kii mountain range, Japan. *Journal of Heritage Tourism*, 11(4), 382–394.

Jones, T. (2012a) A life cycle analysis of nature-based tourism policy in Japan. *Journal of the German Institute for Japanese Studies Tokyo*, 24(2), 179–211.

Jones, T. (2012b) Changing demographics in Japan's national parks: Towards targeted marketing strategy for nature-based tourists. *Tourism and Hospitality Management*, 18(1), 95–109.

Jones, T. (2016).Hokkaido's overlapping protected areas the regional revitalisation: The case of Toya-Usu geopark in Shikotsu-Toya national park. In S. Assmann (Ed.), *Sustainability in Contemporary Rural Japan. Challenges and Opportunities*. New York: Routledge, Chapter 11, 151–166.

Jones, T., Nagata, S., Nakajima, M. and Masuyama, K. (2009) Prefectural branding in Japan: Tourism, national parks and Shinshu brand. *Place Branding and Public Diplomacy*, 5(3), 192–201.

Kim, A., Mayasari , F. and Oh, I. (2013) When tourist audiences encounter each other: Diverging learning behaviours of K-pop fans from Japan and Indonesia. *Korea Journal*, 53(4), 59–82.

Kim, H., Lee, T. and Ko, T. (2016) Satisfaction and subjective well-being of health tourists: The case of Japanese and Korean tourists. *Journal of Travel & Tourism Marketing*, 33(5), 742–756.

Kim, S. (2015) Understanding the historical and geographical contexts of food festival tourism development: The case of the Tatebayashi Noodle Grand Prix in Japan. *Tourism Planning & Development*, 12(4), 433–446.

Kim, S. (2016) Regional development and Japanese obsession with noodles. The Udon noodle tourism phenomenon in Japan. In C.M. Hall and S. Gossling (Eds), *Food Tourism and Regional Development*. New York: Routledge, pp. 135–144.

Kim, S. and Iwashita, C. (2016). Cooking identity and food tourism: The case of Japanese udon noodles. *Tourism Recreation Research*, 41(1), 89–100.

Kim, S. and Prideaux, B. (2012) A post-colonial analysis of bilateral tourism flows: The case of Korea and Japan. *International Journal of Tourism Research*, 14, 586–600.

Knight, J. (1994) Town-making in rural Japan: An example from Wakayama. *Journal of Rural Studies*, 10(3), 249–261.

Knight, J. (1995) Tourist as a stranger? Explaining tourism in rural Japan. *Social Anthropology*, 3(3), 219–234.

Knight, J. (1996) Competing hospitalities in Japanese rural tourism. *Annals of Tourism Research*, 23(1), 165–180.

Knight, J. (2000) From timber to tourism. Recommoditizing the Japanese forest. *Development and Change,* 31, 341–359.

Knight, J. (2010) The ready-to-view wild monkey. The conservation principle in Japanese wildlife tourism. *Annals of Tourism Research*, 37(3), 744–762.

Knight, J. (2011) *Hearing Monkeys to Paradise: How Macaque Troops Are Managed for Tourism in Japan*. BRIL. ebook.

Kominz, L. (1988) The impact of tourism on Japanese Kyogen: Two case studies. *Asian Folklore Studies*, 47(2), 195–213.

Kono, S. and Shinew, K. (2015) Roles of leisure in the post-disaster psychological recovery after the Great East Japan Earthquake and Tsunami. *Leisure Sciences,* 37(1), 1–19.

Korstanje, M. and Tarlow, P. (2013) Disaster, tourism and mobility. The case of Japan earthquake. *PASOA: Revista de Turismi y Patrimonio Cultural,* 11(3), 17–32.

Kubokawa, G., Inoue, T. and Satoh, M. (2014) Evaluation of the Tourism Climate Index over Japan in a future climate using a statistical downscaling method. *Journal of the Meteorological Society of Japan,* 92(1), 37–54.

Kubota, N. (1988) Shikoku pilgrimage. *Japan Quarterly*, 35(2), 171–174.

Larke, R., Kilgour, M. and John, S. (2016) Tourist growth and the implications for retail marketing strategy: Insights from Japan. *Journal of Travel & Tourism Marketing*, 33(5), 658–670.

Lee, C., Song, H. and Bendle, L. (2010) The impact of visa-free entry on outbound tourism. A case study of South Korean travellers visiting Japan. *Tourism Geographies*, 12(2), 302–323.

Lee, K. and Hyun, S. (2016) The effects of perceived destination ability and destination brand love on tourists' loyalty to post-disaster tourism destinations: The case of Korean tourists to Japan. *Journal of Travel & Tourism Marketing*, 33(5), 613–627.

Leelawat, N., Suppasri, A. and Imamura, F. (2015) Disaster recovery and reconstruction following the 2011 Great East Japan Earthquake and Tsunami: A business process management perspective, *International Journal of Disaster Risk Science*, 6(3), 310–314.

Leheny, D. (1995) A political economy of Asian sex tourism. *Annals of Tourism Research*, 22(2), 367–384.

Leheny, D. (2000) By other means. Tourism and leisure as politics in pre-war Japan. *Social Science Japan Journal*, 3(2), 171–186.

Leheny, D. (2003) *The Rules of Play: National Identity and the Shaping of Japanese Leisure.* Ithaca: Cornell University Press.

Leung, R., Au, N. and Law, R. (2015) The recent Asian wave in tourism research: The case of the Journal of Travel & Tourism Marketing. Asia Pacific Journal of Tourism Research, 20(1), 1–28.

Lewis, G. (1996) Beyond the reef: Cultural construction of Hawaii in mainland America, Australia and Japan. *Journal of Popular Culture*, 30(2), 123–135.

Linhart, S. and Frühstück, S. (1998) *The Culture of Japan as Seen Through Its Leisure*. New York: State University of New York.

Love, B. (2007) Fraught fieldsites: Studying community decline and heritage food revival in rural Japan. *Critical Asian Studies*, 39(4), 541–559.

Mak, J., Carlile, L. and Dai, S. (2005) Impact of population aging on Japanese international travel to 2025. *Journal of Travel Research*, 44(2), 151–162.

Marriott, H. (1991) A selected survey of research on Japanese tourism. *Japanese Studies*, 11(1), 70–83.

Martin, D. (2010) Uncovering unconscious memories and myths for understanding international tourism behavior. *Journal of Business Research*, 63, 372–383.

Martin, D. and Woodside, A. (2011) Storytelling research on international visitors: Interpreting own experiences in Tokyo. *Qualitative Market Research: An International Journal*, 14(1), 27–54.

Maruyama, N., Woosnan, K. and Boley, B. (2016) Comparing levels of resident empowerment among two culturally diverse resident populations in Oizumi, Gunma, Japan. *Journal of Sustainable Tourism,* 1–18.

Matsui, K. (2014) *Geography of Religion in Japan. Religious Space, Landscape and Behavior*. New York: Springer.

McMorran, C. (2008) Understanding the heritage in heritage tourism: Ideological tool or economic tool for a Japanese hot spring resort? *Tourism Geographies*, 10(3), 334–354.

McMorran, C. (2015) Between fan pilgrimage and dark tourism: Agendas in overseas field learning. *Journal of Geography in Higher Education*, 39(4), 568–583.

Middeleer, S. (2016) I want to go far away: Discover Japan and Japanese identity tourism in the 1970s. *Studies on Asia,* Series 1(1), 96–121.

Milner, L., Collins, J., Tachibana, R. ans Hiser, R. (2000) The Japanese vacation visitor to Alaska: A preliminary examination of peak and offseason traveller demographics, information source utilisation, trip planning and customer satisfaction. *Journal of Travel & Tourism Marketing*, 9(1/2), 43–56.

Miyakuni, K. (2006) Linking linkage concepts from diverse fields to build a community-based tourism planning framework: The case of Shuri, Japan. *Tourism Geographies*, 8(3), 286–309.

Moeran, B. (1983) The language of Japanese tourism. *Annals of Tourism Research*, 10(1), 93–108.

Moon, O. (1998) Is the *ie* disappearing in rural Japan? The impact of tourism on a traditional Japanese village. In J. Henry (Ed.), *Interpreting Japanese Society: Anthropological Approaches.* London: Routledge, pp. 117–130.

Morris, S. (1990) *The Japanese Overseas Travel Market in the 1990s*. London: Economist Intelligence Unit.

Muscat, B., Nakanishi, H. and Blackman, D. (2015) Integrating tourism into disaster recovery management: The case of the Great East Japan Earthquake and Tsunami in 2011. In B. Ritchie and K. Compiranon (Eds), *Tourism Crisis and Disaster Management in the Asia-Pacific*. Wallingford: CABI, pp. 97–115.

Nakata, M. and Momsen, J. (2010) Gender and tourism. Gender, age and mountain tourism in Japan. *Geografia: Malaysian Journal of Society and Space*, 6(2), 63–71.

Nishiyama, K. (1996) *Welcoming the Japanese Visitors: Insights, Tips, Tactics*. Honolulu: University of Hawaii Press.

Okamoto, T. (2015) Otaku tourism and anime pilgrimage phenomena in Japan. *Japan Forum,* 27(1), 12–36.

Ono, M. (2015) Commoditization of lifestyle migration: Japanese retirees in Malaysia. *Mobilities*, 10(4), 609–629.

Palmer, E. (1988) Planned relocation of severely depopulated rural settlements: A case study from Japan. *Journal of Rural Studies*, 4(1), 21–34.

Park, M. (2000) Social and cultural factors influencing tourists' souvenir purchasing behaviour: A comparative study on Japanese Omiyage and Korean Sunmul. *Journal of Travel & Tourism Marketing*, 9(1/2), 81–91.

Reisinger, Y. and Turner, L. (1999) A cultural analysis of Japanese tourists: Challenges for tourism marketers. *European Journal of Marketing*, 33(11/12), 1203–1227.

Reisinger, Y. and Waryszak, R. (1994) Tourists' perceptions of service in shops: Japanese tourists in Australia. *International Journal of Retail & Distribution Management*, 22(5), 20–28.

Riaz, H. (1975) International tourism and intercultural communication: The case of Japanese tourists in Singapore. *Southeast Asian Journal of Social Science*, 3(2), 25–38

Rimmer, P. (1992) Japan's resort archipelago: Creating regions of fun, pleasure, relaxation and recreation. *Environmental Planning A*, 24(1), 1599–1625.

Ruoff, K. (2014) Japanese tourism to Mukden, Nanjing and Qufu, 1938–1943. *Japan Review*, 27, 171–200.

Ryu, E., Hyung, S. and Shim, C. (2015) Creating new relationships through tourism: A qualitative analysis of tourist motivations of older individuals in Japan. *Journal of Travel & Tourism Marketing*, 32(4), 325–338.

Sasaki, K. and Nishii, K. (2012) Study of blog mining for examination of tourist travel behaviour in Japan. *Transport Research Record*, 2285, 119–125.

Schumann, F. (2010) Historical ties and cultural connections between Guam and Chichijima: Implications for tourism. *Journal of International Business Research*, 9(2), 31–43.

Schumann, F. (2012) The online marketing of Guams overseas weddings in Japan. *International Journal of Hospitality and Tourism Systems*, 5(2), 49–58.

Seaton, P. (2015) Taiga dramas and tourism: Historical contents as sustainable tourist resources. *Japan Forum*, 27(1), 82–103.

Seaton, P. and Yamamura, T. (2015) Japanese popular culture and contents tourism: Introduction. *Japan Forum*, 27(1), 1–11.

Selmi, N, Tur, C. and Dornier, R. (2012) To what extent may sites of death be tourism destinations? The cases of Hiroshima in Japan and Struthof in France. *Asian Business & Management*, 11(3), 311–328.

Shikida, A., Yoda, M., Kino, A. and Morishige, M. (2010) Tourism relationship model and intermediary for sustainable tourism management: Case study of the Kiritappu

Wetland Trust in Hamanaka, Hokkaido. *Tourism and Hospitality Research*, 10(2), 105–115.

Siegenthaler, P. (2002) Hiroshima and Nagasakin in Japanese guidebooks. *Annals of Tourism Research*, 29(4), 1111–1137.

Song, S. and Kim, D. (2016) A pictorial analysis of destination images on Pinterest: The case of Tokyo, Kyoto and Osaka, Japan. *Journal of Travel & Tourism Marketing*, 33(5), 687–701.

Soshiroda, A. (2005) Inbound tourism policies in Japan from 1853 to 2003. *Annals of Tourism Research*, 32(4), 1100–1120.

Stone, P. (2012) Dark tourism and significant other death. Towards a model of mortality mediation. *Annals of Tourism Research*, 39(3), 1565–1587.

Stone, P. (2013) Dark tourism scholarship. A critical review. *International Journal of Culture, Tourism and Hospitality*, 7(3), 307–318.

Stone, P. and Sharpley, R. (2009) *The Darker Side of Travel. The Theory and Practice of Dark Tourism*. Bristol: Channel View Publications.

Sugawa-Shimada, A. (2015) Rekijo, pilgrimage and 'pop-spiritualism': Pop-culture-induced heritage tourism of/for young women. *Japan Forum*, 27(1), 37–58.

Suzuki, T. (2012) Forming an activist bubble in tourism: Peace guiding at Okinawa battle ruins. *Tourist Studies*, 12(1), 3–27.

Suzuki, T. (2016) Bridging between here/now and there/then: Guiding Japanese mainland school peace education tours in Okinawa. *Journal of Cultural Geography*, 33(1), 100–125.

Tsai, P. and Wang, K. (1998) Competitiveness of international tourism in Taiwan: US versus Japanese visitors. *Applied Economics*, 30, 631–641.

Tussyadiah, I. (2006) A gourmet trip: One direction of domestic tourism in Japan. *Tourism Review International*, 9(3), 281–291.

Uzama, A. (2009) Marketing Japan's travel and tourism industry to international tourists. *International Journal of Contemporary Hospitality Marketing*, 21(3), 356–365.

Uzama, A. (2012) Yokoso! Japan: Classifying foreign tourists to Japan for market segmentation. *Journal of Hospitality Marketing & Management*, 21(2), 132–154.

White, M. (2015) Ramen at home and on the road. *Japan Forum*, 27(1), 114–120.

Woosnam, K., Maryuama, N. and Boley, B. (2016) Perception of the other residents: Implications for attitudes of tourism development focused on the minority ethnic group. *Journal of Travel & Tourism Marketing*, 33(5), 567–580.

Wu, C., Funck, C. and Hayashi, Y. (2014) The impact of host community on destination (re)branding: A case study of Hiroshima. *International Journal of Tourism Research*, 16, 546–555.

Wu, L. and Walters, G. (2016) Chinese travel behavior in response to disastrous events: The case of the Japan earthquake. *Journal of China Tourism Research*, 12(2), 216–231.

Yamada, T. (2014) Saving tourism or preventing FMD? Risk discourse in Japan's island community. *Asian Anthropology*, 13(2), 139–148.

Yamaguchi, M., Emenheiser, D. and Reynolds, J. (2000) Meeting the need of Japanese customers: Hotels in Japan and United States. *Journal of Travel & Tourism Marketing*, 9(1/2), 201–210.

Yamaguchi, Y. and Yoshihara, M. (2004) Evolutions of paradise. Japanese tourist discourse and Hawaii. *American Studies*, 45(3), 81–106.

Yamamura, T. (2015) Contents tourism and local community response: Lucky star and collaborative anime-induced tourism in Washimiya. *Japan Forum*, 27(1), 59–81.

Yamashita, S. (2000) The Japanese encounter with the south: Japanese tourists in Palau. *Contemporary Pacific,* 12(2), 437–464.

Yasuda, H. (2010) World heritage and cultural tourism in Japan. *International Journal of Culture, Tourism and Hospitality Research*, 4(4), 366–375.

Yoshida, K., Bui, H. and Lee, T. (2016) Does tourism illuminate the darkness of Hiroshima and Nagasaki? *Journal of Destination Marketing & Management*, 5(4), 333–340.

You, X. and O'Leary, J. (2000) Age and cohort effects: An examination of older Japanese travellers. *Journal of Travel & Tourism Marketing*, 9(1/2), 21–42.

Zhang, R. and McCornac, D. C. (2014) Challenges for the international tourism industry in Japan: An agent for economic recovery and development. *AlmaTourism*: *Journal of Tourism, Culture and Territorial Development*, 5(10), 109–124.

Zhang, Z. (2004) Organising customers: Japanese travel agencies marketing on the internet. *European Journal of Marketing*, 38(9), 1294–1303.

Ziff-Levin, W. (1990) Cultural logic gap. A Japanese tourism research experience. *Tourism Management,* 11(2), 105–110.

3 Urban development and tourism in Japanese cities[1]

Yumiko Horita

Introduction

With the deindustrialization of many cities in developed Western countries, urban tourism has been employed as a means of attracting people to declining inner-city areas, utilizing various resources such as goods, information and people (Law, 1992). Thus, the concept of urban tourism emerged from the tourism development process in areas previously not perceived as destination areas. In other words, urban tourism involves a functional and spatial restructuring process in which an area shifts from a focus on industrial production and its workers' daily lives to the consumption of goods, services and experiences for visitors (Hayllar, Griffen & Edwards, 2008).

Since the beginning of the new millennium, urban tourism has gradually come to be regarded as a key activity in reinforcing the local and national economy in Japan. More specifically, and as explained in Chapter 1, over the last two decades, the Japanese government has actively sought to increase significantly the number of foreign visitors to Japan and, in so doing, has encouraged tourism developments in the expectation that tourism should bring about significant economic benefits. Consequently, influenced not only by the rapid increase in international tourist arrivals but also by the momentum building up towards the hosting of the summer Olympic Games in Tokyo in 2020 (postponed to 2021 following the Covid-19 pandemic), the development of infrastructure and tourism-related facilities has been boosted both in the capital and in other metropolises. At the same time, as tourism is widely considered to have the potential to contribute to the revitalization of regional cities suffering from depopulation and prolonged economic depression (OECD, 2018; Telfer, 2015), various efforts have been made by the public, private and non-profit sectors to develop tourism. This has resulted in many urban spaces in both larger metropolises and medium- and small-sized regional cities being transformed to reflect what visitors desire to see and do.

Urban tourism development is then, at least in Japan, a relatively recent phenomenon, particularly in comparison to Western nations where the transformation of urban spaces can be traced back to the 1980s (Spirou, 2008). However, it can only be fully understood within the context of urban development more generally in Japan following the Second World War, a period during which the country has transformed significantly. The purpose of this chapter, therefore, is to explore how urban tourism space in Japan has been created alongside more general urban development since the end of the Second World War, and then to examine the implications for urban tourism development. In particular, it focuses on two forces that may be in conflict: on the one hand, central government and private enterprises leading urban development as an economic driving force and, on the other hand, citizens seeking to improve and maintain their living space by themselves.

The chapter first reviews briefly the literature on urban tourism in order to identify issues relevant to the relationship between urban tourism and regeneration in general, and in citizen involvement in tourism development in particular, as a framework for considering urban tourism in Japan. It then goes on to explore the formation of urban tourism space in the context of the history of associated urban developments in Japan as well as current trends in urban tourism. Finally, drawing on the representative case of Kyoto, it details urban tourism space development and discusses the implications of the process, identifying some of the characteristics of urban tourism development in Japan. The literature review and surveys are based principally on government websites and policy documents.

Urban tourism and urban space

Urban tourism development: a review

Academic attention has been paid to the phenomenon of urban tourism since the 1980s, coinciding with the development of urban regeneration measures and policies that evolved primarily in industrial cities in developed Western countries (Ashworth & Page, 2011; Law, 1992). Since then, not only has it become an increasingly popular focus of academic attention (e.g., Bellini & Pasquinelli, 2017; Colomb & Novy, 2017; Page & Hall, 2003; Selby, 2004), but the scale of urban tourism itself, in all its manifestations, has also increased significantly. As Popp (2012: 50) puts it, 'urban tourism has been booming since the early 1980s'. In part, this reflects the continuing growth in tourism more generally; annual international arrivals, for example, have increased by more than one billion since 1990. However, transformations in patterns of tourism combined with the emergence of low-cost airlines have also underpinned the ever-growing popularity of cities as tourism destinations (Vera Rebollo & Ivars Baidal, 2009). As a consequence, many urban destinations 'have reached or exceeded their capacity to absorb tourism' (Sharpley & Harrison, 2017) or are suffering from what has come to be referred to as 'overtourism' (Peeters et al., 2018).

A full discussion of this issue is beyond the scope of this chapter although, as considered later, the challenges of overtourism are of particular relevance to Kyoto (see also Chapter 14). However, it is important to point out for the present purposes that what is broadly referred to as urban tourism takes many forms. Indeed, Ashworth and Page (2011: 3) suggest that a variety of other adjectives, such as cultural, shopping or gastronomic, could precede urban tourism and, hence, we should perhaps refer in the plural to urban tourisms. At the same time, a variety of factors determine the extent to which tourism is absorbed and impacts upon urban destinations. Nevertheless, a distinction that can be made, and is of particular relevance to this chapter, is between the 'tourist–historic city' (Ashworth & Tunbridge, 2000), that is, cities to which tourists have long drawn primarily by a critical mass of historic attractions, and former industrial cities that, through a variety of means, 're-invent' themselves as tourist destinations.

With regards to the latter, the advent of the post-industrial period meant that many such cities were no longer significant places of manufacturing production; rather, they were seeking to become management hubs for people, goods and services in an increasingly globalized world (Sassen, 2001). Consequently, urban spaces were restructured as places suitable for the consumption of goods, services, events and so forth. Within with this new role for cities, tourism came to be regarded as an effective conduit to attract consumers, and urban spaces were then shaped as places for consumption. This was reinforced as people started to seek pleasure in consumption, especially the consumption of goods and services that differ from those found elsewhere (Mullins, 1991). Hence, over recent decades, providing a range of goods, services and 'spaces for pleasure' has caused a fundamental restructuring of urban spaces, often along themes that are suitable for such consumption (Bryman, 2004; Hannigan, 2005).

With the expansion of globalization, it is no longer possible for former industrial cities or metropolises to ignore the impacts of tourism. The appeal of urbanity, such as its convenience and pleasant living, in addition to the opportunity to consume urban goods and cultural experiences, are important factors in attracting professionals and intellectual labourers to the service industry (Florida, 2009). In other words, tourism has come to be regarded as having not only significant economic impacts via consumption but also positive social impacts in relation to attracting highly educated and skilled people through the spatial improvement created by tourism promotion. Creativity is a key concept in these spaces, and movements have occurred at both grassroots and policy levels to leverage culture, art, craft and design as a means to restructure urban functions and spaces (Richards, 2014; Sasaki, 2004). Changes in urban spaces and settings can also improve the urban image in addition to attracting investors (Florida, 2009).

Issues in the process of urban tourism development

As many commentators have pointed out, a variety of issues emerge in the process of urban tourism development, some of which may challenge the expectation of positive impacts. First, urban tourism development often generates spatial divides (Judd, 1999). Tourist precincts created by urban regeneration projects in post-industrial cities tend to have been developed away from surrounding disadvantaged areas or by removing pre-existing structures in order to give visitors a sense of security and comfort. These areas feature hotels, conference facilities, restaurants and bars, sports stadiums and shops. They also often feature wide paved walkways with greenery, street furniture and novelty attractions that make tourists feel comfortable. The existing reality of life and culture in the redeveloped area may, thus, be concealed and excluded, being regarded as 'impure', 'dangerous' or 'ambiguous to the urban image' (Judd, 1999). Such developments often entail an overall process of so-called gentrification (Smith, 1982).

In a similar vein, both the state and private enterprises together play a significant role in driving forward urban developments. Harvey (2012) points out that urban space formation is necessary for capitalists to absorb surplus production produced in pursuit of a surplus value. The state and private enterprises conspire in creating urban development visions and setting-up the financial and legislative conditions whereby developers can attract more investment from the global market. As a result, urban developments, including urban tourism development, can distribute wealth to a wealthy class but will not alleviate social and economic inequality and may, rather, result in the destruction of the locality and divide the community through a spatial and class divide (Judd, 1999; Zukin, 2011).

Urban tourism development, however, inevitably requires people who live and work in the city to be involved in the process in order to facilitate its functioning. This is the second issue. In fact, urban tourism development often functions as a driver of inner-city regeneration, which partly aims at improving residents' living environments. Furthermore, most facilities and spaces in urban areas can fulfil roles as tourism amenities, although they are used both by tourists and by non-tourists, that is, those who are typically residents and workers. It can also be said that place marketing to improve the image of the city possibly requires the local community's involvement. Indeed, it has long been suggested that the involvement and support of local residents is fundamental to the success of all tourist destinations, including those in urban areas; as Snaith and Hailey (1999) observe, a 'happy host' is essential to tourism development; hence, 'unhappy' hosts may implicitly threaten the future success of the destination (Lawson et al., 1998). Whether this is in fact the case remains uncertain (Sharpley, 2014); nevertheless, local governments initiate 'place marketing' as a strategy to 'sell' the locality to attract people to the city (Selby, 2004) and, in this process, not only may urban residents and workers prove to be an important element in welcoming visitors, but also they may have the opportunity (if permitted) to think how they might contribute to the management their own area, including visitor spaces.

The third issue to arise in urban tourism development is that of conflict between tourism and residents' daily life, which occurs owing to the incursion of tourists into places on which the tourist gaze has not been previously directed. This is not, of course, a newly emerging issue; the social impacts of tourism development have long attracted significant academic attention (Wall & Mathieson, 2006), as have the attitudes of local communities to tourism and tourists (Sharpley, 2014); however, the nature of these impacts have arguably changed. That is, the behaviours of tourists have diversified; they are now interested in authentic local life and communicating with local people. Some local people also expect to come in contact with tourists. Reflecting this trend, booking accommodation with local hosts is now popular amongst travellers (see Chapter 14). With the emergence of this new service, it is now possible for tourists to easily experience local life while travelling. This also, in principal, increases the opportunity for direct communication between tourists and local host communities.

Overall, then, the process of urban tourism development offers both negative and positive potential: conflicts may occur owing to social and spatial inequality and tourists invading local people's life spaces; at the same time, local people are able to be a part of the tourism development process, with their understanding and consensus. The latter could be a key to overcoming the former negativity. Based on this point, Japanese urban tourism development is now explored in the following sections.

Urban development history and urban tourism development in Japan

The effect of urban development on tourism spaces in Japan can be examined in three contextual periods. The first of these periods is from the post-war era to the end of the 1970s, marked as a rapid economic growth phase. The second period is the time of domestic demand expansion supported by the urban renaissance policy and the collapse of the bubble economy in the 1980s and 1990s. The third period, from 2000 onwards, is the era of expanded globalization and prolonged economic decline. Each period is outlined below.

From post-war to the end of the 1970s

This period of rapid economic growth in Japan corresponds to a time when both urban and rural areas experienced a large wave of development. The government espoused a 'Balanced National Land Development' policy through industrialization, so tourism was viewed as a phenomenon associated with the expansion of industrialization and urbanization. Manufacturing hubs were rapidly constructed across Japan, causing a wave of urbanization over local cities, towns and villages.

In terms of tourist-related development, sightseeing roads and byways were developed by the government, while ski slopes, holiday homes, inns and hotels were mainly developed by private enterprises. Leveraged by mega-events such as the Tokyo Olympic Games in 1964 and the Osaka World Exposition in 1970, the government drove forward the urban development of major cities and transport networks. Consequently, landscapes with historic monuments and cultural heritage sites were often destroyed. As a result of these improved urban infrastructures, however, the Japanese people awoke to the fact that they could enjoy their leisure more easily than before (Ishikawa, 1979).

These waves of development also reached regional historical cities and towns with their well-preserved centres and beautiful townscapes. In some historical cities and towns, lobby groups and residents' conservation groups opposed these developments. This eventually led to activities focusing on the sustainable use of local resources for tourism development from the 2000s onwards.

The 1980s and 1990s: a time of expansion and collapse

The impact of the oil shock at the beginning of the 1970s led to poor economic growth in Japan, whilst the signs of deindustrialization gradually appeared during this period. The transformation of the industrial structure that followed the oil shock seriously affected cities that relied on heavy industries, such as steel manufacturers and ship builders. Cities such as Kobe and Yokohama, which up to this time have relied on these industries and prospered as major port cities, began to explore new economic incentives to attract tourists and visitors. For example, Yokohama enforced a new urban policy that sought to shift its key industry to that of a visitor economy, taking advantage of its cross-cultural character as a harbor city flourishing in a key trade junction. The city implemented the Minato Mirai 21 plan[2] for large-scale redevelopment in 1984, seeking to become a convention city by developing the international convention centre in the waterfront areas (Yokohama Shintoshi Kenkyukai, 1989).

Kobe also developed its waterfront areas as a basis for developing new economic activities and advocated the adoption of the monikers of 'fashion city', 'convention city', 'event city' and 'urban resort city' in their urban policy (Ōta, 2002; Sasayama, 1994). The Kobe Port Island Exposition, which was held in 1981 to celebrate the completion of an artificial island, was the moment that Kobe became a tourist city (Ōta, 2002). These urban policies were enforced with some redevelopment projects under the urban renaissance policy of the Nakasone regime in the 1980s, and formed the foundations of urban tourism.

As for Tokyo, there was a great rush to redevelop its metropolitan districts supported by a monetary easing policy and strategies by both the central government and Tokyo Metropolitan Government to obtain the status of a 'global city'. Aiming to attract a central administration function for multinational enterprises, Tokyo attempted to transform its urban fabric and space into a suitable shape (Sassen, 2001). The redevelopment policy under the Nakasone[3] regime brought a steep rise in property values in metropolitan areas, which also expanded to rural areas. As a result, the restructuring of urban space in Tokyo did not actually make significant progress at this time because of sharp increases in land prices and the bursting of the bubble economy.

It is important to note in relation to the policy that the Act for the Development of Comprehensive Resort Areas (see also Chapter 1) transformed rural natural areas into golf courses, hotels, marinas, ski areas and holiday condominiums under the pretext of regional revitalization. Many municipalities seized the opportunity to develop resort facilities to redress the gap between their and Tokyo's social and economic status. However, the boom in resort development came to an abrupt end when the bubble economy burst. Many businesses and projects fell through or closed, and regional areas were forced to carry a heavier burden than before. Regional decline eventually accelerated.

The 2000s: a time of globalization and urban decline

After the economic bubble burst, Japan faced a long recession. The effects of the urban renaissance policy were limited and urban areas, particularly the Tokyo metropolitan area, suffered from stagnation caused by credit defaults. The Koizumi[4] regime implemented several measures to support private enterprises in disposing of bad debt which, in turn, were part of a broader, drastic series of administrative and fiscal reforms. Tourism-oriented national policy was also introduced later for the implementation of urban regeneration projects.

The primary objective of urban regeneration policies under the Koizumi regime was to relieve the credit defaults that had become a huge barrier to making Tokyo a global city (Cabinet Office, 2002). The central government implemented crucial reforms and deregulation in various fields, such as national land sales, reinforcement of logistics associated with the transportation system and streamlining of development procedures. These reforms led to drastic changes in the existing social system and infrastructure, as they attempted to achieve competitive success for Japan in encouraging investment and consumption in the post-industrial era. Some of the reforms specifically delivered deregulation for urban development, which brought more profits to private enterprises (Igarashi & Ogawa, 2003). For instance, property securitization and a zero-interest-rate policy increased the inflow of capital to redevelopment projects (Toyofuku, 2016). Most urban developments in major cities were implemented based on an especially favourable approach to and legislation for private enterprises. The result was that a more modernistic urban landscape with skyscrapers was formed in the Tokyo metropolitan area and in other major cities.

At the outset, tourism was not directly emphasized in the urban regeneration policy, but redevelopment in urban areas generated spectacular landscapes with sophisticated commercial complexes and comfortable common spaces for both residents and visitors. These kinds of urban regeneration projects were implemented in the central districts of other metropolises, although their volume did not match that of Tokyo. Furthermore, it should be noted that the current sharp increase in foreign tourists has become a driving force in the transformation of the urban space of metropolises. One significant phenomenon is the hotel construction boom in major cities (Isobe & Tauchi, 2016). The expectation of an increase in demand for the Tokyo Olympic Games in 2020 (postponed to 2021), and strong government promotion through deregulation of planning and building standards, are accelerating this trend.

These urban developments were, however, not always favoured by local residents. In Tokyo, for example, strong opposition to the redevelopment project was expressed in one of its unique districts, Shimokitazawa, because people were worried that it would destroy its existing cultural features (Endo & Imanishi, 2010). In other cities, large-scale development of commercial facilities and high-rise buildings often led people to object. In many large cities which have been under strong development pressure, an accumulation of experiences has led to challenges to top-down development projects. Thus, people's awareness of involvement in area management tends to be high (Sorensen & Funck, 2007). Local governments, such as in Kyoto and Kobe, have also developed a system to secure and support people in participating in area management, such as authorizing councils consisting of stakeholders and support for employing consultants, thereby creating a space for deliberation to overcome antagonism and for community empowerment.

Another area management attempt has been striving for consensus among stakeholders in the central business districts (Kobayashi, 2015). Private enterprises and small businesses have voluntarily formed an association to manage the designated area by planting trees and flowers and drawing up guidelines for advertising displays and street and building design. To improve the attractiveness of urban space, it is necessary to solve problems by deliberation and negotiation among stakeholders; it cannot be achieved by competition between them.

Unlike the Tokyo metropolitan area, regional cities have been affected by the Koizumi regime's reductions in local allocation of taxes and government subsidies. As mentioned above, industrialization and the development of its supporting urban infrastructure prevailed in regional cities under previous development policies by the central government. As many regional cities relied on such policies, they were seriously affected by the relocation of manufacturing plants to offshore locations (Okada, 2005). People's lives were also directly impacted by such factors as reduced public services and the hollowing-out of business and commercial districts in regional cities.

Owing to the increased gaps between major and regional cities, the meaning of 'living local' has begun to be re-evaluated, emerging in particular from questions about the uniform urbanization based on industrialization. What has been rediscovered is not something new brought about by development but, rather, the existing resources and values of local living culture within daily life, which may be more easily recognized by outsider perspectives. Based on this movement, activities that utilize local resources have been developed by local people together with outsiders in an effort to enrich local people's daily lives and to create opportunities for communication between local people and visitors. This is part of the 'kankō machizukuri' movement, discussed in more detail below, which became a pillar of tourism-oriented national policy (Nishimura, 2009).

The current situation of urban tourism in Japan

Urban tourism destinations

Tokyo, Osaka and Kyoto are popular urban tourism destinations for both foreign and domestic visitors as shown, for example, by travel reservation data analysis[5]. In addition to such cities with their high international visibility, the cities shown in Table 3.1 are also major urban destinations for domestic travel, including overnight and day trips.

The cities in Table 3.1 can be categorized into four types according to their urban features: historic capital cities, regional major cities, prefectural centre cities and cities depending on tourist facilities. The former two contain internationally well-known cities with large populations, where historic resources and integration of modern urban facilities and cultures attract many tourists. Many of these cities lost tangible heritage because of war damage and urban development during the era of rapid economic growth. Kyoto, as one of the very few cities with remaining rich historic resources, became a major tourist

Table 3.1 Top 10 prefectures attracting large numbers of tourists and day visitors and cities with the largest numbers within their prefectures

Rank	*Prefectures and numbers of tourists and day-visitors in 2014 (000)**		*Cities within the prefecture having the largest number of tourists and day-visitors in 2014 (000)***		*Population of cities (2010)****
1	Tokyo	56,278	–	–	–
2	Chiba	30,274	Urayasu	44,294	164,877
3	Kanagawa	29,248	Yokohama	44,258	3,688,773
4	Shizuoka	28,927	Shizuoka	26,266	716,197
5	Hokkaido	28,819	Sapporo	13,416	1,913,545
6	Osaka	26,975	–	–	–
7	Nagano	23,646	Nagano	9,690	381,511
8	Kyoto	23,069	Kyoto	55,636	1,474,015
9	Hyogo	21,774	Kobe	35,430	1,544,200
10	Aichi	18,039	Nagoya	32,143	2,296,014

Notes

*Data source is Ryoko kanko shohi doko chosa [Travel and tourism consumption survey] in 2014 by JTA.

**Data are collected from each prefecture's statistics on tourist and day visitors in 2014. The source is different from the third column.

***Data come from the 2010 Census.

destination with a large volume of visitors. Tourists are attracted to prefectural centre cities by their natural resources, such as hot springs and views of mountainous areas. They may not be equal to urban tourism destinations, but many regional cities in Japan have developed and promoted tourism using a combination of natural and urban resources. Cities depending on tourist facilities is a special category, into which only Urayasu falls thus far as the city to which the Tokyo Disney Resort attracts a large number of tourists.

Kankō machizukuri movement

With the expansion of tourism development and promotion across the country in the 2000s, some regional cities drew the attention of tourists to their unique activities based on the enduring efforts of local people utilizing their local resources. This range of activities is known as *kankō machizukuri*, a term combining tourism (*kankō*) and community development (*machizukuri*). To put it simply, '*machizukuri*' means the various activities or processes undertaken principally by the residents of an area to improve their living environment. There are expectations that *kankō machizukuri* is an approach to tourism development and management that ensures the preservation of natural and cultural resources, as well as the value of residents' daily lives, for the next generation; in addition, it will contribute to local revitalization (Nishimura, 2009).

Table 3.2 shows the number of cities, towns and villages by population size striving for *kankō machizukuri*. The important point to note in association with urban tourism is that efforts have been made in both metropolises and small- to medium-sized cities to provide opportunities to experience the unique local history, culture and daily life and to communicate with local people through these experiences. At the same time, it shows that the demand for such opportunities has increased.

The formation of urban tourism spaces in Kyoto

This section examines the specific case of urban tourism in Kyoto to identify the significant issues in the formation process of urban tourism spaces. Kyoto is well known as an international and domestic urban tourism destination, and has historically witnessed a disputed relationship over urban development between the central government and enterprises on the one hand, and citizens on the other.

Immediately following the Second World War, a conservation movement developed in the historic landscape of Kyoto formed by grassroots conservation lobbies. Local government developed legislation and bylaws for conservation of its history but, under strong pressure for development, Kyoto has repeatedly destroyed its valuable old buildings and natural resources, leading to a decline in the appeal of its historic landscape. Kyoto Tower in the 1960s

Table 3.2 Kankō machizukuri cities, towns and villages*

Population ('000)	*Numbers of cities, towns and villages and their percentage of the total*		*Names of cities, towns and villages*
Less than 5	14	6.5%	e.g., Kōya, Minami
5~10	17	7.9%	e.g., Hakuba, Hiraizumi, Oguni mach, Aya
10~50	54	25.0%	e.g., Nichinan, Nanao, Usa, Kurayoshi, Noboribetsu, Atami, Usuki, Sakaiminato, Yufu
50~100	45	20.8%	e.g., Hidatakayama, Nikko, Maizuru, Amakusa, Omihachiman, Sahara, Yanagawa, Murakami
100~300	45	20.8%	e.g., Shimonoseki, Hakodate, Echigo, Sasebo, Kure, Matsue, Kamakura, Otaru, Karatsu, Beppu
300~500	19	8.8%	e.g., Kurashiki, Nagasaki, Nagano, Nara, Asahikawa, Kawagoe
500~1000	11	5.1%	e.g., Kitakyushu, Okayama, Himeji, Matsuyama
Over 1000	11	5.1%	e.g., Yokohama, Osaka, Nagoya, Sapporo, Kobe, Kyoto, Fukuoka, Kawasaki, Hiroshima, Kanazawa
Total	216	100%	

Note
*List of kankō machizukuri was retrieved from Japan Tourism Agency: (www.mlit.go.jp/kankocho/page04_000041.html). Population data are from the 2010 census.

and the JR (Japan Railways) Kyoto Station in the 1990s were contentious buildings when planned and constructed. The construction of multi-storey buildings for condominium and hotel development has still continued and these also give rise to opposition.

In contrast, consciousness of the need to respect Kyoto's history and culture has gradually become more widely shared among citizens since the benefits of conservation have clearly emerged with the development of tourism. In 2006, Kyoto City announced a new landscape policy and has tightened regulations since 2007 (Kyoto City, 2007). For example, building heights and the design and posting of billboard advertisements are more strictly controlled than previously. Kyoto City has also encouraged the maintenance of existing traditional townhouses and renovation of vacant houses. In addition, a system has been developed for local residents and enterprises to form a local council to discuss townscape and area management.

Initially, there were major objections when the regulations were tightened; however, a recent survey of residents showed they now regard this regulation favourably (Kyoto Shimbun, 2016). Another survey shows that people in Kyoto tend to feel the landscape is favourably maintained (Kyoto City, 2013). These can provide evidence that a sense of respect for their locality and an understanding of its value have spread among citizens. Furthermore, pavements in the central district were widened, reducing car lanes, so people could walk around more comfortably. Completed in 2016, this project also led to complaints from citizens because of traffic jams; nevertheless, it received some awards from independent institutions related to urban planning and transportation. Considered in the light of past experiences, it is assumed that the citizens' consciousness will change in the near future.

The above tendency has been reinforced by the growth of tourism, as the majority of citizens have realized and accepted the image and evaluation of Kyoto provided by tourists (Muenta, 2009). In fact, many tourists visiting Kyoto expect to experience the atmosphere of an ancient city, including old temples and shrines (Kyoto City, 2014). Consequently, both citizens and visitors can share the image of Kyoto, and it has also been reflected in the formation of urban tourism spaces while ensuring the conservation of the historic townscape.

Here, it is important to point out that such formation of urban tourism spaces has caused changes in the features of the area. For instance, a survey has highlighted the increased use of buildings as souvenir shops and restaurants targeted at tourists and a decrease in the number of dwellings in the areas surrounding Kiyomizu temple, which is a popular tourist destination, since the 1970s (Kim & Munemoto, 2001). The number of traditional townhouses and vacant old houses that have been renovated for use as guesthouses, shops and restaurants has also increased recently (Kyoto City, 2016).

The making of decisions on whether local resources and local identity are ignored and abandoned or maintained could be influenced by their value as tourism products and their appeal to tourists. However, it is impossible to prevent changes in the city and, thus, a more important point is citizens' understanding of the impact of tourism and their active participation in making decisions and taking action in the process of tourism development. In other words, it is the active management of tourism development by citizens, considering their identity and local values that they respect and wish to maintain. It requires people to have high receptiveness towards others, that is, towards visitors and tourists, and to improve their governance for managing the area collaborating with local government and enterprises.

To what degree do the people of Kyoto achieve this? It cannot be to any great extent because people cannot participate sufficiently in decision-making processes or control negative impacts such as congestion, traffic jams and the current tourism development boost. However, citizens' governance power toward urban space is becoming enhanced by taking the expansion of tourism as an opportunity.

Particular note is taken of the emerging consciousness of respecting Kyoto's history and culture and it has brought changes in the formation of urban space. Furthermore, interactions between tourists and residents' daily lives have already occurred on various levels. Tourism products to meet the desire of tourists to experience aspects of local daily life have been provided by grassroots activities and collaboration between businesses and citizens. Noticeable examples of this phenomenon are short tours of local life, experiential programmes and the guesthouses converted from old townhouses. Such interaction between citizens and tourists can create both business opportunities and enjoyment, and especially helps citizens understand how best to welcome tourists, how to create and manage their areas and how to take action for this to happen.

Nevertheless, most recently, this emerging relationship between the local community and tourists in Kyoto is being challenged by the phenomenon of so-called overtourism, which can be defined as 'the situation in which the impact of tourism, at certain times and in certain locations, exceeds physical, social, economic, psychological, and/or political capacity thresholds' (Peeters et al., 2018: 15). Reflecting in part the rapid increase in inbound tourism in recent years, and in part the inevitability of local tourism sector (and community) taking advantage of ever-increasing numbers of tourists being attracted to one of Japan's principle historic city destinations, not only have visitor numbers risen to around 50 million annually, but also there is evidence to suggest that local residents are being affected increasingly negatively by the sheer volume of tourists. Moreover, recent reports suggest that tourists are beginning to avoid the city owing to overcrowding yet, ironically, accommodation is easier to find as yet more new hotels and other lodgings are opened (Baseel, 2019). As such, there is a danger that without appropriate action, Kyoto might follow the path of cites elsewhere, such as Venice and Barcelona, that are suffering severe consequences of excessive tourism (Dodds & Butler, 2019).

Concluding remarks

The foundations for urban tourism space in Japan were formed with expanding urbanization across the country after the Second World War. Along with the progress of deindustrialization, tourism development and promotional activities have gradually expanded among not only metropolises but also regional cities as a measure for tackling economic decline. Particularly in metropolises such as Tokyo, Yokohama and Kobe, the government and private enterprises have promoted and conducted redevelopment projects with transformation of spaces since the 1980s. New urban spaces formed through such redevelopment projects have become urban tourism precincts. Urban tourism developments have been implemented but it has recently become evident that they tend to leverage existing local resources such as history, culture, food and lifestyle. Urban tourism space was often discovered and created by local people around hitherto unrecognized existing local resources. Current urban tourism development in Japan encompasses two types; one is redevelopment for profit and economic growth led by the government and enterprises, the other is re-evaluation and utilization of existing local resources led by local people.

People's consciousness with respect to history and locality has been evoked through the accumulated experiences of counteracting urban development. Respecting local existing values links with regrets regarding the loss of locality as a result of new developments and the search for economic profit, and chance to force people to revitalize their area in the era of shrinking government role and prolonged economic recession. Additionally, it has been influenced by the growth of tourism. Tourism promotes people's mobility and then everywhere can be a tourist place. Tourism creates opportunities to look at other places and to be looked at from the outside. People have become more respectful of their local history, resources and lifestyles through tourism experiences. At the same time, such consciousness leads to active participation in the process of managing their living places and places to which they feel attached as their own place. Local area management activities by local people have been encouraged and could become part of the process of tourism development.

Another important point here is that there has been a significant shift from dependence on urban development to collaborative management or partnership among stakeholders, including local people, following the growth of tourism, although a number of conflicts remain. Thus, it can be said that most cities in Japan have been formed with two conflicting forces: development-oriented force based on economic growth and collaborative management force based on valuing the locality with the prospect of active citizen participation in local management. The future direction of these tendencies will be worth watching, particularly given the most recent experience of Kyoto, where the former appears to be supressing the latter as the city becomes a victim of overtourism.

Acknowledgement

This work was supported by JSPS under KAKENHI [15K1945].

Notes

1 A previous version of this chapter appears in *Tourism Planning & Development*, *15*(1), 26–39.
2 This is a name of an urban development plan that in English translates to 'Future Port of the 21st Century'.
3 Nakasone Administration (1982–1987).
4 Koizumi Administration (2001–2006).
5 Expedia. Available at: https://welove.expedia.co.jp/press/14735/ (Accessed 10 September 2016).

References

Ashworth, G. and Page, S. (2011) Urban tourism research: Recent progress and current paradoxes. *Tourism Management*, 32(1), 1–15.

Ashworth, G. and Tunbridge, J. (2000) *The Tourist-Historic City: Retrospect and Prospect of Managing the Historic City.* Oxford: Pergamon.

Baseel, C. (2019) Japanese travelers avoiding Kyoto as number of foreign visitors continues to grow. *Japan Today*, 11 March. Available at: https://japantoday.com/category/features/travel/japanese-travelers-are-avoiding-kyoto-as-the-city%E2%80%99s-number-of-foreign-visitors-continues-to-grow (Accessed 16 July 2019).

Bellini, N. and Pasquinelli, C. (2017) *Tourism in the City: Towards an Integrative Agenda on Urban Tourism*. Switzerland: Springer International Publishing.

Bryman, A. (2004) *The Disneyization of Society*. London: Sage Publications.

Cabinet Office (2002) Toshi saisei kihon hōshin (Basic principle for urban renaissance) [html]. Available at: www.kantei.go.jp/jp/singi/tiiki/toshisaisei/kettei/020719kihon.html

Colomb, C. and Novy, J. (2017) *Protest and Resistance in the Tourist City*. London: Routledge.

Dodds, R. and Butler, R. (Eds) (2019) *Overtourism: Issues, Realities and Solutions*. Berlin/Boston: Walter de Gruyter.

Endo, A. and Imanishi, K. (2010) Citizen-led community planning through the collaborative partnership with the local government; the case of the shimokitazawa redevelopment project. *Summaries of technical paper of Annual Meeting Architectural Institution of Japan, F-1, Urban Planning, building economics and housing problems 2010,* 187–188. Available at: www.aij.or.jp/scripts/transac/trans_eR.htm

Florida, R. (2009) *Who's Your City? How the Creative Economy Is Making Where to Live the Most Important Decision of Your Life.* New York: Basic Books.

Hannigan, J. (2005). *Fantasy City: Pleasure and Profit in the Postmodern Metropolis*. London: Routledge.

Harvey, D. (2012) *Rebel Cities: From the Right to the City to the Urban Revolution*. New York: Verso.

Igarashi, T. and Ogawa, A. (2003) *Toshisaisei wo tou* (*Question the urban regeneration*). Tokyo, Japan: Iwanami Shoten.

Ishikawa, (1979). Yoka no Sengoshi [Postwar history of leisure]. Tokyo: Tokyo Shoseki.

Isobe, M. and Tauchi, K. (2016, July 2). Hōnichikyaku jyuyo waku ginza [Boiling foreign visitor demand at Ginza]. Asahi: Shimbun, p. 35.

Judd, R. (1999) Constructing the tourist bubble. In R. Judd and S. Fainstein (Eds), *The Tourist City*. New Haven, CT and London: Yale University Press, pp. 35–53.

Kim, H. and Munemoto, J. (2001) A study on the relationship between the change in building use and ownership transfer in Sanneizaka Preservation District for groups of historic buildings. *Journal of Architecture, Planning, Environment and Engineering,* 545, 215–221.

Kobayashi, S. (Ed.) (2015) *Saishin eria manejimento* [Area Management]. Kyoto, Japan: Gakugei Publishing.

Kyoto City (2007) *Kyoto City Landscape Policy: Forming Timeless and Radiant Kyoto Landscapes* [pdf]. Available at: www.city.kyoto.lg.jp/tokei/cmsfiles/contents/0000062/62129/HP-English.pdf

Kyoto City (2013). *Kyotoshi keikanhakusho dētashu heisei 25 nendo* (Landscape white paper data collection 2013. Available at: www.city.kyoto.lg.jp/tokei/cmsfiles/contents/0000146/146590/H25de-tasyuu.pdf

Kyoto City (2014) *Kyoto kankō sōgō chōsa* (Kyoto tourism comprehensive Survey) Available at: https://kanko.city.kyoto.lg.jp/chosa/image/kanko_chosa26.pdf
Kyoto City (2016) *Ryokangyo shisetsu ichiran* (List of Licensed accommodation). Available at: /www.city.kyoto.lg.jp/hokenfukushi/page/0000193116.html
Kyoto Shimbun (2016, August 14) Kyoto shimin wa ochituita iro wo konomu: Kanban jyōrei de keikan kaizen (Kyoto citizens prefer muted colours: Advertising display bylaw for improved landscape). Kyoto Shimbun. Available at: www.kyoto-np.co.jp/politics/article/20160814000079
Law, C. M. (1992). Urban tourism and its contribution to urban regeneration. *Urban Studies*, 29 (3/4), 599–618.
Lawson, R., Williams, J., Young, T. and Cossens, J. (1998) A comparison of residents' attitudes towards tourism in 10 New Zealand destinations. *Tourism Management*, 19(3), 247–256.
Muenta, Y. (2009) *Sōzōtoshi notameno kankōshinkō* (*Tourism promotion for creative cities*). Kyoto, Japan: Gakugei Publishing.
Mullins, P. (1991) Tourism urbanisation. *International Journal of Urban and Regional Research*, 15(3), 326–342.
Nishimura, Y. (Ed.) (2009) *Kankō machizukuri: machi jiman kara hajimaru chiiki manejimento (Tourism and Community Development: Local Management Stemmed from Town Pride*). Kyoto, Japan: Gakugei Publishing.
OECD (2018) *OECD Tourism Trends and Policies 2018*, OECD Publishing, Paris. Available at: http://dx.doi.org/10.1787/tour-2018-en (Accessed 11 July 2019).
Okada, T. (2005) *Chiikizukuri no keizaigakunyumon; Chiikinai saitoushiryoku* (Introduction to Economics for Local Development: Reinvestment within Local Areas). Kyoto, Japan; Jichitaikenkyusha.
Ōta, S. (2002) Kobe no toshiseisaku no kiseki to sono kadai; Kobe burando machizukuri no sōzō hasshin (Urban policy in Kobe and its issues: Creation and dispatch of Kobe brand city development). In S. Ōta and K. Nakashima (Eds), *Kobe toshigaku wo kangaeru* (*Considering Kobe Urbanology*). Tokyo, Japan: Minerva, pp. 28–103.
Page, S. and Hall, C.M. (2003) *Managing Urban Tourism*. Harlow: Pearson Education.
Peeters, P.M., Gössling, S., Klijs, J., Milano, C., Novelli, M., Dijkmans, C.H.S., Eijgelaar, E., Hartman, S., Heslinga, J., Isaac, R. and Mitas, O. (2018) *Research for TRAN Committee-Overtourism: Impact and Possible Policy Responses*. European Parliament, Directorate General for Internal Policies, Policy Department B: Structural and Cohesion Policies, Transport and Tourism.
Popp, M. (2012) Positive and negative urban tourism crowding. *Tourism Geographies*, 14(1), 50–72.
Richards, G. (2014) Creativity and tourism in the city. *Current Issues in Tourism*, 17(2), 119–144.
Sasaki, M. (2004) The role of culture in urban regeneration. *Forum Barcelona*. Available at: www.fundacioforum.org/b04/b04/www.barcelona2004.org/esp/banco_del_conocimiento/docs/PO_22_EN_SASAKI.pdf (Accessed 12 July 2019).
Sasayama, Y. (1994) Kobe shi: āban rizōto toshi (Kobe City: Urban resort city). In K. Niino (Ed.), *Āban rizōto no tanjyō; 21 seiki no toshi senryaku* (Birth of an Urban Resort: Urban Strategy for the 21st Century). Tokyo, Japan: Bunshindo, pp. 69–73.
Sassen, S. (2001) *The Global City: New York, London, Tokyo*. New Jersey, NJ: Princeton University Press.
Selby, M. (2004) *Understanding Urban Tourism: Image, Culture and Experience*. London: I. B. Tauris & Co. Ltd.

Sharpley, R. (2014) Host perceptions of tourism: A review of the research. *Tourism Management*, 42(1), 37–49.

Sharpley, R. and Harrison, D. (2017) Conclusion: Mass tourism in the future. In D. Harrison and R. Sharpley (Eds), *Mass Tourism in a Small World*. Wallingford: CABI, pp. 232–240.

Smith, N. (1982) Gentrification and uneven development. *Economic Geography*, 58(2), 139–155.

Snaith, T. and Hailey, A. (1999) Residents' opinions of tourism development in the historical city of York. *Tourism Management*, 20(5), 595–603.

Sorensen, A. and Funck, C. (Eds). (2007). *Living Cities in Japan: Citizens' Movements, Machizukuri and Local Environments*. London: Routledge.

Spirou, C. (2008) The evolution of the tourism precinct. In B. Hayllar, T. Griffen and D. Edwards (Eds), *City Spaces – Tourist Places: Urban Tourist Precincts*. Oxford: Butterworth Heinemann, pp. 19–38.

Telfer, D. J. (2015) Tourism and regional development issues. In R. Sharpley and D. J. Telfer (Eds), *Tourism and Development Concepts and Issue*, 2nd Edn, Bristol: Channel View Publications, pp. 140–177.

Toyofuku, Y. (2016) Mainasukinriseisaku to Fudōsansijyō no kanetsu (Zero-interest-rate policy and overheating property market), *Keizai* (*Economy*), 254, 66–75.

Vera Rebollo, J. and Ivars Baidal, J. (2009) Spread of low-cost carriers: Tourism and regional policy effects in Spain. *Regional Studies*, 43(4), 559–570.

Wall, G. and Mathieson, A. (2006) *Tourism: Change, Impacts and Opportunities*. Harlow: Pearson Prentice Hall.

Yokohama Shintoshi Kenkyukai (1989) *Yokohama shin-toshi no mirai* (Future of new city, Yokohama). Tokyo, Japan: Bunshindō.

Zukin, S. (2011) *Naked City: The Death and Life of Authentic Urban Places*. Oxford: Oxford University Press.

4 Transition of forest tourism policies in Japanese national forest management[1]

Yumi Oura

Introduction

The majority (67 percent) of the natural land area of Japan is covered by forest. Here, the term 'forest' is used to refer to the non-built environment, ranging from urban peripheral fields to forested areas of mountains and wilderness. According to the Public Opinion Survey on Life and Forest conducted in December 2011, 72.2 percent of respondents stated that they had visited mountains and/or forests within the last year in order to either 'enjoy the landscape and scenery' or 'refresh mentally and physically through forest walks' (Cabinet Office, 2012). In Japan today, forests represent a significant recreational and tourism resource, a major transformation from the time when they were simply a place of timber production.

This transformation is evident in the history of the national forests (NF), which account for approximately one-third of the total forested areas of Japan. Following the establishment of NF by the Meiji Government in the late 1800s, their primary aim was to contribute to economic development through timber production. Over time, however, the administration of NF went through a number of restructuring and management changes owing to fluctuating production demands and fiscal pressures whilst, more recently, NF management and financial considerations have again reached a critical point. Nowadays, NF play a major role in conservation and tourism, and they also promote a more participatory management style and community engagement. Corporate social responsibility activities and volunteer tourism are part of such engagement.

This chapter critically analyses the historic development of the NF management[2] and policies in Japan and discusses the relationship with tourism, focusing in particular on the transition from the post-war production era to the emergence of recreational and tourism-focused developments incorporating concepts such as 'forests for people'.[3] To provide the background to the discussion, however, it is first necessary to define concepts such as *forest recreation* and *forest tourism* as they relate specifically to the NF of Japan.

National forests and forest tourism

The characteristics of the NF management in relation to tourism in terms of meeting public demand and financing forest management can be summarized as follows:

First, the basis for the NF was established in the Meiji era (1868–1912). In contrast to the cases of the US or the UK, the main reason for the nationalization of forests in Japan was to overcome feudalism. At the time of their establishment, however, the Government had no clear purposes or objectives for forest management and so the NF management policies, in effect, evolved in response to particular demands at particular times, specifically, either for timber production/exploitation or conservation (Oka, 1990). As a consequence, although NF were and remain the common property of the people now managed by the Forestry Agency (FA), the emphasis was primarily on profit-making, with insufficient opportunities being given to people to participate in their management and use (Shioya, 1997). This meant that forest management responded primarily to government policy and the needs of business communities as well as the social issues of the time, as is clearly reflected in forest tourism policies.

Second, given that NF were formed from the amalgamation of vast common agricultural forests, the FA proposed to 'revitalize rural areas through the utilization of NF' as one of the missions of NF management. Until the 1950s, the FA promoted the agricultural use of NF mainly for timber production, firewood and grazing pasture. From the 1960s to the 1970s, however, rapid economic growth in Japan contributed to the widening economic gap between cities and rural areas and, hence, tourism development became an important part of national rural development policy. With drastic changes in the socioeconomic status of rural areas, the agricultural use of NF decreased and, instead, tourism came to be seen as more important as a means of achieving rural area development (Oura, 2004). This lead to the establishment of recreational forests, though more strongly reflecting supply conditions (rural agricultural areas or government) than the user demands.

The third factor is the effect of economic issues associated with NF management. Post-war NF management was positioned as a national enterprise, with a Special Account[4] being established with the aim of developing a financially self-sustaining system, and forest management was expected to become self-sufficient. With strong demand for timber following the end of the Second World War (from the 1950s to the early 1970s), there was a surplus of funds in this Special Account. However, during the mid-1970s, when the system became dependent on imported timber, this surplus turned into a chronic deficit and, subsequently, considerable improvements became necessary in NF management. In particular, a forest tourism policy was identified as a measure to secure self-revenue though an expected increase in land use revenue in connection with the resort development boom at the time. Thus, forest tourism policy changed to actively providing NF for large-scale development projects and the NF were, therefore, at the mercy of national policies and economic problems.

Fourth, changes in NF policies after 2000 saw a new development. Despite the improvement plans that had spanned the previous 20 years, NF management had become saddled with significant accumulated debts and, in 1996, the Board of Audit declared the NF to be bankrupt. In 1998, a new policy called 'radical reform' was introduced, developed around the central concept of 'forests for people'. Under this new concept, collaboration and cooperation between the FA, the local authorities and citizens became the flagship policy which, with the recreation forest at its centre, strengthened its role in incorporating various leisure activities.

Following the 2013 law reform, the 'corporatized' NF management based on the Special Account was amalgamated back into the general account system, in essence becoming an entity. This required forest management to become citizen-centred and to be seen to be operating to the benefit the public. Under the current NF management system, however, the achievement of such an ideal continues to remain doubtful, for a number of reasons. In particular, difficulties have been experienced in implementing the mission in daily operations owing to the increase in areas managed by each member of staff. After the 'Radical Reform', the number of workers continued to be cut, which nearly doubled the area assigned to each member of staff. Indeed, by the early 2000s, workforce shortages had already been observed in some of the District Forest Offices (Zhang, Noguchi, Koike & An, 2009).

In addition, workers' skills, local knowledge and experience accumulated over a long period of time are being lost; owing to management restructuring, outsourcing and the reduction in workforce, such a legacy is not being passed on to the next generations (Komaki, 2008). As Yamamoto (2008) points out, foresters did not enjoy sufficient opportunities to gain an accurate grasp of the forest condition owing to workforce shortages. In short, then, it is evident from this discussion that the NF management was placed in a difficult situation as the consequence of financial problems occurring over a number of years. The question is: to what extent do such problems influence forest tourism in a national forest?

National forests, recreational forests and forest tourism

For the Japanese, irrespective of whether they live on rural farms, or in mountainous or urban areas, opportunities for outdoor recreation tend mostly to be on forested lands. 'Forests' in Japan, therefore, refer broadly to natural areas and, consequently, concepts such as 'forest recreation' or 'forest tourism' may be defined broadly as visits and activities taking place in natural surroundings. In Japan, the idea of long-term holidays or leave is still not widely accepted, and trips are typically short-term, including day visits. This is certainly the case with 'forest tourism', although recreational activities in

NF may take the more conventional form of long-term tourism, as these sites tend to be in remote mountainous regions.

The basic development framework for the use of certain sections of NF as, for example, ski areas, campgrounds and hiking trails was established in the 'Forest Park System' in 1973 (Oura, 2004). This classified Recreational Forests into six use categories, including nature education, outdoor sports and hiking. In 2019, there were approximately 300,000 ha (4 percent of the total NF area) and 727 designated Recreational Forests in Japan (Forestry Agency, 2019), including designated World Natural Heritage Sites such as Yakushima Island and Shiretoko,[5] and well-known attractions such as Arashiyama in Kyoto, Miyajima in Aki, Hiroshima and Mount Takao near Tokyo, and a variety of forest types ranging from pristine areas to urban forests.[6]

Transition in post-war national forest management and forest tourism policies

The late 1950s to mid 1960s: the stage of passive approaches

From 1955 onwards, Japan's economy began to grow significantly and, during the 1960s, Japanese people enjoyed increases in both income and leisure time. In addition, with the development of new transport networks, the demand for tourism and recreation gradually increased. At this time, however, NF management was focused primarily on establishing a timber production system following the policies in the 1957 National Forest Productivity Enhancement Plan. Supported by the strong domestic demand for timber and forest land development in the backcountry, this plan reached its peak in the middle of 1960s (see Figure 4.1).

Under these circumstances, the first move to offer opportunities for tourism in natural (forested) areas was the establishment of so-called 'natural parks', including national parks. The government valued the development of tourism within its post-war economic growth strategy, and tourism in national parks was considered an important element of this strategy. In 1952, the 'Special Measures Concerning Road Construction and Improvement' were enacted and, as a consequence, the construction of toll roads in mountainous areas improved access to national parks. By 1970, 64 such roads had been constructed nationally (Murakushi, 2015). In addition, the construction of ski areas, golf courses and accommodations facilities began. In 1960, there were approximately 140 million visits to national parks and quasi-national parks, a figure which had increased to 400 million by 1965 (Ministry of the Environment, 2016). This growth in demand, however, also brought threats, such as damage to the forests, accidents involving visitors and fires. Consequently, the FA established the National Ski Area System in 1959 and, in 1966, the National Camp Sites System, and also placed emergency shelters in mountainous areas. During this period, recreational activities became popular in Japanese forested lands, especially around national

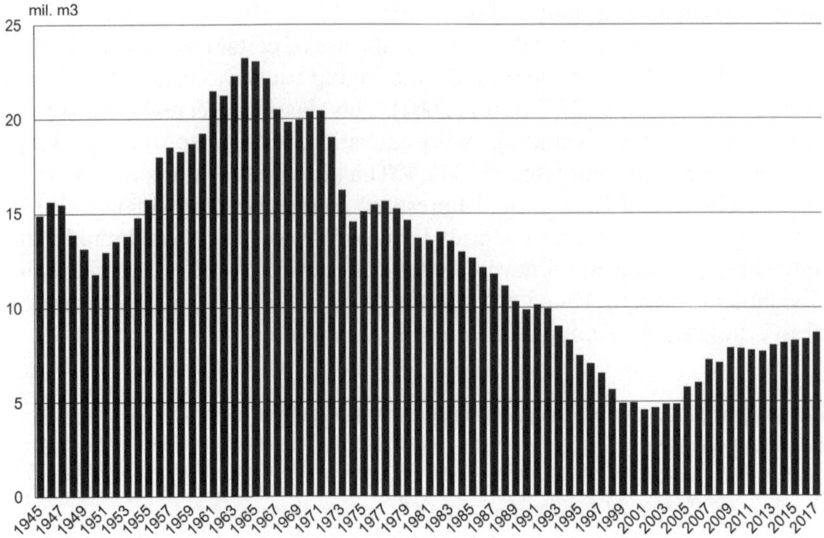

Figure 4.1 Volume of wood production of national forests (1945–2017).

Source: Kokuyurinya jigyou toukeisho [Statistics of the national forest management]. The rising trend from 2005 is strongly influenced by a change in the way statistics were collected and collated from that year onwards.

parks and forests. Broadly speaking, management systems during this period focused on zoning measures to divide recreational and forestry areas rather than actively encouraging recreational use.

The late 1960s to 1970s: establishment of the forest park system in national forest management

As noted above, increases in the rate of economic growth during the 1960s were reflected in a commensurate growth in levels of wealth in Japanese society. However, the consequences of rapid urbanization and industrialization, such as pollution and the destruction of natural areas, became rapidly evident; at the same time, urban–rural income disparity, along with the increased demand for a labour force in urban areas, accelerated a massive population flow from rural to urban areas.

The Comprehensive National Development Plan (1962) and the Act of Measures for Promoting Mountain Villages (1965) were therefore established to solve this 'regional gap'. A comprehensive development plan for rural municipalities was designated as the method to develop tourism recreation in rural areas, and private forest areas were developed as resorts and golf courses. This was, however, often undertaken in an uncontrolled manner, resulting in serious problems arising from both the overuse and the destruction of natural resources. Hence, in the late 1960s, a government council began to examine measures to mediate these issues. The government demanded the active and prompt development of a system that promoted planned and harmonious development along with the protection of natural recreation areas for NF that would qualitatively and quantitatively provide the majority of the natural recreation areas.

At the same time, the development of NF, especially those in backcountry areas, that had been heavily exploited during the period of strong economic growth had led to problems including loss of mature forest ecosystems and rare plants, and frequent occurrence of disasters. Local communities had been strongly opposed to these developments and, with the establishment of a new system dependent on imported timber, the demand for the development of domestic forest resources weakened. Consequently, NF management was obliged to adapt its policies from focusing on timber production to a more public function for the forests. With the introduction of New Forest Management in 1973, a streamlined management policy was presented which sought a reduction in timber harvesting, clearance-cutting and staffing levels – in short, an overall reduction in the business itself.

Subsequently, the Natural Recreation Forest System was established in 1967, the aim of which was the controlled use of NF for both nature conservation and public recreational use, while making adjustments to the forestry management. In 1972, this system was further developed into the Forest Park System as a comprehensive system that incorporated traditional uses of ski areas and campgrounds. In so doing, the overall approach to forest tourism policy in NF experienced a fundamental transformation. In other words, in a departure from the preceding passive approach, a critical stance against uncontrolled development of forested areas was adopted and, at the same time, there was a proactive response to the public's needs for forest recreation while protecting natural resources. This policy was established as a unique approach to managing to NF that contributed to the revitalization of rural areas.

The 1980s: expansion of the scale of forest tourism development following the financial difficulties facing national forest management

As described above, NF management transitioned toward decreasing deforestation but, because of low timber prices affected by the imported timber system and the large investment costs of overharvesting, the project fell into deficit by the mid-1970s. The FA had no option but to improve its management and, with the establishment in 1978 of a Special Measures Law with the purpose of enhancing the management of the NF, a management improvement plan was implemented with the aim of achieving income and expenditure equilibrium within 20 years. In the following decade, and through four revisions, costs were reduced by streamlining management through staff reductions, by the elimination and consolidation of organizations and the contractor system, and though anti-deficit measures such as the sale of unnecessary land and forested areas in order to secure self-revenue. The promotion of the management improvement plans had a significant impact on many forest tourism policies including: (i) the withdrawal of the FA as the leading management position in forest parks; (ii) a decreased financial burden on forest parks management following the introduction of facility user fees; (iii) the awarding of contracts to local municipalities and private businesses; and (iv) beginning in the late 1980s, the active promotion of large-scale resort development with the hope of increasing land-use fees.

The latter policy, promoting the development of large-scale resorts, reflected national policies at the time. While ostensibly seeking to increase opportunities for the use of people's leisure time, the true nature of national policy was to target rural agricultural, forestry and fishery areas for investment by large companies and the public as part of an economic structural adjustment policy that would change the Japanese economic structure from being external demand (export) dependent to internal demand-focused. This, in turn, sought to decrease the pressure from the US regarding the imbalance in Japan–US trade. Meanwhile, in rural agricultural areas, population decline again became an issue in the 1980s, with de-industrialization and severe challenges facing the agricultural and forestry sectors further accelerating out-migration form rural areas. Combining the desperate hope of rural areas for revitalization through resort development and the expectations of large companies to find new areas of investment, the trend for large-scale resort development swept through the nation, stimulated in no small measure by the Comprehensive Resort Area Development Law (aka Resort Law) in 1987 (see also Chapter 6). In tandem with this trend, NF developed a so-called Human Green Plan that allowed for the development of NF into large-scale resorts. This was actively promoted, and forest tourism policy for NF went through a major qualitative change from providing services to citizens to a business that focused on self-revenue as part of its fiscal reconstruction programmes.

In contrast with policy changes within forest tourism, and in addition to traditional activities such as skiing, driving and hiking, other activities that focused on the forests themselves began to emerge, the most notable being 'forest bathing'. Since it was first proposed to the FA in 1982, the concept of walking in forests in order to promote mental and physical health rapidly became accepted and adopted by society as health awareness increased. In the mid-1980s, nature experience programmes and guided tours were also introduced into each area whilst, at the end of the decade, bureaucrat-led guide training projects, such as forest instructor training, were also introduced. Consequently, in recreational areas in NF, the forest recreation experience gradually became as diverse as the forest tourism policy, including programmes such as regular forest bathing events organized by local forestry offices and fee-paid guided trips offered by trained staff.

The 1990s and later: failure of the large-scale resort development project and bankruptcy of the national forest system

At its peak, the private sector-led large-scale resort development trend that swept through the nation from the late 1980s onwards resulted in some 20 percent of national land being included in the project areas. With the collapse of the bubble economy in the early 1990s, however, private businesses started to withdraw one-by-one, and projects were suspended or discontinued. By 1992, out of 77 priority sites designated under the Resort Law, 39 percent had not begun construction, while 83.8 percent of 2,046 planned sports and recreation facilities had not even begun (Nihon Keizai Shinbun, 1994). Overall, the large-scale resort project was a complete failure as a development programme; in particular, the decline in the popularity of skiing had a major effect on forest tourism. Many of the ski resorts had been developed from the late 1980s into the early 1990s in mountainous regions that usually experienced heavy snow. However, following a peak in 1992–1993, visitor numbers almost halved within 5 years and, in the 2000s, fell to between a third and a quarter of the peak level, driving many ski areas into bankruptcy, see Figure 4.2. In addition to the economic downturn, the diversification of leisure activities and the increased time that younger people spent on ICT-based leisure activities,

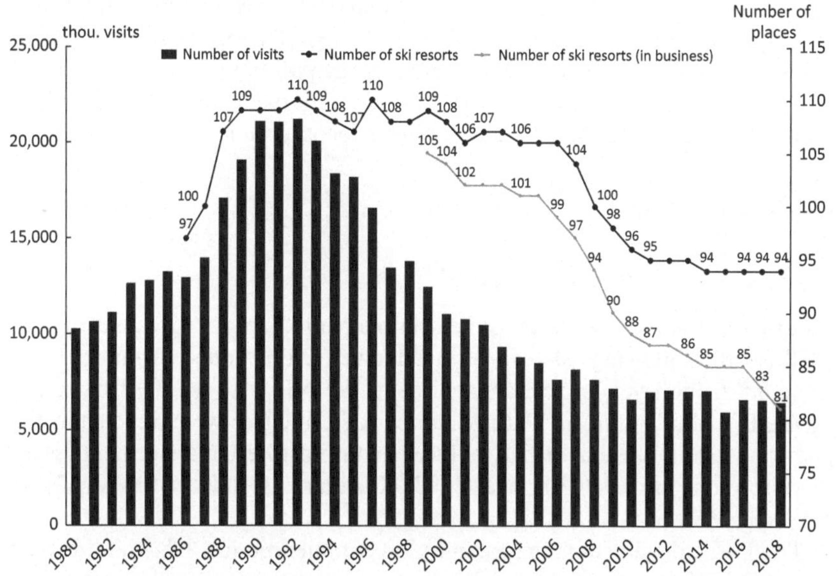

Figure 4.2 Number of ski resorts and its visitors in Nagano prefecture (1980–2018).

Source: Adapted from Suki jou tou genkyo chousa kekka [Report of present state of Ski resorts] and Suki suketo jou riyousya toukei chousa [Statics of visitors of ski/skate areas] in each fiscal year by Nagano prefecture.

such as utilizing social media or playing on game consoles, were contributing factors. Inevitably, the decline in skiing had major negative economic impacts on many of the mountain villages that had depended on ski areas for their winter revenue.

Meanwhile, with regards to forest administration from the late 1980s, public participation in forest management became the main thrust. For both national and private forests, as the profitability of forestry rapidly decreased, the cost and burden of maintaining the maturing artificial forests (approximately 40 percent of the forested land) became an issue. This burden was shared by rural agricultural areas, municipalities and residents of downstream urban areas, all of whom benefitted from these forests. The government and forestry-related organizations developed a system for downstream urban municipalities to contribute to forest maintenance and to organize fund-raising projects, while actively holding educational events and volunteer training to encourage public participation. In the 1990s, forestry activities as public-led initiatives rapidly increased in number, reflecting the increasing public participation in forest management. Indeed, although forest volunteer activities originated as citizens' movements to solve various problems, such as the increased number of uncared for forests and the lack of manpower resulting from the forestry recession, government outreach and awareness activities increased the number of participants in what has now become a socially acknowledged activity. A series of forestry activities were traditionally categorized as work came to be considered as enjoyment and established as a form of forest recreation that actively engage people with the forests. Along with this change in attitude, public participation in forest management also became the mainstream policy for the national forest tourism policy, taking the place of the facility development-focused policy.

New developments

Appearance of the 'forests for people'

As discussed above, NF management implemented an improvement plan in 1978 for the following 20-year period. However, this ended in failure; with some 3.8 trillion yen in accumulated debt, the NF were declared bankrupt by the Board of Audit in 1996. A number of criticisms were directed at the plan, most notably that it had reduced national forest management issues to simply one of cost (Arinaga, 1997; Kasahara, Kouda & Shioya, 2008; Noguchi, 1999).

The Radical Reform management plan commenced in 1998. In return for using 2.8 trillion yen of taxpayers' money to pay-off the accumulated debt, the NF were declared as forests for people. Under this new concept, the national forest policy changed direction towards a new 'collaboration and cooperation' system with goals of: (i) further promoting cooperation with private forests in forest development and forestry production; (ii) creating and improving a system for developing diverse activities, such as citizen-led forest development projects, corporate social action programmes and forest environment education programmes; and (iii) increasing cooperation with citizens in forest conservation management activities, such as patrolling the forest. Since 2004, this has been regarded as a model project that focuses on shared management of NF by the FA, local municipalities, local resident organizations and non-profit organizations. This collaboration set out to achieve a balance between nature conservation and forest development within the wide expanses of NF (ca. 10,000 ha), and was an attempt to implement a new form of national forest management within the forests for people concept. In addition, as a facilitation body, forest interaction promotion centres were installed in nine locations to support citizen-led activities and environment education programmes.

Overview of forest park reform measures

Following the radical reform referred to above, national forest tourism policy was acknowledged as a model framework toward actualizing 'forests for people'. Existing recreation areas and their surroundings were actively promoted as places for citizens to voluntarily undertake forestry activities, and for school and social education programmes. However, a variety of factors contributed to management challenges, such as the economic and population decline of rural areas and the rise of new tourism in the 1990s, the large number of ageing and unused facilities and the unstructured nature of management organizations in many areas. The FA therefore established the Committee on Forest Parks in National Forests in 2004 and, in the following year, the Forest Park Reform Measures (Reform Measures) were introduced.

The Committee addressed several problems in recreation forests, including: decreasing business opportunities owing to low economic growth and the diversification of leisure activities; quantitative saturation from the increased number of similar facilities in private forests; decreasing appeal owing to aging facilities; and stagnation in large-scale facility development, such as ski areas. Meanwhile, various activities responding to public interests, such as guided tours with interpretive materials, became necessary. In addition, awareness about nature conservation in response to environmental degradation and the need for biodiversity conservation was increasing. Consequently, the Committee concluded that contributions should be made toward the solution of global environmental issues that are also policy issues for the government and that public participation in the forest management approach should stimulate efforts to actualize the 'forests for people' concept. As a future measure, a policy change from quantitative sufficiency to qualitative improvement that suits user needs was proposed as the overall concept with the primary implementation goals to: (i) review the settings for forest parks; (ii) examine new zoning measures to allow for different recreational opportunities; (iii) promote cooperation with local municipalities and citizens in management and operation projects; and (iv), improve activity programmes and information provision and promote safety measures.

The greatest progress made from a national perspective was the review of forest parks. Between 2005 and 2012, about 13 percent of existing projects and 5 percent of the areas were discontinued, most of which were related to large-scale facilities such as ski areas. To promote cooperation in management and operation, the Council was revised to improve its function by including members from the local municipality, businesses and citizens' groups at the core and a new agreement with the District Forest Offices was made. Diverse partnerships occurred in many locations, allowing for unique management, but the degree of success varied from region to region owing to economic conditions and extensive amalgamation of municipalities.

Characteristics of the new developments

In the areas where measures were possible, cooperation between local municipalities and citizens' groups progressed, with the user becoming more active in some situations. Some case examples are now provided.

The first case involves local municipalities valuing NF for the purpose of community building. In the Poroto Forest Park in Shiraoi town, Hokkaido, following the introduction of a project to renew traditional indigenous people's (the Ainu) living spaces in the 2000s, efforts were made to use resources within local municipalities, such as the training of nature guides, forest development to support Ainu culture and the implementation of environmental education programmes that incorporated the study of Ainu culture. This encouraged the local people to acknowledge NF as local resources and, in 2012, a partnership council was developed. Nature guide organizations led by citizens and the Ainu Museum joined in this partnership council, along with the town-affiliated organizations and, through user surveys, a new direction for the use of Poroto Forest Park was developed.

The second example highlights a case in which the activities of citizen groups that enjoy forestry activities are actively accepted. Minoh Forest Park in Minoh City, Osaka Prefecture, is familiar to many local people as it is a green space near an urban area, with about 2.6 million annual visits. In the 2000s, in parts of this park, volunteer forestry activities were welcomed, allowing new kinds of collaboration in forest management and making the management system more people-centred. Currently, nine citizens' groups are working together, using their expertise to provide environmental education opportunities for children, while installing management and operation programmes that incorporate biodiversity and landscape conservation.

The third case provides an example of efforts associated with the creation of stay-type tourism that utilizes so-called 'forest therapy'. Forest therapy takes forest bathing one step further to scientifically elucidate the physiological effects of forests, and utilizes the findings to improve mental and physical health. Along with the changes in the medical system of Japan from being disease to health oriented (i.e., a shift in emphasis from disease treatment to prevention), studies on the physiological activation function of forests were gradually conducted in the early 2000s. In 2006, a certification system began. The main theme was recuperation which was facilitated by long-term stays in health resorts, the use of local food, and the creation of new jobs such as medical staff. Forests are garnering attention as the core for the stay-type tourism that comprehensively utilizes local resources. These points are valued in the Reform Measures as well, and NF actively provide the field space.

In the Akasawa Forest Park in Agematsu town, Nagano Prefecture, a forest therapy programme was introduced in 2005. The council, consisting of the municipality, medical staff, tourism staff and a group of local residents, was established in 2005. The council offered health consultation tours in cooperation with medical organizations, weekly physician-led tours of forest park facilities, three times a week placement of resident nurses and public health nurses, and tours accompanied by instructors. This programme catered to tourists while also valuing medical measures. Therefore, health consultation in recreation facilities became available for local residents. The physician-led tours of the recreation forest facilities included a tour of backcountry settlements, thereby supplementing the medical care in these mountain settlements. With these efforts, repeat visitors and overnight guests gradually increased and, furthermore, through therapy guide training projects, the number of residents involved in forest park increased, hence enhancing cooperation with local residents.

A fourth example illustrates the use of the area outside of the existing forest parks. Instead of establishing specific areas and facilities, maintenance roads and forest roads that were originally used for timber production and dam management were restored within the NF for use as recreation trails. Shin-etsu trail, located along the border of Nagano and Niigata Prefectures, is an 80 km-long trail that was opened in 2008. The majority of the route through the mountains is within NF boundaries. From around 2000, the trail was conceptualized because of the decline in the ski industry and the transition to green tourism. A feasibility study was conducted with nine surrounding municipalities and, in 2003, the NPO Shin-etsu Trail Club was established to maintain trails, perform natural environment surveys, conduct guided tours and provide information for users. Trail maintenance was performed mainly by local residents, and many volunteers from both within and outside of the prefecture participate. Nowadays, trail maintenance work by volunteers is undertaken more than 20 times a year.

In addition to such efforts, in the project-type cooperation that has been implemented on a trial basis since the mid-2000s, tourism is considered to be a part of sustainable area development. In the AKAYA Project in Minakami town, Gunma Prefecture, visitors are able to enjoy nature through engaging in educational activities such as water resource and forest conservation, joining a nature observation club and undertaking biological surveys. In addition, activities that aim to build a conservation/management/utilization system of the natural environment through sustainable use, and projects that restore old roads for use as trails for guided tours, are being implemented. To restore biodiversity in the target areas, anyone who agrees with the goal of the project can participate. Currently, about 50 people have registered, many of whom live outside of the prefecture. Activities including monitoring and surveying animals and plants or maintaining boardwalks and forests are performed once a month over a two-day and one-night period.

Discussion

With the development of mass tourism in the 1960s, forest tourism policy came into full swing. Until the early 1990s, it progressed based on quantitative expansion, with the principal focus of facility construction. However, because the development of these forest tourism facilities had a strong public work aspect as regional development in rural areas, development was strongly influenced by not only the demand side but also the supply side. As a result, there was excessive facility construction. Furthermore, the impact of financial issues on NF management further accelerated such excess.

In the late 1990s, decentralization of management began to reflect new developments in tourism and challenges in location issues as local municipalities took the lead. In addition, as a result of accepting diverse usages under the concept of forests for people, new stakeholders appeared and new management and service providers became involved. The fact that voluntary forest maintenance and boardwalk maintenance by citizens have increased indicates that Japan is moving from a stage of simply using forest recreation facilities as public works provided from top-down to a stage where citizens want to be involved with the creation and conservation of favourable forest tourism environments. In addition, as it is the age of leisure diversification, conservation activity itself began to be considered as a leisure activity, incorporating the participation of both tourists and local residents. This not only indicates a change in forest tourism policy, but also changes in the historical relationship between NF and local residents. Through a downturn in agriculture and a rapid decline in NF management, the relationship between NF and local residents through traditional agricultural use weakened. Meanwhile, a new relationship based on forest tourism was born. Currently, such a change in the use of forests is providing an opportunity for local municipalities, local residents and even urban residents to gain awareness about NF and to participate in their management.

While the establishment of partnerships promoted by the FA demonstrates a progressive stance, it also has a passive directionality of streamlining the programmes of the FA by supplementing the significant decline in NF management system owing to financial difficulties. For example, the number of staff in the national forest management was 7,588 in 1999, but declined to 4,060 in 2018 (Forestry Agency, 2000, 2020) and, as mentioned before, it is near the limit of its ability to fulfil its essential management functions (Zhang

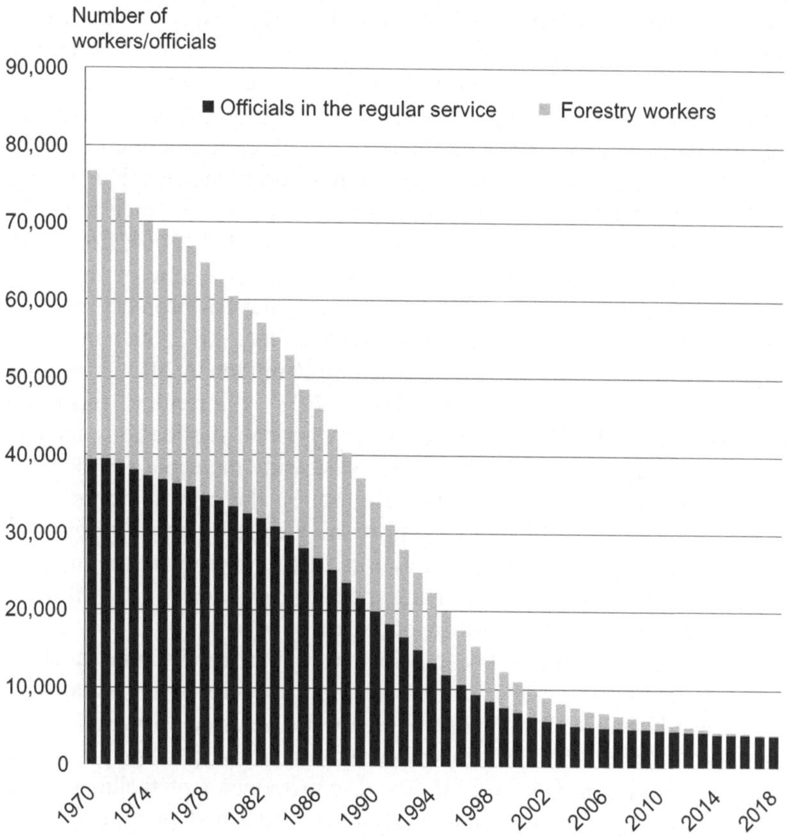

Figure 4.3 Number of workers in NF management (1970–2018).

Source: Adapted from Kokuyurinya jigyou toukeisho [Statistics of the national forest management] in each fiscal year by Forestry Agency.

et al., 2009) – see Figure 4.3. In addition, Japanese foresters at the District Forest Office level are generally reshuffled within several years, quite different from Western countries where foresters are assigned to the same area over a long time to ensure adequate management with the full local knowledge (Ota, 2015). Such frequent changes in personnel is one of major factors that brought stagnation and confusion in many of the collaborative projects (Oura, 2009).

Therefore, in areas where it is difficult to obtain the cooperation of local municipalities, management is declining owing to the recent declining local economy and to factors such as staff shortages and ageing facilities. In addition, the ageing of citizens' group members who played an important role in management and service provision is becoming an issue in various locales, posing a threat to building a stable management system. The introduction of guided tours and interpretation made great progress through the popularization of ecotourism since the late 2000s. However, plans for developing understanding of usage condition, monitoring the impacts on the natural environment and monitoring from a social science perspective, such as satisfaction level of users (which are all essential in understanding the maintenance of a good forest tourism environment), and the introduction of a user management system based on such surveys, have been delayed. This is particularly because the facility development-focused policy in which management focused on the user experience was insufficient (Aikoh, 2014; Ito, 2003).

Since the early 2000s, Japan has positioned tourism as one of the main pillars for the country's economic growth strategy, and has been aggressively working to attract international inbound tourists. Since 2011, the number of inbound tourists has rapidly increased, reaching 10 million in 2013, 20 million in the following 2 years, and over 30 million in 2018. As a result of this rapid growth, overtourism has become a concern in popular destinations such as Tokyo, Kyoto and Osaka. In an effort to redistribute the economic benefits of inbound tourism to rural regional economies, the central government established the 'Tourism Vision to Support the Future of Japan' [*Ashita-no nihon o sasaeru kanko bijon*], with the aim of further increasing the number of inbound tourists, as well as diverting them to rural areas (Council for a Tourism Vision to Support the Future of Japan, 2016).

New tourism destinations became increasingly attractive and actively promoted, including the rich natural landscapes of reforested recreational areas within NF. Between 2017 and 2019, the FA selected 93 model forest parks deemed to have particularly exceptional forest landscapes, and promoted them through multilingual information and improvements of facilities. As a result, the budget for forest recreational areas, which had previously been declining when compared to the total expenditure of the FA, began to significantly increase (Ministry of Internal Affairs and Communications Kanto Regional Administrative Evaluation Office, 2018).

Although this was seen as an opportunity for forest recreational areas, the maintenance of which had been stagnant owing to budget cuts in the FA and local governments, the results of the 2018 administrative evaluation of the project implementation conducted in the Kanto region were poor (Ministry of Internal Affairs and Communications Kanto Regional Administrative Evaluation Office, 2018). Among the seven selected areas in the region evaluated, only two were found to have conducted facility inspections in accordance to the safety measures guidelines. In addition, a number of management deficiencies were highlighted, including cases of trees falling on the sidewalk, inadequate trail signs that may cause visitors to become lost, insufficient warnings about dangerous wildlife, abandoned huts left crumbling, the illegal dumping of garbage overlooked and incorrect access information causing trouble to the visitors.

In this way, the inadequacy of the management systems in some of the recreation areas, contrary to the government's intention to attract international inbound tourists to rural forested recreation areas, potentially reduced its value as a tourism resource and in some instances could be considered a threat to visitor safety. To address these issues in the future it is essential to allocate persons equipped with specialized knowledge about tourism to these recreational forest areas.

Conclusion

The concept of 'forests for people' is the development of a new public sphere around the governance of NF. In addition, various interactions with tourism as a medium are contributing greatly by promoting the establishment of social relationships between forests, rural villages and urban residents. This is a socially significant characteristic part of present-day tourism.

To achieve the ideal of forests for people, the most important point is to strengthen and stabilize the system of the whole NF management. Consequently, the current trends of significant reduction of and frequent changes in personnel should be reconsidered. Then, it is important to keep promoting the active involvement of local municipalities and citizens in the management of recreation areas. At the same time, the FA should be actively involved in resolving issues such as support for sites where such promotion is difficult, development of a thorough management system for conservation of biodiversity and measures against excessive uses, and training and placement of experts to realize these goals.

Notes

1 An earlier version of this chapter appears in *Tourism Planning & Development*, 15(1), 40–54.
2 The term NF management (kokuyurin-keiei) refers not only the management of the timber business but also the public service management in national forests.
3 A study supported by JSPS under KAKENHI [No. 23614008, and No. 15K01946].
4 The Special Account (tokubetsu-kaikei) for NF Management introduced after the Second World War was an autonomous accounting system independent of the national budgetary system.
5 95% of the four World Natural Heritage listed areas in Japan are located within the national forests.
6 They receive approximately 120 million visits annually.

References

Aikoh, T. (2014) Kokuritsukouen no keikaku to kanri no kadai: Daisetsusan kokuritsukouen wo jirei to shita kensho [Issues in planning and management in the Japanese national park system: Perspectives from Daisetsuzan National Park], *Journal of Forest Economics,* 60(1), 14–21.

Arinaga, A. (1997) Shinrin kokuyu to kokuyurin mondai: syakaiteki kisei to rin'ya no shoyu to kanri [State-ownership of forest and today's problems concerning national forest in Japan: social control over ownership and management of the forest], *Journal of Forest Economics,* 43(1), 3–14.

Cabinet Office (2012) Mori to seikatsu ni kansuru yoronchousa [Public opinion survey on life and forest]. Retrieved from: http://survey.gov-online.go.jp/h23/h23-sinrin/index.html

Council for a Tourism Vision to Support the Future of Japan (2016) Asu no nihon wo sasaeru kanko bijon [Tourism vision to support the future of Japan]. Retrieved from: www.mlit.go.jp/common/001126598.pdf

Forestry Agency (2000) Dai 52 ji heisei 11 nen kokuyurinyajigyou toukeisho (heisei 10 nendo) [51th statistics of the national forest management (Fiscal year 1999)] Retrieved from: www.rinya.maff.go.jp/j/kokuyu_rinya/tokei/b_n/1999/pdf/20014230_06.pdf

Forestry Agency (2019) Dai 71 ji reiwa 1 nen kokuyurinyajigyou toukeisho (heisei 30 nendo) [71th statistics of the national forest management (Fiscal year 2018)] Retrieved from: www.rinya.maff.go.jp/all_71.pdf

Forestry Agency (2020) Japan's Forests with Breathtaking Views: Selections from the Tourist Destinations in National Forest. Retrieved from: www.rinya.maff.go.jp/e/national_forest/recreation_forest/index.html

Ito, T. (2003) Nichibei hikaku ni yoru shinrin rekurieshon kenkyu no kensho [A comparative review of forest recreation studies in Japan and the United States], *Journal of the Japanese Forest Society,* 85(1), 33–46.

Kasahara, Y., Kouda, T. and Shioya, H. (Eds) (2008) *Dousuru kokuyurin* [How the national forest should be?]. Tokyo, Japan: Liberta Shuppan.

Komaki, T. (2008) Kokuyurinya no yakuwari to kaikaku no yukue [The role of national forest and direction of devolution], *Forest Economy*, 61(3), 11–13.

Ministry of Internal Affairs and Communications Kanto Regional Administrative Evaluation Office. (2018) Rekurieshion no mori no riyou kanri tou ni kansuru gyousei hyouka kanshi kekka houkokusho [Results report of Administrative Evaluation and Inspection on utilisation and management of forest parks in national forests] Retrieved from: www.soumu.go.jp/main_content/000590860.pdf

Ministry of the Environment. (2016) Shizen kouen no kouen su, nenkanriyosyasu no suii [Trends in number of natural parks and of visits per year]. Retrieved from: www.env.go.jp/park/doc/data/natural/naturalpark_6.pdf

Murakushi, N. (2015) Koudoseichoki ni okeru kokuritsu kouen no kajoriyo to sono heigai (jo); koudoseichoki kokuritu kouen seido no kenkyu(4) [The harmful effect of the overuse of the national parks in Japan during the high economic growth period (1)], *Keizai shirin,* 81(2–4), 317–354.

Nihon Keizai Shinbun. (1994, January 17). Shisetsu no 8 wari ga michakko: Zusan na keikaku rotei [80% of construction works of planned facilities hadn't started yet: a faulty plan came out to light], 3.

Noguchi, T. (1999) Shinrin kokuyu no igi to nihon no kokuyurin mondai [Meanings and problems of state-ownership of forest and the national forest in Japan]. In S. Fukao (Ed.), *Ryuiki ringyo no totatsuten to tenkaihouko* [Achievement and direction of river-basin forestry]. Fukuoka, Japan: Kyushu University Press, pp. 239–255.

Oka, K. (1990) Kokuyurin keiei [National forest management]. In R. Handa R. (Ed.), *Rinseigaku* [Forestry Policy]. Tokyo, Japan: Bun-eido shuppan, pp. 141–156.

Ota, I. (2015) Wagakuni ni okeru kokuyurin no sonzaiigi ni kansuru ichi kousatsu [Purpose of national forest management in Japan], *Journal of Forest Economics,* 61(1), 3–14.

Oura, Y. (2004) Kokuyurinya ni okeru shinrin rekurieshon jigyo to chiiki syakai [Recreational enterprises in national forests and their influence on local society], *Nagoya University Forest Science,* 23, 19–98.

Oura, Y. (2009) Kokuyurinyajigyou ni okeru 'moderu purojekuto' ni kansuru ichikousatsu: Miyazaki ken aya no shoyojurin purojekuto wo jirei to shite [Present state of model project in national forest management: A case study in aya no shouyoujurin project in Miyazaki Prefecture], *Tourism Studies,* 0, 1–13.

Shioya, H. (1997) Kokuyurinya no 'koukyosei' to hoseidojou no kadai [Public interests of national forest and problems of laws], *Journal of Forest Economics,* 43(1), 23–30.

Yamamoto, H. (2008) Kokuyurin ni okeru shinrinkanrigijutsu no kadai [Technical problems of national forest management], *Forest Economy,* 61(4), 17–19.

Zhang, X., Noguchi, T., Koike, M. and An, B. (2009) Kokuyurin ni okeru suigenkan'yohoanrinseibi no jittai to mondai ten; naganokennai no ukeoijigyotai wo jirei to shite [Actual condition and problems of protected forest management as source of water supply in national forest; a case study on private contract enterprises in Nagano Prefecture], *Journal of Forest Economics,* 55(1), 99–108.

5 A systematic review of sport tourism research in Japan[1]

Eiji Ito and Tom Hinch

Introduction

Japan will host the Rugby World Cup in 2019, the Tokyo Olympic and Paralympic Games in 2020 (postponed to 2021) and the World Masters Games Kansai in 2021. These 3 years have been called the 'Golden Sport Years' (Mano, 2015) and the announcement of these events has resulted in sport tourism receiving much attention in Japan. Moreover, recognition of its potential significance has also been reflected in an increase in the Japanese government's promotional activities with regards to sport tourism. For example, in 2011 the Japan Tourism Agency proposed the Basic Policy to Promote Sports Tourism in order to promote the coordination of sport and tourism. In the following year, the Japan Sport Tourism Alliance was launched to help bridge the spaces between public, private and academic stakeholders whilst in 2015, the Japan Sport Agency was organized under the Ministry of Education, Culture, Sports, Science and Technology. This agency announced the Sport Basic Plan in 2017, which emphasized sport tourism promotion. More specifically, one of the aims of this plan was to increase the number of inbound sport tourists to 2.5 million and sport tourism related consumption to 380 billion yen by 2022. In addition, for the Rugby World Cup and the Tokyo Olympic and Paralympic Games, many municipalities have entered into agreements to host visiting national teams' pre-training camps and associated participant-based events. It is also expected that up to 30,000 domestic and 20,000 international athletes will take part in the World Masters Games Kansai in 2021. Hence, the prospect of all three of these events provides a strong incentive for sport tourism research in Japan; however, it is important to ensure that this scholarly effort is sustained after the sporting events have taken place.

Sport tourism refers to 'sport-based travel away from the home environment for a limited time where sport is characterized by unique rule sets, competition related to physical prowess, and a playful nature' (Hinch & Higham, 2001: 49). This concept covers a variety of sport activities ranging from recreational to elite sports. Higham and Hinch (2018) propose the following four types of sport tourism.

1 Active sport tourism: tourists engage directly in sport activities (e.g., skiing) at a destination.
2 Spectator events sport tourism: tourists watch high performance sport events (e.g., Tokyo Olympic Games) at a destination.
3 Participant events sport tourism: tourists/recreational athletes participate in a sport event (e.g., World Masters Games 2021 Kansai) at a destination.
4 Heritage sport tourism: tourists visit a sport heritage site (e.g., sport hall of fame) at a destination.

Of these, heritage sport tourism has been researched the least (Higham & Hinch, 2018), although Ramshaw and Gammon have been making substantive contributions to this literature (e.g., Hinch & Ramshaw, 2014; Ramshaw & Gammon, 2005). The lack of literature in this realm may be because some researchers (e.g., Kudo & Nogawa, 2002) have argued that sport tourism only occurs when sport involvement is the primary purpose of travel. However, Higham and Hinch (2018) demonstrate that some tourists travel to a destination for sport involvement as a main purpose (primary), whereas others travel to a destination for multiple reasons that include sport involvement as a (secondary) purpose. Moreover, some tourists may even not be aware of the opportunity for sport involvement until they are at their destination. Once there, however, they may engage in sport which may provide them with a memorable sport tourism experience (tertiary).

The relationships between sport and tourism have attracted academic attention since the mid- to late 1990s (Gibson, 2016) whilst, in particular, the specialized *Journal of Sport & Tourism*, which enjoyed its 20th anniversary in 2016, has made a substantial contribution to the accumulation of sport tourism knowledge. Sport tourism research has also been published in the field of sport sciences, such as in sport management publications, as well as in tourism journals. Recently, Hinch, Higham and Sant (2014) reviewed sport tourism articles published in five journals (*Journal of Sport & Tourism, Journal of Sport Management, European Sport Management Quarterly, Tourism Management, Journal of Travel Research*) from 2007 to 2011 in order to examine the major trends in sport tourism literature in the West. They identified 102 sport tourism articles, with 67 articles of these being published in *Journal of Sport & Tourism*, 23 published in sport management journals (16 in *Journal of Sport Management* and 7 in *European Sport Management Quarterly*), and 12 published in tourism journals (10 in *Tourism Management* and 2 in the *Journal of Travel Research*). Of these articles, 55 percent focused on the spectator events dimension followed by active (20 percent), participants' events (12 percent), and heritage (2 percent) sport tourism. Most of these articles were empirical (86 percent) with 12 percent being conceptual and 2 percent being reviews of the literature. In terms of the methodological approaches employed in the empirical articles, 59 percent were quantitative and 36 percent were qualitative in nature, whilst the remaining 5 percent of the articles featured mixed-method approaches. In terms of methods, self-completed survey questionnaires were employed the most frequently (46 percent) followed by interviews (14 percent), case studies (10 percent), secondary data analysis (6 percent), modelling and scaling (6 percent), content analysis (4 percent), ethnography (4 percent) and other, such as focus groups (9 percent). From a geographical perspective, the empirical articles were largely dominated by Europe (30 percent) and North America (25 percent), while Asia only accounted for 11 percent. Hinch et al. (2014) also reported that there have been some advances in the research of spectator events sport tourism, such as the FIFA World Cup and the Olympic Games, whereas heritage sport tourism was largely overlooked in the sport tourism literature with some exceptions (e.g., Ramshaw & Gammon, 2005). It was also reported that few studies have addressed the role of sport tourism in small communities and peripheral areas. Hinch et al.'s (2014) findings are of particular importance to the field by revealing the trends in the sport tourism literature in the West.

In the context of this chapter, it is also important to recognize that some sport tourism studies in Japan have been published in English in Western-based journals. For example, Lakshman (2008) and Manzenreiter (2008) examined the economic impact of the 2002 FIFA World Cup in Japanese contexts. Shimizu (2014) focused on the bidding stage of another mega event in Japan (the 2020 Tokyo Olympics) reporting that: (a) a strong link between sport and national identity was reflected in the process, and (b) more public discussion was needed about this relationship and how it is formalized in policy and governance. In terms of social-psychological dimensions, Nishio and his colleagues (Nishio, 2014; Nishio, Larke, van Heerde & Melnyk, 2016) investigated Japanese sport tourist behaviours by focusing on their constraints and motivations. However, as Hinch and Ito (2018) report, English language publications focusing on sport tourism in Japan are limited in number and, similar to the other sport tourism literature published in the West, they tend to focus on major spectator events.

In Japan, sport tourism research has been published in the sport sciences field, such as in sport management and sport sociology journals, as well as in publications focused on tourism studies. However, the equivalent of a specialized journal such as the *Journal of Sport & Tourism* does not exist in Japan. It is, therefore, important to build on and integrate existing knowledge published in sport sciences and tourism studies journals whilst in particular, understanding the state of sport tourism knowledge in Japan is important if the intent is to manage sport tourism in a sustainable way. One way to consider whether sport tourism research in Japan has a strong foundation is to conduct a systematic review of the current state of this body of literature (Ito, Walker & Liang, 2014). More specifically, examining trends in research methods and themes will help to determine future directions that sport tourism should take if it is to contribute to sustainable development in Japan. Thus, the purposes of this chapter are to: (a) review sport tourism research articles published in Japanese journals; (b) identify trends in research methods and thematic patterns; and (c) propose avenues for future research.

Methods

The benefit of conducting a systematic review is that the current state of knowledge in a particular area can be articulated (Ito et al., 2014). More specifically, our systematic review presented in this chapter attempts to clarify 'what we know', 'what we do not know' and 'what we need to know' (Weed, 2014: 1) in terms of the literature on sustainable sport tourism development in Japan. In our systematic review, we reviewed sport tourism articles published in Japanese journals up to May 2016. The two major academic search engines, CiNii (Articles) and J-Stage, were used to identify relevant sport tourism research articles published in Japanese journals. In order to ensure a comprehensive search of published articles within this domain, the following two combinations of keywords were used: (a) 'sport (スポーツ)' and 'tourism (ツーリズム)' and (b) 'sport (スポーツ)' and '*kankou* (観光: a Japanese indigenous word for tourism)'. The CiNii search located 108 articles for the former and 244 articles for the latter terms, whereas the J-Stage search located 96 articles for the former and 1,117 articles for the latter. Further criteria for inclusion of an article in this systematic review were that it contained 'sport' and 'tourism (or *kankou*)' in its title, keyword, abstract or journal name. The abstracts for these articles as well as the article itself were then reviewed to confirm whether it addressed sport tourism in a substantive way. Articles in which sport tourism was a peripheral topic were dropped from the review. Because we searched for the articles by using an online database, this review did not include unpublished works (grey literature). A decision was also made not to include columns/commentaries and conference proceedings and abstracts.

As a result, 52 articles met the criteria for inclusion in the comprehensive review. As per Hinch et al. (2014), the analysis focused on the following characteristics: (a) the number of articles; (b) the year of publication; (c) journal fields; (d) types of sport tourism; (e) geographical focus; (f) methodological approaches; (g) empirical methods; and (h) research themes. In addition, the articles were categorized according to their focus on the three dimensions of the triple bottom line of sustainable development (social/cultural, economic and environmental) (Hall & Lew, 2009) as Higham and Hinch (2018) argue that all three of these aspects need to be considered to achieve sustainable sport tourism. The social and cultural dimension included various characteristics of sport tourism destinations including participation, planning, education, health and visitor satisfaction, whilst the economic dimension involved characteristics of sport and tourism industries, such as financial viability and economic impact. The environmental aspect related to natural resources including environmental impacts, resource degradation, resource values and the balance between natural and built facilities. These three dimensions tend to be inseparable and integrated (e.g., the economic and environmental aspects), so it was sometimes challenging to categorize a publication as having a primary focus in one dimension or another (Higham

& Hinch, 2018). To ensure the reliability of the categories, the first author and a Japanese sport tourism researcher who did not know about the research purposes coded the articles individually using the same criteria. Inter-rater agreements (Tinsley & Weiss, 1975) or Cohen's kappa was 0.69 in this study. A kappa coefficient of 0.61 to 0.80 is considered to be indicative of 'substantial agreement' (Landis & Koch, 1977); therefore, it is reasonable to state that the first author and the Japanese research assistant were substantially consistent with their coding results. Any remaining rater disagreements were then resolved through discussion between them.

Results and discussion

The number of articles and years of publication

Table 5.1 reports the results of the systematic review. Between January 1991 and May 2016, 52 sport tourism articles were published in Japan. Hence, by simple calculation, two research articles have been published on average every year for the past 25 years. This number amounts to about 10 percent of Hinch et al.'s (2014) count in the five targeted Western-based journals. However, given that 9, 10 and 33 articles were published in the 1990s, 2000s and 2010s, respectively, it is reasonable to state that sport tourism research has been gaining momentum in Japan. Indeed, as anticipated in the introduction to this chapter, the rapid increase in the number of sport tourism publications in the early 2010s coincides with the announcement of the Basic Policy to Promote Sports Tourism in 2011 and the establishment of the Japan Sport Tourism Alliance in 2012. Based on this record, it is very likely that sport tourism research will become even more prevalent in Japan with the three major international sport events scheduled from 2019 to 2021.

Of the 52 sport tourism articles, nine were published in sport science journals (e.g., *Japan Journal of Physical Education, Health and Sport Sciences*, *Journal of Japanese Society of Lifelong Sport*) and three articles were published in tourism journals (*The Tourism Studies* and *International Tourism Review*). Three articles were published in other fields (*Geographical Review of Japan* and *City Planning Review*) and the remaining 37 were published in bulletins commonly published by universities, colleges or faculties/departments in Japan, including *Kanoya Taiiku Daigaku Gakujyutsu Kenkyuu Kiyou* and *Osaka Kankou Daigaku Kiyou*. The dominance of articles published in the university bulletins is noteworthy as the peer review process for university bulletins is commonly conducted internally, if at all. It is, therefore, widely acknowledged that articles published in university bulletins tend to be of lower quality than those published in regular journals which tend to have a rigorous review process. One explanation for university bulletins being the dominant outlet for sport tourism research in Japan may be the lack of accumulated sport tourism knowledge at this exploratory stage of the field. That is, it may be difficult to publish sport tourism articles in established academic fields in

Table 5.1 Results of the systematic review (sorted by year)

Lead author	*Year*	*Type*	*Region*	*Approach*	*Method*	*Theme*
Kureha	1991	Active	Japan	Empirical	Case	Environmental
Nogawa	1992	Participant Event	America	Empirical	Survey	Social/Cultural
Kano	1993	Active	Japan	Review	NA	Environmental
Nogawa	1994	Participant Event	Japan	Empirical	Survey	Social/Cultural
Nogawa	1995	Participant Event	Japan	Empirical	Survey	Social/Cultural
Matsubara	1996	Spectator Event	Japan	Review	NA	Social/Cultural
Kudo	1998	Participant Event	Japan	Empirical	Survey	Environmental
Maeda	1998	Active	Japan	Empirical	Case	Environmental
Tomiyama	1999	Participant Event	Japan	Empirical	Survey	Social/Cultural
Kudo	2002	Other	NA	Conceptual	NA	Other
Okamoto	2002	Participant Event	Scotland	Empirical	Observation	Social/Cultural
Kang	2004	Spectator Event	Korea	Empirical	Case	Environmental
Ito	2007	Spectator Event	Japan	Empirical	Case	Social/Cultural
Usui	2008	Active	Japan	Empirical	Interview	Social/Cultural
Amano	2009	Participant Event	Japan	Empirical	Survey	Social/Cultural
Ito	2009	Spectator Event	Japan	Empirical	Survey	Social/Cultural
Matsunaga	2009	Active	Japan	Review	NA	Social/Cultural
Ninomiya	2009	Other	Japan	Review	NA	Other
Sato	2009	Active	Japan	Empirical	Survey	Social/Cultural
Eto	2010	Participant Event	Japan	Empirical	Interview	Social/Cultural
Kato	2010	Participant Event	Japan	Empirical	Case	Environmental
Miyara	2010	Other	Japan	Empirical	Case	Social/Cultural
Murata	2010	Active	Japan	Empirical	Mixed	Environmental
Hayashi	2011	Active	Japan	Empirical	Case	Social/Cultural
Nishio	2011	Spectator Event	New Zealand	Empirical	Secondary	Environmental
Okamoto	2011	Other	Japan	Review	NA	Environmental

Table 5.1 Results of the systematic review (Continued)

Lead author	*Year*	*Type*	*Region*	*Approach*	*Method*	*Theme*
Sugitani	2011	Active	Japan	Empirical	Case	Social/Cultural
Murata	2012	Other	NA	Review	NA	Other
Oshimi	2012	Active	Japan	Empirical	Interview	Social/Cultural
Sugitani	2012	Other	Japan	Empirical	Case	Economic
Takizawa	2012	Participant Event	Japan	Empirical	Case	Social/Cultural
Emi	2013	Active	Japan	Empirical	Case	Social/Cultural
Nishio	2013	Participant Event	America	Empirical	Survey	Social/Cultural
Okayasu	2013	Active	Canada	Empirical	Case	Social/Cultural
Sato	2013	Active	Japan	Empirical	Case	Environmental
Watanabe	2013	Participant Event	Japan	Empirical	Interview	Social/Cultural
Takizawa	2013	Participant Event	Japan	Empirical	Case	Social/Cultural
Akiyoshi	2014	Spectator Event	Japan	Empirical	Mixed	Social/Cultural
Emi	2014	Other	Japan	Empirical	Survey	Social/Cultural
Ohashi	2014	Spectator Event	NA	Review	NA	Social/Cultural
Motoshima	2014	Active	Japan	Empirical	Survey	Social/Cultural
Yamazaki	2014	Other	Japan	Review	NA	Economic
Kitahama	2015	Other	Japan	Empirical	Case	Social/Cultural
Okamoto	2015	Other	Japan	Review	NA	Social/Cultural
Park	2015	Spectator Event	Japan	Empirical	Survey	Social/Cultural
Shiotsuki	2015	Participant Event	Japan	Empirical	Survey	Social/Cultural
Tanno	2015	Participant Event	Japan	Empirical	Survey	Social/Cultural
Tsujimoto	2015	Active	Japan	Empirical	Case	Social/Cultural
Uchida	2015	Active	Japan	Empirical	Case	Social/Cultural
Yumita	2015	Participant Event	Japan	Empirical	Survey	Social/Cultural
Ogasawara	2015	Active	Korea	Empirical	Case	Social/Cultural
Enokido	2016	Spectator Event	NA	Review	NA	Social/Cultural

Japan, including sport sciences and tourism studies, simply because they do not have a track record for publication in relevant journals. Furthermore, it is common that journal article submissions are based on whether or not the author(s) has a membership with a particular association in Japan. A lower incidence of sport tourism focused conferences may also have contributed to the relatively low number of sport tourism publications in established journals. Another possible reason for the tendency to publish in university bulletins rather than peer reviewed journals could simply be the fact that there is not a direct Japanese equivalent to *Journal of Sport & Tourism* which has a mandate to publish sport tourism research. For all of these reasons, the tendency in Japan has been for sport tourism articles to be published in university bulletins although, as the field transitions through this exploratory phase, it is likely that a greater proportion of sport tourism articles will be published in peer reviewed journals.

Of the 52 articles, the majority of the sport tourism articles focused on active sport tourism (32.7 percent: $n = 17$) and participant-based sport events (30.8 percent: $n = 16$) followed by spectator-based sport events (17.3 percent: $n = 9$) and other (19.2 percent: $n = 10$). This finding contrasts with Hinch et al.'s (2014) findings which reported that 55 percent of the sport tourism articles in their sample were spectator-based sport events. While Hinch et al. (2014) argued that more sport tourism research was needed in the areas of participant-based events and active tourism, it appears that Japanese researchers have already adopted this focus. In other words, sport tourism research in Japan to date has typically focused on the situations in which tourists are physically engaged in sport activities, that is, on active and participant-based sport events. This result suggests that Japanese sport tourism researchers have adapted tourism frameworks or concepts to sport participants rather than focusing on spectators as in the case of the international field in general. For example, a series of pioneering sport tourism studies conducted by Nogawa (Nogawa, 1992; Nogawa & Yamaguchi, 1994; Nogawa et al., 1995) examined participant behaviours at participant-based sport events (e.g., a walking event) from the perspective of sport tourists and excursionists. This focus on participant-based events and active sport tourism is seen as a positive characteristic of the field in Japan; however, with the exception of sport tourism articles published in relation to spectator-based events like the FIFA World Cup 2002, such as Kang (2004), research into spectator events is not well-represented in Japan. One of the obstacles to conducting research at such events relates to the empirical nature of much of the sport tourism literature. For example, it may have been difficult for researchers to obtain permission from stakeholders, such as event organizations or sponsor companies, to conduct surveys at the mega sport events that have been hosted in Japan. In order to overcome this obstacle, Japanese sport tourism researchers need to position their work in terms of win–win outcomes that benefit both researchers and major event stakeholders. Public funding grant agencies must also be convinced of the benefits of this research in relation to the public interest. The major public investments associated with hosting the upcoming mega events offer a good opportunity to argue the need for more research in this area.

Similar to Hinch et al.'s (2014) finding that there was a lack of heritage sport tourism research in the West, no such research publications were identified in this systematic review. This omission is a glaring gap in this body of literature in Japan as even at an intuitive level, it is clear that heritage sport-based travel is a common travel motivation. Examples of heritage-based sport travel range from sport-induced travel patterns associated with Sumo through to travel to pay homage to the Japanese baseball great Ichiro Suzuki who played in a Major League Baseball game between the Seattle Mariners and the Oakland Athletics hosted Tokyo in 2019. Heritage sport tourism research is likely to emerge as a key sport tourism promotion strategy after the three upcoming major sport events (e.g., Olympic stadium and museum tours) as the host cities strive to prolong the benefits of these events. Accumulating research evidence that provides insight into this type of sport tourism should therefore be a priority.

Finally, it is noted that ten articles did not match any of the four types of sport tourism research. These articles mostly explored conceptual foundations (e.g., Kudo & Nogawa, 2002; Murata, 2012) and reported states and challenges of sport tourism in Japan (e.g., Emi, 2014; Kitahama, 2015). This conceptual focus reflects the relatively early phase of this field of research in Japan and will serve the field well as it builds on this foundation. Similarly, it is not surprising to find that 78.8 percent of the articles (n = 41) had a Japanese context as a geographic focus. What is surprising is that just over 20 percent of the publications addressed sport tourism in America, Canada, New Zealand, Scotland and South Korea. This external focus may have been a function of Japanese scholars studying abroad as doctoral students or visiting scholars, or more generally collaborating with international colleagues. As Ito et al. (2014) stated in terms of leisure studies, further cross-national/cultural research is warranted in the study of sport tourism.

Research methods

The vast majority of the articles (78.8 percent: n = 41) were empirical in nature with 19.2 percent (n = 10) being reviews of the literature and 1.9 percent (n = 1) being conceptual. This strong emphasis on empirical approach aligned with Hinch et al.'s (2014) findings. Empirical studies included: case studies (43.9 percent: n = 18), questionnaire surveys (39.0 percent: n = 16), interviews (7.3 percent: n = 3), observations (2.4 percent: n = 1) and secondary analysis (2.4 percent: n = 1). Only two articles (Akiyoshi et al., 2014; Murata, 2010) employed mixed-method approaches. A similar dominance of case studies and questionnaire surveys was also reported in Hinch et al.'s review, such as articles focusing on one or a small number of sport events and sport tourism destinations.

However, the quality of the case studies in the systematic review of the literature in Japan was a major concern. Only a few of the case studies, such as Maeda (1998) met the demanding standards of a case study to 'illustrate a problem; indicate a means for solving a problem; and/or shed light on needed research, clinical applications, or theoretical matters' (American Psychological Association, 2009: 11). Similarly, in terms of the quantitative approaches, some researchers used advanced statistical techniques such as multiple group analysis (Sato et al., 2009) and structural equation modelling (Nishio et al., 2013; Park, 2015; Yumita & Harada, 2015) but most of the others (e.g., Emi, 2014; Watanabe et al., 2013) presented only descriptive statistics, such as means, or proportions, in their findings. Although Nogawa's sport tourism publication series (e.g., Nogawa, 1992) also relied primarily on descriptive statistics, personal computers and statistical software were not easily accessible during this era. While this level of analysis has considerable value in terms of describing the state of sport tourism, it falls short of providing the explanatory power that more advanced statistical techniques can now provide. As Gibson (1998) argued two decades ago, discussions based on the results of sophisticated statistical analyses are needed for future quantitative research. In addition to this limitation, issues of measurement validity and reliability have tended to be overlooked. One example of this omission includes the failure to back-translate measurement instruments developed in another language (Brislin, 1970). These instruments should also be reviewed by experts to consider their applicability in a Japanese cultural context if they originated in a foreign cultural context. Conceptual and measurement invariance need to become important topics of discussion in cross-cultural sport tourism research. Lastly, the use of random, rather than convenience sampling is desired to enable a broader generalization of study results.

Qualitative research approaches, such as interviews or observation, were less common than quantitative approaches. As Ito et al. (2014) recommended, qualitative research helps researchers to provide the type of 'thick description' desired by anthropologists (Geertz, 1973) and cultural psychologists (Shweder, 1991). Qualitative research results in different types of insights than quantitative research with both types being necessary for the further development of sport tourism research (Gibson, 1998). However, even in the context of the qualitative studies that were found, only a handful of articles, including Oshimi et al. (2012) explicitly reported helpful details such as the nature of the interview questions (structured, semi-structured, unstructured) and the process used to develop the question lists. Similarly, the articles that purported to use observation-based methods failed to explain their procedures in detail (e.g., what behaviours the researchers tried to observe or how they recorded the data). While the advantages of triangulation are broadly recognized in research (Flick, 2017), the publications reviewed tended to focus on a single approach, failing to incorporate multiple theoretical approaches, methods and types of data.

Nishio (2011) was the only researcher who employed secondary data analysis as the primary method in his study. This is a missed opportunity as this research approach would open a major line of inquiry related to 'big data' collected by governments and companies. Cultural and institutional practices in Japan have long valued record keeping, which suggests that there are likely to be relevant databases that may allow sport tourism researchers to explore the big picture inclusive of spatial and temporal dimensions of sport tourism in Japan. Lastly, no examples were found in the review of experimental research, even though this approach has long been recommended for sport tourism research (Gibson, 1998). Although experimental methods have limitations in terms of the generalizability of the results, this approach has the major advantage of allowing researchers to measure the effect of the manipulation of an independent variable on dependent variables of interest (Vaske, 2008). It is, therefore, important for sport tourism researchers in Japan to move expand their repertoire beyond case studies and questionnaire surveys. Methodological innovation will contribute to the accumulation of sport tourism knowledge in Japan.

Research themes

The majority of the reviewed articles addressed social and cultural themes (71.2 percent: $n = 37$), suggesting that sport tourism research in Japan tends to be examined from a sociological perspective. Murata (2012) argued that Japanese sport tourism research has primarily been focused on the political and industry functions, such as the impacts of sport tourism on local communities. Sport tourism research in Japan has also been characterized by social–psychological perspectives, such as studies that have examined the intention of sport tourists to return to the destination in the future (Sato et al., 2009), motivations and constraints (Nishio et al., 2013) and destination image (Yumita & Harada, 2015). These social–psychological experiences are likely to influence sport tourists' spending behaviours; therefore, these publications have valuable practical implications for the sport tourism industry. Although these articles were categorized into a social/cultural aspect, they are closely related to the economic realm.

Seven articles (13.5 percent) investigated economic aspects of sport tourism with an emphasis on economic impact (e.g., Kang, 2004; Kato & Yeh, 2010; Kudo, 1998; Nishio, 2011). This research theme, embracing issues such as employment creation, aligns with the Basic Policy to Promote Sports Tourism proposed by the Japan Tourism Agency in 2011. Although the series of sport tourism research projects conducted by Nogawa (Nogawa, 1992; Nogawa & Yamaguchi, 1994; Nogawa et al., 1995) was not included in this category, they did examine the spending behaviours among Japanese sport tourists as a subtheme within their studies. For example, Nogawa (1992) found that Japanese outbound sport tourists at the Honolulu Marathon spent less money while visiting Hawaii compared to general tourists owing to the former's lack of discretionary time to shop and due to their age cohort (20s and 30s) with its associated discretionary income constraints. While not included as part of the sample of academic articles in this review, the Saitama Sport Commission (n.d.) reported eight billion yen as the economic impact of hosting sport events in their sport commission business from April 2015 to March 2016. This 'grey' literature merits further investigation and suggests that some municipalities and sport commissions may have valuable data that could be productively utilized for secondary data analyses.

Only five (9.6 percent) of articles in the review addressed environmental research topics, although this research theme is particularly important in the context of sustainable sport tourism development (Higham & Hinch, 2018; Murata, 2010). One of the challenges of sport tourism in the environmental domain is the speed at which new sports are developed and diffused, often leading to mass participation with all of its associated impacts (Higham & Hinch, 2018). For example, the number of golfers and golf courses in Japan increased rapidly from the 1980s into the 1990s, with substantial implications in terms of land use and the environmental impacts of the traditional maintenance practices in these dedicated green spaces. More recently, trail running seems to be following the same trend, with the number of trail running events increasing by 20 percent from 2014 to 2015, reaching a total of 276 events (Yamakei Research Institute, 2016). It was calculated that the participant population was 200,000 trail runners in 2014 but that the population with the 'intention' to participate was estimated at 700,000, indicating a major growth potential for this sport (Japan Management Association Research Institute, 2014).

Both sports involve the use of natural resources. The likely negative impact on the environment brought about by golf and trail running courses has been noted at a basic level but more insight is needed. The management of the impacts of sport tourism in natural areas is a complex task because: (a) measuring the impact of tourism is fraught with difficulty; (b) baselines rarely exist from which to assess change in the natural environment; and (c) the impact of tourism can seldom be disaggregated from the direct or indirect impact of other human activities (Higham & Hinch, 2018). Given that the Japan Sports Agency promotes outdoor sport tourism as a way of taking advantage of Japan's abundant natural resources, further development of advanced insights into the relationship between sport tourism activities and environment should be one of the priorities for research. Similarly, the impact of seasonality should be examined because sport tourism, particularly outdoor sport tourism, is strongly characterized by seasonal variations (Hudson & Hudson, 2015). While these seasonal variations may be challenging from a business perspective, they may be beneficial in terms of the sustainability of the natural resources. Furthermore, it is not just wilderness and rural settings that need to be studied in terms of the environmental aspects of sport tourism; urban settings also have important environmental implications that should be considered (Higham & Hinch, 2018). In particular, the three major sport events in Japan may give rise to negative impacts on local communities (e.g., increases in rates and rents, community displacements), security concerns and vandalism. Sport tourism research related to the urban environment is likely to be a significant line of inquiry in Japan for the next 3 years given the schedule of international events.

Lastly, most of the studies in this systematic review were descriptive. Gibson (1998: 57) made a similar observation of the broader field of sport tourism 20 years ago, stating that 'there is a need to expand the use of theory to all realms of sport tourism so that the dynamics underlying sport tourism might be better understood and explained'. Theory provides a systematic explanation of observed events and 'it extends the generalizability of the findings; improves the rigor and confidence in the research; provides a structure for integrating and building upon previous findings; and moves beyond the endless, repetitive cycle of purely descriptive research' (Vaske, 2008: 17). By taking advantage of the characteristics of interdisciplinary fields, sport tourism phenomena in Japan should be systematically analysed based on relevant theories found in other fields.

Conclusion

The purposes of this chapter were to: (a) review sport tourism research articles published in Japanese journals; (b) identify trends in research methods and thematic patterns; and (c) discuss the avenues for future research, for sustainable sport tourism research in Japan. This systematic review located 52 sport tourism articles and revealed that sport tourism research has become increasingly popular since the 2010s. In terms of 'what we know', there is a solid foundation of literature related to active and participant-based sport events, empirical research featuring case studies and questionnaire surveys and research focused on social and cultural themes. In terms of 'what we do not know', this systematic review indicated that future research should address gaps related to heritage sport tourism, qualitative methodologies, cross-cultural/national approaches, environmental topics and theory-driven research. Lastly, in terms of 'what we need to know', this study indicates that the integration of interdisciplinary knowledge in tourism, sport and other related studies is the key in helping to encourage sustainable sport tourism development after the events. For example, as Hinch and Ito (2018: 99) state, 'there are significant gaps in the literature including research that examines the potential of mega sport events to leverage active sport tourism initiatives that coincide with *chiiki saisei* (local revitalization) and *machizukuri* (place-based community development)'. These topics should be tackled not only by sport tourism research but also by related research fields, such as geography, sociology or psychology. Additionally, more research is needed in the area of heritage sport tourism which is likely to become increasingly relevant in the aftermath of the three major international sport events scheduled for Japan from 2019 to 2021. To this end, it is important to take a process-oriented focus rather than a goal-oriented focus (Higham & Hinch, 2018) in research associated with the forthcoming major international sporting events to be hosted in Japan. The 'Golden Sport Years' will serve as a stepping stone for sustainable sport tourism development in Japan as long as researchers adopt a continuous, rather than a temporary interest in this topic.

As with all systematic reviews, this one has some limitations. A major one concerns the manner in which the articles included in the review were identified. One of the criteria for the inclusion of an article in this study was that it contained 'sport' and 'tourism (or *kankou*)' in its title, keyword, abstract or journal name. However, it is possible that this criterion might miss some key sport tourism articles. For example, although this study's search did not identify it, Jung's (2002) article, which examined the golf course development in Cheju-Island, South Korea, could be regarded as sport tourism research from the environmental perspective. In addition, the three categories employed for the analysis of research themes, that is, social/cultural, economic, environmental, should be refined. The social/cultural category proved to be a much boarder category than the economic and environmental categories. This may be one of the reasons why this study identified a predominance of social/cultural publications (71.2 percent). The overlap between these three categories also proved challenging as they were not mutually exclusive (Higham & Hinch, 2018).

Despite these limitations, however, this review was successful in building on the approach used by Hinch et al. (2014) in their earlier review, which simply described the research topics of the publications under review rather than categorizing them into research themes. In contrast, this systematic review addressed this issue and, as a result, identified a strong emphasis on social and cultural research trends in the Japanese sport tourism literature. Lastly, unlike Hinch et al.'s review (2014), this study included articles published in university bulletins, thereby better reflecting the body of research literature that exists in a Japanese context. Given Torrance's (2017) emphasis on the importance of reviewed articles in terms of quality, future reviews should consider excluding articles published in universities. Similarly, given the projected boost in sport tourism research in Japan owing to high profile international events being hosted over during the 'Golden Sport Years' (Mano, 2015), another comprehensive review of the literature will soon be merited. In particular, it will be interesting to see if the research focus on these major spectator events spills over into the other realms of sport tourism and whether such interest is maintained. More fundamentally, and as Weed (2005) has called for at a global level, a comparison of the findings from comprehensive reviews separated by 5–10 years will provide an indication of whether the body of sport tourism literature in Japan has continued to progress in an integrated and insightful manner. Accumulating additional knowledge about the sport tourism dimensions of the major international sporting events during these 'Golden Sport Years' in Japan is crucial to the advancement of this research field in Japan. Progress in this area will not only contribute to sport tourism scholarship and practice in Japan but will also contribute to the understanding of sport tourism development at a global level.

Note

1 This chapter is a modified version of the research note published in Japanese (Ito, E. & Hinch, T. [2017]. Kokunai supotsu tsurizumu kenkyu no keitouteki rebuy [A systematic review of sport tourism research in Japan]. *Taiikugaku Kenkyu, 62,* 773–787). This work was supported by Japan Society for the Promotion of Science [grant numbers 16K21172 and 19K20568].

References

Akiyoshi, R., Yamaguchi, Y. and Inaba, S. (2014) Okinawaken no jyumin to puroyakyu kansensha kara mita supotsu tsurizumu no suishin ni kansuru kenkyu [A study on the promotion of sport tourism with local residents and spectators of professional baseball camps in Okinawa], *SSF Supotsu Seisaku Kenkyu,* 3, 64–71.

Amano, K. (2009) Supotsu ibento no soushutsu to kankou shinkou ni kansuru kenkyu: Suponichi Sado ronguraido 210 o jirei ni [Study of creation of sports event and sightseeing promotion: A case of SADO long-ride 210]. *Surugadai Daigaku Bunka Jouhougakubu Kiyou,* 16, 35–52.

American Psychological Association. (2009) *Publication Manual of the American Psychological Association* (6th Ed.). Washington, DC: American Psychological Association.

Brislin, R. W. (1970) Back-translation for cross-cultural research. *Journal of Cross-Cultural Psychology*, 1(3), 185–216.

Emi, R. (2013) Sougougata chiiki supotsu kurabu ni yoru chiiki kasseika no kanousei: Sasebo-shi o jirei to shite [The potential of general community- based sports clubs for local area revitalization: The case of Sasebo city]. *Nagasaki Kokusai Daigaku Kokusai Kankou Gakkai Kankougaku Ronshu*, 8, 55–63.

Emi, R. (2014) Supotsu ni yoru chiiki kasseika ni kansuru kenkyu: Nagasaki-ken Hasami-chou o jirei to shite [Community revitalization through sports: The case of Hasami town, Nagasaki prefecture]. *Nagasaki Kokusai Daigaku Kokusai Kankou Gakkai Kankou Ronbunshu*, 9, 85–94.

Enokido, K. (2016) Orinpikku to toshi soshite kankou no rikai ni mukete: Supotsu kankou to toshi kankou ni okeru gakusaiteki kenkyu to sono kadai o saguru [Toward an understanding of the Olympics, the city, and tourism: Exploring issues in Olympic tourism as an interdisciplinary study of sports tourism and urban tourism]. *Nihon Toshikeikaku Gakkai,* 65, 48–51.

Eto, M. (2010) 'Tsuru do Okinawa' sankasha zoukayouin ni kansuru kenkyu: Okinawa chanpuru moderu [Factors causing the increase of participants in the 'Tour de Okinawa': Okinawa chanpuru model], *Shobi Gakuen Daigaku Sougou Seisaku Kenkyu Kiyou,* 19, 29–50.

Flick, U. (2017) Triangulation. In N. Denzin and Y. Lincoln (Eds.), *The Sage Handbook of Qualitative Research*. Thousand Oaks, CA: Sage Publications, pp. 444–461.

Geertz, C. (1973) *The Interpretation of Cultures.* New York: Basic Books.

Gibson, H. J. (1998) Sport tourism: A critical analysis of research. *Sport Management Review*, 1(1), 45–76.

Gibson, H. J. (2016) Tourism. In G. J. Walker, D. Scott and M. Stodolska (Eds.) *Leisure Matters: The State and Future of Leisure Studies*. State College, PA: Venture Publishing, pp. 85–92.

Hall, C. M. and Lew, A. (2009) *Understanding and Managing Tourism Impacts: An Integrated Approach*. London: Routledge.

Hayashi, T. (2011) Kangaku renkei ni yoru paburishitei o katsuyou shita, supotsu tsurizumu no puromoshon ni kakawaru jirei kenkyu: Atsuma-chou no safu tsurizumu o jirei to shite [Promotion of sports tourism through publicity generated by universities and local government: A case study Atsuma's 'surfing tourism' as a case example]. *Sapporo Kokusai Daigaku Kiyou,* 42, 143–167.

Higham, T. and Hinch, J. (2018) *Sport Tourism Development* (3rd Ed.). Bristol: Channel View Publications.

Hinch, T. and Higham, J. (2001) Sport tourism: A framework for research. *International Journal of Tourism Research,* 3(1), 45–58.

Hinch, T., Higham, J. and Sant, S. (2014) Taking stock of sport tourism research. In A. Lew, C. M. Hall and A. Williams (Eds.) *The Wiley Blackwell Companion to Tourism*. Chichester: John Wiley & Sons, pp. 413–424.

Hinch, T. and Ito, E. (2018) Sustainable sport tourism in Japan. *Tourism Planning and Development,* 15(1), 96–101.

Hinch, T. and Ramshaw, G. (2014) Heritage sport tourism in Canada. *Tourism Geographies*, 16(2), 237–251.

Hudson, S. and Hudson, L. (2015) *Winter Sports Tourism*. Oxford: Goodfellow Publishing.

Ito, R. (2007) 2002 nen FIFA warudokappu no kyanpuchi yuuchi to chiikikasseika ni kansuru ichikousatsu: Oita-ken Nakatsue-mura o jirei to shite [Induction of training camp facilities of the 2002 FIFA World Cup and local revitalization: A case study of Nakatsue Village, Oita Prefecture]. *Kankougaku Ronshu*, 2, 39–49.

Ito, T., Fuzimoto, N. and Shen, M. (2009) Puro supotsu o jiku toshita chiiki okoshi no torikumi to sono kadai: Hiroshima Toyo Kapu Yuu renshujou no kankyaku chosa o chushin ni [Regional vitalization with a professional sport and the challenges it is facing: With a focus an audience survey at the Yuu stadium of the Hiroshima Toyo Carp], *Gendai Shakaigaku,* 10, 101–114.

Ito, E., Walker, G. J. and Liang, H. (2014) A systematic review of non-Western and cross-cultural/national leisure research. *Journal of Leisure Research,* 46(2), 226–239.

Japan Management Association Research Institute. (2014). Toreiru ranning ni kansuru jittai chousa [Field survey on trail running]. Retrieved from: www.jmar.co.jp / job/ sports/data/trailrun-release20140905.pdf

Jung, S. (2002) Reja supotsu kaihatsu to 'naihatsuteki hattenron': Kankoku Cheju tou no gorufujyo kaihatsu no jirei kara [The leisure sport development and endogenous development: The case of golf course in Cheju-Island, Korea]. *Supotsu Shakaigaku Kenkyuu,* 10, 86–100.

Kang, I. (2004) 2002 Kannichi warudokappu to supotsu kankou [2002 FIFA World Cup Korea Japan and sport tourism]. *Momoyama Gakuin Daigaku Sougou Kenkyujo Kiyo,* 29, 57–67.

Kano, K. (1993) Supotsu shakaigaku ni okeru sukijou kaihatsu kenkyuu no ichizuke ni kansuru ichishikaku: Kankou chirigaku o sankou ni shite [An approach for the ski area development in sport sociology: A view from tourist geography]. *Musashigaoka Tankidaigaku Kiyou,* 1, 7–16.

Kato, K. and Yeh, T-M. (2010) Suki intahai ga kaisai chiiki ni motarasu keizai kouka no suikei: Akita-ken Kazuno Hanawa intahai o jirei toshite [Estimating the impact of the inter-high school ski championship on the host city's economy]. *Supotsu Manegiment Kenkyu,* 2, 47–56.

Kitahama, K. (2015) Chihou chuushin toshi ni okeru supotsu tsurizumu no torikumi no kentou [Consideration of effort toward developing sport tourism in regional cities]. *Tokai Daigaku Tankidaigaku Kiyou,* 49, 19–26.

Kudo, Y. (1998) Supotsu tsurisuto no kankou koudou to keizai kouka ni kansuru kenkyu [A study of touristic activities and economic impact of Japanese sport tourists]. *Jouchi Daigaku Taiiku,* 31, 15–26.

Kudo, Y. and Nogawa, H. (2002) Supotsu tsurizumu ni okeru kenkyu wakugumi ni kansuru kenkyu: "Supotsu" no toraekata ni chakumoku shite [A study of framework for sport tourism research: How seize an idea of 'sport' at sport tourism?]. *Juntendo Daigaku Supotsu Kenkou Kagaku Kenkyu,* 6, 183–192.

Kureha, M. (1991) Gunma-ken Katashina-mura ni okeru suki kankou chiiki no keisei [Development of a ski-tourist region in Katashina-mura, Gunma Prefecture]. *Chirigaku Hyouron Series A,* 64, 818–838.

Lakshman, C. (2008) Conditions for hosting mega-sporting events in Asia: Comparing Japan and India. *Asian Business & Management,* 7(2), 181–200.

Landis, J. and Koch, G. (1977) The measurement of observer agreement for categorical data. *Biometrics,* 33, 159–174.

Maeda, K. (1998) Kanu kurabu ni yoru ryuiki nettowaku keisei to sono kanousei: Aratana kasen kankyou seibi no tameno ba no keisei ni mukete [Networking in the river basin by canoe clubs and its possibility: For forming the arena for new way of river improvement]. *Supotsu Shakaigaku Kenkyuu,* 6, 17–29.

Mano, Y. (2015) *Kiseki no 3 nen 2019, 2020, 2021: Goruden supotsu iyazu ga chihou o kaeru* [The miraculous three years, 2019, 2020, and 2021: The Golden Sport Years change rural areas]. Tokyo: Tokuma Shoten.

Manzenreiter, W. (2008) The 'benefits' of hosting: Japanese experiences from the 2002 Football World Cup. *Asian Business & Management,* 7(2), 201–224.

Matsubara, Y. (1996) Supotsu ibento to shiteno Tokyo Orinpikku to kankou: Wagakuni ni okeru kankou kokusaika no shuppatsuten [Tourism and Tokyo Olympics as a sport event], *Ouyou Shakaigaku Kenkyu,* 38, 67–76.

Matsunaga, K. (2009) Nihon ni okeru supotsu herusu tsurizumu no genjou to kadai [On the current state and issues of sports and health tourism in Japan]. *Ryukoku Daigaku Keieigaku Ronshu,* 48, 126–140.

Miyara, T. (2010) Supotsu tsurizumu kosu no kanousei: 'Tougougata chiiki supotsu kurabu' no katsuyou o teigensuru. [The potential of sports tourism course: A proposal for the use of 'integrated community sports club']. *Kankougaku Ronshu,* 5, 29–33.

Motoshima, N., Murashima, T., Nabeyama, T. and Sakairi, Y. (2014) Kendou jugyou ni yoru chuugakusei oyobi daigakusei no imeji no henka: Supotsu tsurizumu ni okeru budou puroguramu no kentou [Kendo trial lesson and its effect on prior image in junior high school and college students: Budo program and sport tourism]. *Kankougaku Ronshu,* 9, 73–84.

Murata, S. (2010) Eko supotsu ni yoru kankou kaihatsu no seitouka to sono riron: 'Seikatsu no umi' no jyuusouteki riyou o meguru gyomin no taiou [Justification of tourism development by eco-sport, and its logic: Dealings of fishermen against overlapping activities on 'seikatsu no umi']. *Soshioroji,* 55, 21–38.

Murata, S. (2012) Supotsu tsurizumu kenkyu no gendaiteki saikousei [Contemporary reconstitution of sports tourism studies]. *Taiikugaku Kenkyu,* 57, 471–482.

Ninomiya, H. (2009) Nihon ni okeru supotsu tsurizumu no shosou: Supotsu tsurizumu douteki moderu no kouchiku [The aspects of the sport tourism in Japan: Construction of the dynamic model is sport tourism]. *Doshisha Supotsu Kenkou Kagaku*, 1, 9–18.

Nishio, T. (2011) Supotsu tsurizumu no keizai bunseki: Nyuzirando ni okeru mega supotsu ibento ni yoru tsurisuto heno inpakuto no kenkyu. *Nihon Kokusai Kankou Gakkai Ronbunshu*, 18, 69–74.

Nishio, T. (2014) The constraint factors of potential international sports fan tourists. *Journal of Vacation Marketing*, 20(2), 137–147.

Nishio, T., Okamoto, J. and Ishimori, M. (2013) Sankagata kaigai supotsu ibento ni okeru autobaundo tsurisuto no kenkyu: Honoruru marason sankasha no douki to seiyaku youin ni tsuite [Study of Japanese outbound sports tourists in an international sports event: Motivations and constraints of Honolulu Marathon participants]. *Supotsu Sangyougaku Kenkyuu*, 23, 75–88.

Nishio, T., Larke, R., van Heerde, H. and Melnyk, V. (2016) Analysing the motivations of Japanese international sports-fan tourists. *European Sport Management Quarterly*, 16(4), 487–501.

Nogawa, H. (1992) Supotsu tsurizumu ni kansuru kenkyu: Honoruru marason no jyuudanteki kenkyu [A study of sport tourism: A longitudinal research study on Honolulu Marathon]. *Kanoya Taiiku Daigaku Gakujyutsu Kenkyu Kiyou*, 7, 43–55.

Nogawa, H. and Yamaguchi, Y. (1994) Kokunai supotsu tsurizumu ni kansuru kenkyu: Touki supotsu ibento o jirei toshite [A study of domestic sport tourism: In the case of winter sport event]. *Kanoya Taiiku Daigaku Gakujyutsu Kenkyu Kiyou*, 11, 103–113.

Nogawa, H., Yamaguchi, Y. and Hagi, Y. (1995) Syougai supotsu ibento ni okeru supotsu tsurizumu no jisshouteki kenkyu [An empirical research study on sport tourism in sport-for-all events: Comparisons between single-night multiple-night tourists]. *Kanoya Taiiku Daigaku Gakujyutsu Kenkyu Kiyou*, 13, 1–17.

Ogasawara, M. and Nakajima, K. (2015) Minkan hieiri dantai ga sousetsu shi unei kanri suru Jejutou shuukai choukyori toreiru 'Jeju Orure' toho ryokou bumu no jittai [An analysis of the popular walking tours of the Jeju Olle long-distance walking trails around Jeju island founded and managed by a non-profit organization]. *Supotsu Sangyougaku Kenkyu*, 25, 61–73

Ohashi, S. (2014) Supotsu tsurizumuron no tenkai: Supotsu kankyaku no tsurizumu mondai o chuushin ni [Problems of sport (sports) tourism: Focused on the issues of sports spectatorial travel]. K*eizai Riron*, 377, 43–67.

Okamoto, J. (2002) Sukottorando no kankou sangyou ni okeru dentou supotsu minzoku buyo [A traditional sport and folk dance in Scotland tourism industries]. *Kenkyu Nenpou (Hitotsubashi Daigaku Supotsu Kagaku Kenkyushitsu)*, 36–42.

Okamoto, J. (2011) Chiikikasseikasaku to shite no supotsu tsurizumu no kanousei [Possibility for sport tourism as policy of local revitalization]. *Hitotsubashi Daigaku Supotsu Kenkyu*, 30, 61–66.

Okamoto, J. (2015) 'Supotsu kankou' no manazashi [The 'ports' tourist gaze]. *Hitotsubashi Daigaku Supotsu Kenkyu*, 34, 30–35.

Okayasu, I., Ito, E. and Yamaguchi, S. (2013) Kanada arubatashuu no shougai supotsu reja to chiiki no kanrensei ni tsuite [Relationship between lifelong sport/leisure and community in Alberta, Canada]. *Shougai Supotsugaku Kenkyu*, 9, 49–55.

Oshimi, D., Harada, M., Sato, S. and Ishii, J. (2012) Supotsu chimu no gassyukuchi senkou ni okeru ishi kettei purosesu no kentou: Koukou daigaku supotsu chimu ni tsuite [Destination decision making process in selecting sports training camps: Focusing on high school and university sports teams]. *Supotsu Sangyougaku Kenkyuu,* 22, 9–27.

Park, Y. (2015) Okinawa-ken supotsu kankou ni kansuru kenkyu: Supotsu kankou no ishiki chousa to moderu kenshou [A study on the impact of sports tourism in Okinawa prefecture: Survey of attitudes and model-based test on tourism among residents of Nago city]. *Osaka Keizai Houka Daigaku Chiiki Sougou Kenkyuujo Kiyou,* 7, 43–67.

Ramshaw, G. and Gammon, S. (2005). More than just nostalgia? Exploring the heritage/sport tourism nexus. *Journal of Sport Tourism,* 10(4), 229–241.

Saitama Sport Commission. (n.d.). Heisi 27 nendo saitama supotsu komisshon ni kakawaru supotsu ibento kaisai ni tomonau nenkan keizai kouka (suikei) [The annual economic impact of sport events hosted by the Saitama Sport Commission]. Saitama Supotsu Komisshon.

Sato, D. (2013) Kindai no Kamisuwa ni okeru onsen supotsu fukugougata kankouchi no keisei to kohan riyou [Formation of hot spring resort area and change in coastal area use with introduction of modern sports around Lake Suwa]. *Rikkyo Daigaku Kankougakubu Kiyou,* 15, 29–45.

Sato, S., Harada, M. and Ohnishi, T. (2009) Supotsu tsurisuto to supotsu ekusukashonisuto no saihouito: Niseko no raihousha no chikakukachi ni chakumoku shite [Sport tourists' and sport excursionists' intention to revisit: Focusing on the perceived value of Niseko visitors]. *Supotsu Manejimento Kenkyuu,* 1, 19–31.

Shweder, R. (1991) *Thinking Through Cultures: Expeditions in Cultural Psychology.* Cambridge, MA: Harvard University Press.

Shimizu, S. (2014) Tokyo: Bidding for the Olympics and the discrepancies of nationalism. *The International Journal of the History of Sport,* 31(6), 601–617.

Shiotsuki, R., Watanabe, R. and Arakawa, M. (2015) Okinawa no supotsu tsurizumu to hosupitaritei ni kansuru hikaku kenkyu: 'ECO Supirittoraido & Uoku in Nanjyou-shi' to 'GREAT EARTH Ishigakijima Raido' ni okeru jirei kara [A comparative study on sports tourism and hospitality in Okinawa: Case studies on 'ECO Spirit Ride & Walk in Nanjyo City' and 'GREAT EARTH Ishigakijima Ride]. *Kankoukagaku,* 7, 1–20.

Sugitani, M., Aoki, M., Ishikawa, Y., Misono, S. and Sugiura, T. (2011) Supotsu tsurizumu no kanousei o saguru: Kokusai rizoto o mezasu Hokkaido Niseko chiiki no jirei kara. [A study on possibility of sport tourism: Case of Hokkaido Niseko, aiming to international resort] *Touhou Gakushi,* 40, 1–15.

Sugitani, M. (2012) Okinawa kankou ni okeru supotsu tsurizumu no genjyou to kadai [On the current state and issues of sports tourism in Okinawa]. *Toho Gakushi,* 41, 47–64.

Takizawa, H. (2012) Fuji-shi ni okeru supotsu kankou ni tsuiteno ichikousatsu: Aruteimetto no zenkokutaikai shouchi o rei ni [A study of sport tourism in Fuji-city]. *Fuji Tokoha Daigaku Kenkyu Kiyou,* 12, 225–247.

Takizawa, H. (2013) Fuji-shi ni okeru supotsu kankou ni tsuiteno ichikousatsu: Aruteimetto no zenkokutaikai shouchi o rei ni (Dai 2 hou) [A study of sport tourism in Fuji-city (The second report)]. *Fuji Tokoha Daigaku Kenkyu Kiyou,* 13, 183–211.

Tanno, T., Watanabe, R. and Shiotsuki, R. (2015) Okinawa ni okeru supotsu tsurizumu to chiikishigen no katuyou: ECO supirittoraido & uoku in Nanjou-shi no juudanteki hikaku [Use of sports tourism and local resources in Okinawa: Longitudinal comparison in ECO spirit ride & walk in Nanjo]. *Atomi Gakuen Joshi Daigaku Manejimento Gakubu Kiyou,* 20, 85–110.

Tinsley, H. E. and Weiss, D. J. (1975) Interrater reliability and agreement of subjective judgments. *Journal of Counseling Psychology*, 22, 358–376.

Tomiyama, K. (1999) Supotsu tsurisuto no kankou koudou: Kitakyushu-shi no kankou syukyaku shisaku heno teigen [Tourist behaviors among sport tourists: A suggestion for tourism policy in Kitakyusyu city]. *Kitakyusyu Daigaku Bungakubu Kiyou,* 6, 51–62.

Torrance, H. (2017) Evidence, criteria, policy, and politics: The debate about quality and utility in educational and social research. In N. Denzin and Y.Lincoln (Eds.), *The Sage Handbook of Qualitative Research* (5th ed.).Thousand Oaks, CA: Sage Publications, pp. 766–795.

Tsujimoto, C. (2015) Supotsu kankou ni yoru chiiki kasseika ni kansuru ichikousatsu, Niigata-ken Tookamachi-shi: Kenkoukei tsurizumu ni yoru chiiki kasseika no youin kyotenron [Study on regional activation by sport tourism: Factor-based theory of regional activation by health-based tourism]. *Osaka Kankou Daigaku Kiyou,* 15, 23–32.

Uchida, K. (2015) Supotsu tsurizumu ni yoru kokusai kouryuu: Honkon bareboru chimu o jirei toshite [International exchanges through sport tourism: A case study of the Hong Kong volleyball team]. *Kyoto Notoru Damu Jyoshi Daigaku Kenkyuu Kiyou,* 45, 45–57.

Usui, F. (2008) Supotsu, tsurizumu, bunka no 3 youso no shinketsugo ni yoru chiikikasseika senryaku: Kushiro-shi Akanko onsen ni okeru Kushiro-shi kokusetsu Akan kohan sukijyou o meguru chiiki katsudou no kenkyu [Regional revitalization strategy by synthesizing three factors; sports, tourism and culture: A study of regional activities around Kokusetsu Akan Lakeside Ski Area at Lake Akan hot springs]. *Kankou Souzou Kenkyu,* 3, 1–21.

Vaske, J. J. (2008) *Survey Research and Analysis: Applications in Parks, Recreation and Human Dimensions*. State College, PA: Venture Publishing.

Watanabe, R., Shiotsuki, R. and Tanno, T. (2013) Chiikishigen o ikashita supotsu tsurizumu no arikata ni tsuite: Jitensha ibento ni okeru kenkou anzen kanri to suibunseshu ni tsuite no kentou [A study of sport tourism in regional resources: Health, safety and drink management in cycling event]. *Atomi Gakuen Joshi Daigaku Bunkabu Kiyou,* 48, 211–231.

Weed, M. (2005) Research synthesis in sport management: Dealing with 'chaos in the brickyard'. *European Sport Management Quarterly*, 5(1), 77–90.

Weed, M. (2014) After 20 years, what are the Big Questions for sports tourism research? *Journal of Sport & Tourism,* 19(1), 1–4.

Yamakei Research Institute. (2016) *Tozan hakusho 2016 [The White Paper of Mountain Climbing*]. Tokyo: Yama-kei Publishers.

Yamazaki, Y. (2014) Supotsu tsurizumu to ryokou bijinesu: Ryokou sangyou ni okeru supotsu tsurizumu no kanousei [Sports tourism and travel business: Possibility of the sports tourism in the travel industry]. *Nihon Kokusai Kankou Gakkai Ronbunshu,* 21, 283–289.

Yumita, E. and Harada, M. (2015). Supotsu ibento sankasha no desutineshon imeji ga hyouka, manzokudo, koudouito ni oyobosu eikyou: Sankagata supotsu ibento no supotsu tsurisuto ni chakumoku shite [The influence of sport event participant's destination image on evaluation, satisfaction, and behavioral intentions: Focusing on sport tourists of a sport event]. *Kankou Kenkyuu,* 27, 101–113.

6 Mobilizing stoke

A genealogy of surf tourism development in Miyazaki, Japan[1]

Adam Doering

Introduction

The International Olympic Committee's (IOC) decision on 1 June 2016 to introduce surfing as a demonstration sport in the Tokyo 2020 (now 2021) Olympics has put Japan on the mainstream media map as a world class surf destination. Yet, the idea that Japan even has surf, let alone being a surf destination capable of hosting the sport's first Olympic appearance, has come as something of a surprise to the general public both within Japan and internationally. Nevertheless, as the 11-time World Surfing Champion Kelly Slater recently commented, 'There's a pretty strong surf culture there actually. People might not realize there is great surf in Japan' (Bensinger, 2016).

For non-surfers, Japan does not easily conjure up an image of a surfers' paradise. Furthermore, despite the fact that the international surfing community has long recognized Japan alongside America, Hawaii, Australia and Brazil as a well-established surf nation (O'Brien & Ponting, 2013), many surfers know surprisingly little about what comprises a 'strong Japanese surf culture'. Indeed, until now, Japan's surf scene has largely existed outside the gaze of the global mainstream media. However, the introduction of surfing into the Tokyo 2020 Olympics marks an attempt by the Japanese Government, the IOC, International Surf Association, World Surf League and the associated global surf industry to capitalize on this opportunity.

Nowadays, surfing means big business. Estimates suggest that the population of surfers worldwide increased from a total of 25 million in 2001 to 35 million in 2011 (The Economist, 2012), whilst more recent economic analyses indicate that surfing waves potentially contribute around US$50 billion to global economic activity each year (McGregor & Wills, 2016). The surf cultural industry alone, which includes sales of functional surf items such as wetsuits, surf fashion and accessories sold to non-surfers, and related surf experiences, is said to be a US$15 billion global market (Lazarow, Miller & Blackwell, 2008). These figures are, however, a point of contention. Carroll (2014), for example, maintains that such statistics are excessively optimistic and are inflated at times by the surf industry in order to expand transnational commercial interests. Nevertheless, while there is no consensus over the figures, there can be little doubt surfing is an influential global industry capable of shaping people and places in significant ways. However, as Wheaton (2010) points out, despite its diffusion across the globe, few studies have explored the diversity of surf cultures, histories and economies existing outside Western understandings and experiences.

Surf tourism research has largely been confined to the established markets in the USA, Europe and Australia (Lazarow et al., 2008; Martin & Assenov, 2012). Indeed, Martin and Assenov (2012) suggest that only 10 percent of those countries of the world possessing coastal surfing resources have been studied. More specifically, although there is an emerging body of literature analysing the implications of surfing (Cheng & Tsaur, 2012; Evers, 2017; Tien-Ming & Chiang-Chuan, 2015), action sport (Thorpe, 2014; Thorpe & Wheaton, 2011) and lifestyle sports (Evers & Doering, 2019; Wheaton, 2013) in East Asia, there are only a few sporadic contributions dedicated to analysing the complexities of surf destinations in Japan (Doering & Evers, 2019; Konagaya, 2009; Manzenreiter, 2013; Mizuno, 2002, 2007, 2015; Murata, 2017). However, although scholarship has yet to devote empirical attention to the social construction of surf destinations in Japan in particular, there is extant literature exploring the historical production of seaside tourism development (Funck, 2006; Tada, 2015), the relationship between leisure and consumer culture in the Japanese context (Manzenreiter & Horne, 2006), and the social construction of surf spaces in the international literature (Laderman, 2014; Ponting & McDonald, 2013; Preston-Whyte, 2002; Thompson, 2011). In this chapter, I build on this emerging literature to extend the geographical reach of surf tourism research into East Asia and Japan in particular, while also contributing to empirical analyses dedicated to unpacking how surf destinations are socially constructed over time and the ways in which contemporary rural seaside communities negotiate the increasing exposure to the transnational surf industry.

More specifically, drawing on 2 month's ethnographic fieldwork conducted between July and August 2016, this chapter traces the historical production of one of Japan's most prominent domestic surfing destinations, Miyazaki Prefecture. Inspired by Foucault's (1977) notion of genealogy, I combine secondary historical research with participant observation and opened-ended conversations with Japanese surfers, 'surf legends,[2] local surf shop owners, surfing association representatives at the national and regional levels, and the Miyazaki Prefecture Convention and Visitor Bureau, to provide insight into how surf tourism came to be mobilized in Miyazaki as an important development tool for revitalizing its rural economies. The aim of this historical and ethnographic account is to unpack the 'conditions of possibilities' (Foucault, 1977) – that is, the economic, political, material and ideological assemblages that enable the formation of surf tourism destinations in Japan and its emergent subjectivities. I am less concerned here with defining surf tourism or how to best manage surf tourism destinations (see, e.g., Buckley, 2002; O'Brien & Ponting, 2013) than with interrogating the worldmaking force of surf tourism development in making and remaking peoples, places and pasts (Hollinshead, 2009). The chapter is guided by three questions: Through what lineages and processes did surf tourism development come to be mobilized as a key discursive and promotional tool for revitalizing the rural Japanese seascape? What social and cultural transformations does surf tourism development enable in the present? And, how might the development of a surf tourism destination change the local experience of place and the sea?

The chapter begins by examining the continuity and change of the Miyazaki seascape between 1930 and 2016. The genealogy then traces the convergence of surfing history and tourism development from its unlikely beginnings from the 1930s to the1950s that manufactured a southern tropical aesthetic, to the reimaging of the beach as a leisure space during the honeymoon boom of the 1960s and 1970s, to the 'Japan Made' surf culture emerging in the 1990s and the present branding of Surf City Miyazaki today. The narrative shifts to the present, drawing attention to three notable transformations currently underway in Miyazaki's surf destination development: the feminization of surf capital; global cultural gentrification of surf space; and, the ongoing geopolitical authority of urban ideologies within rural Japanese space. I conclude by speculating on the ways in which surf tourism development may (dis) connect local surfing communities in Japan and offering initial insights into potential implications a new era of Olympic surfing invites.

Drawing a picture on the land: manufacturing Miyazaki's surf aesthetic (1930–1950s)

Miyazaki's surf scene begins with its geography. Miyazaki is located on the southernmost part of Kyushu facing the Pacific Ocean to the east, and is surrounded by the Kyushu mountain range in the northwest and the Kirishima range in the south. The mountains and sea act as a natural barrier against the winter winds, creating a mild climate year round. This inaccessibility combined with one of the nation's lowest population densities (Statistics Bureau, Japan, 2016) has also prevented the development of high-speed train connections, making the region less accessible than most areas of Japan. These natural barriers make Miyazaki Prefecture feel like an island within an island.

Early tourism development in the 1930s sought to enhance Miyazaki's natural geography by manufacturing a 'southern tropical paradise' (Funck & Cooper, 2013; Tada, 2015). Shotaro Iwakiri, the founder of the Miyazaki Bus Company and 'father of Miyazaki tourism' is said to have had one main mission: 'to draw a picture on the land' [*daichi ni e wo kaku*] (Sato, 1997). Iwakiri began reconfiguring the built environment by planting rows of Phoenix Palm (Phoenix canariensis) trees in the area, starting from the beach resort area of Aoshima and travelling south along a 100-km stretch of coastline called the Nichinan Coast Road. The idea was to transform the coastline into what Iwakiri referred to as a 'moving frame', offering a dynamic visual experience of sea, sky and palm trees to tourists as they drive the coastal road

(Tada, 2015). The distinctive foreign and tropical feel of Miyazaki's coastline was further enhanced with the establishment of *Shaboten Koen* [Cactus Park] in 1937 and *Kodomo no kuni* [Children's Country] in the Aoshima beach area in 1939. The southern tropical motif was urbanized in October 1954 when Iwakiri planted 51 Phoenix palm trees alongside the Oyodo River in front of the Miyazaki Tourist Hotel (MCTA, 1997).

By the end of the 1950s, Miyazaki's southern tropical imagery was well established: Iwakiri's picture had its frame. Looking at Miyazaki retrospectively, this era marked the beginnings of the surfing aesthetic that dominates representations of Miyazaki today. One would be hard pressed to find a tourism promotion of Miyazaki without one of Iwakiri's palm trees in it. The Phoenix palm, a native of the Canary Islands, would eventually become the prefecture's official tree. The Nichinan Coast Road, a product of the desire for the picturesque, would create easy accessibility to the numerous surf breaks along Miyazaki's rocky stretch of coast. However, it would take a few more decades before surfers and surf tourism would be integrated into Miyazaki's southern tropical seascape.

Japan's Hawai'i: the honeymoon boom and surfing's emergence (1960–1970s)

The southern tropical imagery positioned Miyazaki well for the post-war domestic tourism market seeking something different, foreign even, but without leaving Japan. The 1960s marked the first major tourism development in Miyazaki: the honeymoon boom. This boom began when a couple from the imperial family travelled to Miyazaki for their honeymoon in 1960. Two years later, the then reigning Emperor Akihito and Empress Michiko visited Miyazaki, sparking a honeymoon boom that would last nearly two decades (MCTA, 1997). Miyazaki came to be referred to as Japan's Hawai'i.

Miyazaki's honeymoon destination brought together mass tourism infrastructure and mythical history with a foreign island aesthetic. The honeymoon boom reached its peaked in 1974 when a total of 370,000 newly married couples visited Miyazaki, representing almost one-third of all married couples in the nation (Funck & Cooper, 2013). Graburn (1983) describes this era of Japanese tourism accurately as 'pay, play and pray'. Large hotels were constructed on the seaside of Aoshima, a rural fishing village and tourism hub 10 minutes' drive south of Miyazaki City, to accommodate the rapidly increasing demand. Tourism promotion during the honeymoon boom prominently featured the Aoshima Shrine and the Nichinan Coastline road as its main attractions. Aoshima beach acted as a departure point for guided tours of the Nichinan coastline, travelling to the sea cliff shrine of Udo Jingu and the numerous sightseeing spots developed in the early years by Iwakiri.

During this time, the image and meanings of the seascape gradually shifted towards becoming a leisure space as hotel patrons returning from their day trips had immediate access to Aoshima's swimming beach. By 1963,

Miyazaki's tourism campaign began to capitalize on this trend by developing the motto, 'Let's play with the sun' [*taiyo to asobou*]. In 1968, Miyazaki city hotels worked with the Miyazaki City Bus Company to establish the 'Bathing suit bus' [*mizugi basu*] to take people from downtown to Aoshima beach in their bathing suits (Miyazaki Kotsu, 1997; Tada, 2015).

Miyazaki's southern tropical imagery and seaside leisure ideology had been established, but the practice of surfing was just starting to emerge. It is frequently assumed surfing in Japan began at the Yokosuka naval base, located just 65 km south of Tokyo, shortly after the end of the Second World War (Funck, 2006; Konagaya, 2009; Sakaguchi, Yoneda & Sato, 2010). However, this version of history is contested. Others argue that Japan's traditional wave riding culture was present as early as the 1800s (see Ohkawa, n.d.). The earliest record of wave riding dates back to 1821 when Haiku poet Dokurakuan Kanri wrote of a beach scene at Yunohama beach in Yamagata Prefecture where a group of young boys were playfully skimming along the waves on *itago*, small wooden bodyboards shaped out of the removable floor planks found on traditional Japanese-style wooden boats (Ohkawa, n.d.). Conversations with representatives from the Wakayama Surf Association[3] and Yoron Island Surf Association[4] also sought to decentralize this dominant Tokyo-centric narrative by suggesting surfing could also have started around the same time in Okinawa, and potentially even Osaka. Nonetheless, the dominant framing of Japan's surf history argues that, following the war, American soldiers based out of Yokosuka would surf fibreglass longboards on the breaks scattered around the base during their downtime. This eventually led to the establishment of Japan's first 'surf town' along the Shonan area beaches in Kanagawa Prefecture.

According to the Miyazaki surf legends, surfing started to find its way to Miyazaki Prefecture in the late 1960s.[5] In these early years, surfing was limited to a very small group of dedicated surfers. Those first- and second-generation surf legends now act as local custodians of the waves around Miyazaki. For nearly four decades, they have governed certain breaks through surf shop affiliations and informal surf break localism. By the 1970s, Miyazaki's surfing scene began to officially organize when it joined the Nippon Surfing Association (Konagaya, 2009). As we will now explore, the end of the millennium saw Miyazaki become a recognizable surf destination, resulting in an influx of surf lifestyle migrants from urban areas which started to destabilize this localized control.

'Japan made': Seagaia Ocean Dome, Miyazaki Pro and typhoon Orchid (1980–2000s)

The honeymoon boom came to an abrupt halt in the 1970s when travel to the real Hawai'i and more remote island areas of Japan, such as Okinawa, became more affordable and appealing for the emerging middle classes (Tada, 2015). With increased disposable income, the development of the new leisure class and cheaper international flights, the domestic Japanese tourism market was internationalized (Harada, 1994; Manzenreiter & Horne, 2006; see also Chapter 1). In 1964, only 13,000 people travelled abroad; by 1990, that number exceeded more than 10 million (Harada, 1994). Honeymooners and surfers alike could now travel to the real Hawai'i rather than its Japanese replica. The honeymoon boom went bust, with annual visits to Miyazaki being reduced to one-third of the peak visits. Aoshima's tourism development entered into a slow decline for nearly three decades, which came to be referred to as 'the lost decades' (Manzenreiter & Horne, 2006). Moreover, the rapid economic development of the 1980s eventually led to the depopulation of rural Japan, making rural communities like Aoshima bear a disproportionate amount the costs of industrialization.

This left the Japanese government needing to consider the best way to distribute the economic benefits more equally amongst the peripheral regions of Japan, and it turned to large-scale tourism development as the solution. The Law for the Development of Comprehensive Resort Areas, in short the 'Resort Law', was introduced in 1987 with the stated aim 'to revive peripheral regions while at the same time boosting domestic demand' (Funck, 1999: 333). This ambitious plan called for large-scale, multifunctional resort areas offering everything from action sport activities to business conferences (Funck, 1999).

Resorts require cheap, large and mostly unused real-estate, which is not easy to find even in rural areas of Japan owing to its mountainous terrain. The Resort Law consequently had serious implications for Miyazaki as it shifted its tourism development efforts away from the already existing and deteriorating tourism infrastructure in Aoshima to the lesser developed area of Hitotsuba located 45 min north at the other end of the large bay. Miyazaki's previous tourism development experience ensured it had institutions and infrastructure in place to respond quickly to the new Resort Law (Funck, 1999). The initial public–private resort proposal to develop the Seagaia Resort was accepted for approval in 1988 and opened for business in 1993, making Miyazaki Prefecture the first to take advantage of the Resort Law.

Seagaia Resort is a complex of hotels, a golf course, a tennis court and a shopping centre and at one time was home to the world's largest indoor wave pool, the Seagaia Ocean Dome. The Ocean Dome marked Miyazaki's first attempt at intentional surf tourism development, albeit via an artificial wave pool. The Polynesian-style indoor wave pool was located within 300 m of an actual surf break, and comprised a fire-spitting volcano, a 100-m wide white sand beach as well as employing foreign hula dancers and surfers for the evening shows (Figure 6.1). The Ocean Dome's indoor wave pool made immediate international headlines (Pollack, 1993) and, with the revival of wave pool technology today, the Ocean Dome continues to be a topic of conversation (SURFING Magazine, 2015). However, the Ocean Dome was expensive to operate and the demand for indoor waves could not meet expectations. Visitors to the Ocean Dome peaked in 1995 at 1.25 million visits and slowly declined

Figure 6.1 Seagaia Ocean Dome first week of demolition. 1 September 2016. Photo: Adam Doering.

until its eventual closure on 1 October 2007 (Mainichi Newspaper, 2016). Although short-lived, the Ocean Dome introduced Miyazaki's surf scene to international audience.

Miyazaki's surf culture began to assert itself by hosting a series of large international surf events. In April 1990, Miyazaki became the first Asian country to host the World Amateur Surfing Championships, which welcomed 230 surfers from 20 countries to Kizaki beach (MCTA, 1997). From 1991 until 1993, Miyazaki Prefecture would also host the 'Miyazaki Pro' as part of the Association of Surfing Professional World Tour Event, the world's leading surf organization at the time (now called the World Surf League). The Miyazaki Pro included some of surfing's biggest stars of all time: Sunny Garcia, Tom Carroll, Kelly Slater and Tom Curren. However, the biggest legacy to emerge from these events occurred outside of the official competitions. In October 1991, Typhoon Orchid hit Miyazaki's shoreline on the fourth and fifth days of that year's Miyazaki Pro, stopping the competition but putting Miyazaki's surfing on the world stage. Typhoon Orchid produced up to 15-m waves and, with the world's best surfers in town, a group headed south to explore reef breaks down the Nichinan Coastline road. It was then that professional surfer Tom Curren paddled out into a large typhoon swell breaking over the appropriately named 'Devils Washboard'. What were referred to at the time as 'the greatest waves in the history of surfing in Japan' (Surfn' Life, 1992, n.p.) would be recorded on film for the first time. Written alongside the images and names of the foreign surfers was the proud tag line 'Japan Made'. Subsequently, the surf spot would be named 'Curren's Point' in Tom Curren's honour, leaving a lasting legacy within the Miyazaki surf culture and developing a sense of pride that the Japanese waves are world class.

Hosting a series of large-scale international surfing events achieved two things. First, in terms of domestic tourism development, these events solidified Miyazaki as premier surf destination in Japan. Beginning in the 1990s, Miyazaki would come to be referred to as Japan's surf 'Mecca' (MCTA, 1997). Second, these events helped legitimize surfing's image within mainstream Miyazaki society. As Keison, an Aoshima surf instructor in his 50s described it, 'People became proud there were international events being held here'.[6] The legitimization of surfing through its professionalization opened up new potential for sports tourism development in the region. The ideological groundwork had been established for a surf tourism destination to emerge.

Surf City Miyazaki: the convergence of surf culture and tourism (2010–the present)

Today, Miyazaki is a surf tourism destination. In 2011, the *Namitabi Miyazaki* [Miyazaki Wave Tour] initiative began, marking the prefecture's first official attempt to align surf culture with tourism development. Surfing's importance is evident in the fact that the first surf tourism economic analysis for Miyazaki Prefecture was conducted in 2015 (Miyagin Economic Research Institute, 2016). Initial figures from the report indicate the economic benefit from inbound surfers to Miyazaki is an estimated US$15.5 million (Miyanichi e-Press, 2015). This is a much needed boost within an otherwise struggling rural economy; however, identifying the economic benefits of surf tourism is only the starting point for examining the cultural and social formations that today's surfing tourism industry enables (Stranger, 2010). A closer look at the recent developments around Aoshima beach, specifically Surf City Miyazaki which opened in April 2016 and the Aoshima Beach Park which started a year earlier in 2015, offer initial insight into the contested nature and cultural politics associated with surf tourism development in Miyazaki.

Surf City Miyazaki is a locally owned outdoor fitness centre but is designed, managed and 'produced' by the Kamakura-based company, Beach Town Corporation. Beach Town was established in June 2008 when it opened the Beach Hayama Outdoor Fitness Club. It currently operates 16 outdoor fitness centres in prime beachside locations throughout Japan and as the Creative Director recently described, there were four more projects scheduled to open in 2016. The aim was to have as many outdoor fitness in operation prior to the Tokyo 2020 Olympics as possible.[7] The business model has been such a success that Beach Town has even been celebrated at the national level for its innovative approach to leisure and ability to develop the female outdoor fitness market (Japan Productivity Centre, 2016).

Surf City Miyazaki and the concept of outdoor fitness developed in response to 1990s 'indoorization' (see also Spielvogel, 2003) of fitness: going to the gym, running on a treadmill, riding a stationary bike.[8] Travelling to Southern California during his holidays, the CEO of Beach Town explained how impressed he was by the way Californians used the outdoors as an exercise space and wanted to introduce this concept to the Japanese market. So despite its name, Surf City Miyazaki is about encouraging a broader concept of outdoor fitness, with surfing being just one component. As the motto on the homepage reads, 'Yoga, Surf, SUP, Run: A new lifestyle through nature begins in Miyazaki' (Surf City Miyazaki, n.d.). The main source of income is yoga, with customers often trying out surfing as an aside (Figure 6.2).[9] In addition, Surf City Miyazaki hosts monthly 'Roxy Fitness' programmes, Roxy being Quiksilver's brand targeting young women, and international events

Figure 6.2 Surf City Miyazaki's yoga studio and Roxy standup paddle boards for rent. August 2017.

Photo: Adam Doering.

like the Hawai'i-based movement *The Butterfly Effect* (2012) where women participants are described to be 'asserting their femininity during sports'. Surf City Miyazaki is further supported by the transnational surf industry through a Quicksilver affiliated surf school and surf tour program (Surf City Miyazaki, n.d.). Capitalizing on corporate sponsorship, global branding and surf imagery, nearly 95 percent of Surf City customers who try surfing are 'absolute beginners' who just want to try a new experience.[10] In short, they are surf tourists.

The complexities surrounding local ownership, out of prefecture operation and management and reliance on the transnational surf industry raises some interesting questions for Miyazaki's surf tourism development. What new ideologies, values, imaginaries and subjectivities emerge when the seascape is framed through as an outdoor fitness centre? Who is in control of this development? How will the already well-established Miyazaki surf culture respond to the production of a newly imported beach style? Although it is beyond the scope of this introductory analysis to fully respond to these questions, it is important to offer some initial insights into three transformations that will help direct future academic inquiry on Japanese surf culture and surf tourism destinations in the lead up to the Tokyo 2020 Olympics.

Gendered development: the feminization of surf capital

The first reconfiguration to note is the ways Miyazaki is able to tap into what Lanagan (2002) calls 'surf capital'. Lanagan (2002) describes surf capital as the employment of surf culture and associated imagery to develop broader commercial interests. More specifically, surf capital enables Surf City Miyazaki to attract and develop the highly sought after Japanese female market aged 20–40, characterized as single and professional women who are adventurous and self-reliant (Hashimoto, 2000; Yang, Khoo-Lattimore & Arcodia, 2017). According to the economic impact analysis on surf tourism in Miyazaki, female surfing tourists are more likely to stay longer, to stay in hotels rather than their cars, and spend more money overall (Miyagin Economic Research Institute, 2016). National statistics on leisure also indicate that interest in outdoor fitness is increasing and that females aged 25–45 comprise the largest market segment (Japan Productivity Centre, 2016). The gender ratio of outdoor fitness leisure participation is estimated to be 3:7 in favour of females, making this the most attractive domestic tourism and leisure market in Japan today (Japan Productivity Centre, 2016). Surf capital and its associated discourses of female empowerment, adventure and independence, has played a critical role in reshaping the experiences, meanings and aesthetics of the Miyazaki seascape.

Gender research in lifestyle sports has been critical of how the transnational surf industry frequently co-opts a feminist discourse for neoliberal ends (Comer, 2010; Evers, 2017; Wheaton, 2010). As Evers (2017) describes in his analysis of surf tourism development in China's Hainan Island, politicians, property developers and the global surf industry frequently employ the 'hetero-sexy surfer-girl ... to generate consumer interest, promote consumer femininities, and initiate and widen the marketplace in new locations'. Wheaton (2010: 1061) also writes of how in many lifestyle sports, 'the female consumer has become one of the most valuable and targeted commodities'. Building on a similar transnational feminization of surf capital, Surf City Miyazaki is able to align Miyazaki's rich surfing past with the feminization of surf culture for the purposes of encouraging broader forms of consumption, namely, the outdoor fitness market. The feminization of surf capital is revitalizing the local economy and developing new sense of agency and belonging for some, but future research should also remain attentive to the ways Japan's female empowerment discourse may also be implicated in the globalization and commercialization of the Japanese seascape.

Global cultural gentrification: surf style for the new transnational economy

A second related transformation concerns the cultural gentrification of the seascape in Miyazaki through a combination of the transnational surf industry and domestic urban–rural geopolitics. Anderson (2014: 246) refers to this contested space as a 'trans-local node located within a mobile surf space'. Miyazaki's seascape has become increasingly globalized through surf tourism development. During the Roxy Fitness events, for example, the Aoshima seascape becomes imprinted with corporate sponsorship with 'Roxy Fitness' sketched into the sand, Red Bull tents providing shade for participants, Roxy sportswear adorning the bodies of the female participants, and Roxy stand-up paddle boards blanketing the sea. The transnational aesthetic is furthered by the establishment of Aoshima Beach Park, which opened a year prior to Surf City Miyazaki in 2015. Located 5 minutes down the beach from Surf City, Aoshima Beach Park is an area to hang out, drink and eat. The stated aim of the beach park is to establish a 'beach culture with style' taking as its model the strong beach cultures of California, Australia and Hawai'i (Miyahara, 2016, n.p). This has been interpreted as a surf beach aesthetic comprising hammocks, pop-up clothing shops, American style burger joints and a pizzeria, all operating within a fully English linguistic landscape. Combined, these establishments illustrate how Miyazaki's surf styled destination development is trending towards becoming a productive mechanism in the new transnational economy.

Japan's domestic geopolitics: surf tourism reinforcing rural–urban disparities

Surf tourism development also offers insight into Japan's domestic geopolitics. The power and influence of Japan's urban centres are being reposition away from centralized governmental control and decision-making, as it did in the 1980s with the Resort Law, towards a decentralized small-scale, urban-stylized and microenterprizing, free-market economic and cultural expansionism. A similar soft power shift of tourism destination development is evident in the discourses and practices of the lifestyle migrants to Miyazaki. In addition to the transnational surf space is a creative collective of lifestyle migrants from Japan's urban centres who become artisanal surf equipment designers, local shapers, manufacturers and retailers. Lifestyle migrants have introduced a highly stylized surf culture, referred to as 'shabby chic'[11] and/or 'modern retro',[12] and have actively sought to reconstruct the seascape of Miyazaki through a discourse of change is good. One of the first signs seen walking from Miyazaki Surf City to the Aoshima Beach Park is the bright red neon sign of the burger stand reading, 'Change is good'. The surf shop and café down the road even goes so far as to exclaim, 'We are making heaven'. The discourse of change is reinforced in surf films such as Good Morning Miyazaki (Westcott, 2013) where the tagline reads, 'This is not your average Japan'. But what has changed, whose heaven is being made, and for whom?

The 'change is good' discourse is contested by several of the surf legends,[13] local tourism businesses[14] and recreational organizations[15] in Miyazaki who feel the overly managed, planned and commercialized management/production style misalign with the slow, organic and rural mode of life that has traditionally defined Miyazaki (for a comparative mountain lifestyle example, see Knight, 2003). Mado, a first-generation surfer and surf entrepreneur in his 50s, expressed his frustration that the Miyazaki surf community is not just divided, it has been blown to pieces and will not likely be repaired.[16] The introduction of a 'new' beach culture to Miyazaki, where according to the promotional narrative there previously was none (Miyahara, 2016), is designed by urban lifestyle migrants who have the promotional skills, desire and finances to produce a new surf destination. Representatives from the Miyazaki Convention and Visitors Bureau have even stated that it is the role of the prefectural government to promote the surf/sports destination of Miyazaki, but the investment and creative direction largely comes from the lifestyle immigrants arriving from the urban areas of Tokyo and Osaka.[17] It could be argued, then, that Miyazaki's surf tourism development reinforces rather than remedies the longstanding geopolitical disparities between Japan's urban and rural spaces. As Murata's (2017) study on surf tourism in the Kanto region suggests, surf destination place-making is political and riddled with conflicting values. In an emerging post-industrial and post-Fordist economy, one would be wise to question the broader political and economic structures that require urban lifestyle migrants to rural areas to be *the* tool for revitalization and economic redistribution.

Surf tourism development in Japan: where next?

The aim of this chapter was to extend the geographical reach of surf tourism research into Northeast Asia more broadly and Japan more specifically. It also sought to contribute to the empirical analyses concerning how surf destinations are socially constructed over time and the ways in which surf destination development is negotiated today as rural seaside communities become increasingly exposed to the transnational surf industry. The genealogy first drew attention to the ways local meanings and understandings of the seascape shift over time: from the picturesque, to a leisure space, an indoor phenomenon and surf 'Mecca', towards the outdoor fitness ideology and practice emerging today. By mapping out the convergence of Miyazaki's surf and tourism histories, the chapter then examined how surf tourism development came to be mobilized as an important tool for revitalizing rural coastal economies like Miyazaki. In doing so, it is hoped the chapter offers a more nuanced understanding of the conditions of possibilities that enable surf tourism development and the associated subjectivities that emerge as a result.

The case of Miyazaki emphasizes how surf tourism development is about expressing alternatives and revitalizing rural communities using the scarce resources available. The lifestyle migration encouraged by surf tourism imagery and development has also brought people out of the cities and into the countryside, affording rural communities like Miyazaki an opportunity to reinvent and reimagine themselves in difficult economic times. The once rural fishing village and honeymoon hotspot of Aoshima has become Surf City Miyazaki. However, like most tourism development, Japan's surf tourism is a double-edged sword. Reimaging Miyazaki's seascape as an outdoor fitness centre and surf tourism destination may open new expressions of agency and self-determination for some, but may also lead to the co-option of surf capital that encourages transnational cultural gentrification and reinforces domestic geopolitical disparities between Japan's urban and rural spaces.

The introduction of surfing as a demonstration sport in the Tokyo 2020 (2021) Olympics has equal potential to heighten a sense of connection with surf culture and surf communities in Miyazaki and/or further disconnect these communities through further commercialization, cultural gentrification and reinforcing longstanding urban–rural disparities. The fact that Tokyo has hosted the Olympic Summer Games twice speaks to the Tokyo-centricity that already pervades the Olympic discourse in Japan. And while many Japanese surfers, surfing association representatives and destination management organizations have expressed hope that the Olympics may further legitimize surfing and demonstrate the ways surfing contributes positively to the local community, culture and environment, there is limited evidence to suggest that local communities have been able to leverage the Olympics to support development beneficial to the wider population (Reis, de Sousa-Mast & Gurgel, 2014).

How surf tourism destinations like Miyazaki will negotiate their Olympic hopes with Olympic realities is yet to unfold, and it is not the aim here to provide solutions. As surf tourism development in Japan has yet to be explored, the analysis presented here is only the first step for informing a more robust theoretical analysis of Japan's contemporary surf destinations leading up to the Tokyo 2020 (2021) Olympics. The intersection of surf culture and tourism development, the cultural politics negotiated within these emerging surf destinations, their multiple histories and points of conflict, new formations of 'community' and attachments to place that surf culture invites, and the Olympic legacies that will emerge from introducing surfing into the

Tokyo Olympics, are all rich areas of research deserving of further academic attention.

Acknowledgements

I would like to thank Clifton Evers (Newcastle University, UK) and Debra Occhi (Miyazaki International College, Japan) for their insights and comments on earlier drafts of this chapter.

Notes

1 A previous version of this chapter appears in *Tourism Planning & Development*, *15*(1), 68–81.
2 'Surf legend' is a term used in Japan to describe the first-generation local surfers and surf shop owners in their 60s, and 70s that formally and informally manage surf breaks throughout Japan.
3 Kaisers Surf Shop, Wakayama City, interviewed by Adam Doering, 27 May 2016. Machu (anonymized), Wakayama Surfing Association, interviewed by Adam Doering, 23 March 2016.
4 Yoron Island Surf Association, interviewed by Adam Doering, 4 September 2016.
5 Ryu (anonymized), Miyazaki surf shop owner and surf legend, interviewed by Adam Doering and Clifton Evers, 16 August 2016.
6 Keison (anonymized), personal communication with author, 2 July 2016.
7 Aki Matsumura. Creative Director, Beach Town Co., interviewed by Adam Doering, 1 September 2016.
8 Takashi Kurono, CEO, Beach Town Co., interviewed by Adam Doering, 2 July 2016.
9 Ibid.
10 Ibid.
11 Aki Matsumura, Creative Director, Beach Town Co., interviewed by Adam Doering, 1 September 2016.
12 Shiro (anonymized), interviewed by Adam Doering, 1 September 2016.
13 Mado (anonymized), in post-surf conversation with the author, 27 August 2016, and Ryu (anonymized), 16 August 2016.
14 Sora (anonymized), surf accommodation provider, personal communication with author, 30 June 2016.
15 Lei (anonymized), Hula dancer, personal communication with author, 2 July 2016.
16 Mado (anonymized), 27 August 2016.
17 Itsuki (anonymized), Representative from the Miyazaki Convention and Visitors Bureau, interviewed by Adam Doering, 1 September 2016.

References

Anderson, J. (2014) Surfing between the local and the global: Identifying spatial divisions in surfing practice. *Transactions of the Institute of British Geographers*, 39(2), 237–249.

Bensinger, G. (2016) *In Depth with Graham Bensinger: Kelly Slater* [TV Series], 16 March. Los Angeles, CA: GB Entertainment.

Buckley, R. (2002) Surf tourism and sustainable development in Indo-Pacific Islands. I. The industry and the islands. *Journal of Sustainable Tourism*, 10(5), 405–424.

Carroll, N. (2014) Lies, damned lies and statistics. *Surfing Life*, May 9.

Cheng, T-M. and Tsaur, S-H. (2012) The relationship between serious leisure characteristics and recreation involvement: A case study of Taiwan's surfing activities. *Leisure Studies*, 31(1), 53–68.

Comer, K. (2010) *Surfer Girls in the New World Order*. Durham: Duke University Press.

Doering, A. and Evers, C. (2019) Maintaining masculinities in Japan's transnational surfscapes: Space, place and gender. *Journal of Sport and Social Issues*, 43(5), 386–406.

Evers, C. (2017) Surfing and contemporary China. In D. Hough-Snee and A. S. Eastman (Eds.), *Critical Surf Studies Reader*. Durham: Duke University Press, pp. 386–402.

Evers, C. and Doering, A. (2019) Lifestyle sports in East Asia. *Journal of Sport and Social Issues*, 43(5), 343–352.

Foucault, M. (1977) *Discipline and Punish: The Birth of the Prison*. New York, NY: Vintage.

Funck, C. (1999) When the bubble burst: Planning and reality in Japan's resort industry. *Current Issues in Tourism*, 2(4), 333–353.

Funck, C. (2006) Conflicts over space for marine leisure: A case study of recreational boating in Japan. *Current Issues in Tourism*, 9(4–5), 459–480.

Funck, C. and Cooper, M. (2013) *Japanese Tourism: Spaces, Places and Structures*. New York, NY: Berghahn.

Graburn, N. (1983) *To Pray, Pay and Play: The Cultural Structure of Japanese Domestic Tourism*. Aix-en-Provence: Cahiers du Tourisme, Centre des Hautes Études Touristiques.

Harada, M. (1994) Towards a renaissance of leisure in Japan. *Leisure Studies*, 13(4), 277–287.

Hashimoto, A. (2000) Young Japanese female tourists: An in-depth understanding of a market segment. *Current Issues in Tourism*, 3(1), 35–50.

Hollinshead, K. (2009) The 'worldmaking' prodigy of tourism: The reach and power of tourism in the dynamics of change and transformation. *Tourism Analysis*, 14(1), 139–152.

Japan Productivity Center. (2016) *Rejya-Hakusho* [White paper on leisure]. Tokyo: Seisansei Shuppan.

Knight, J. (2003) Organic farming settlers in Kumano. In A. Waswo and N. Yoshiaki (Eds.), *Farmers and Village Life in Twentieth-Century Japan*. New York, NY: Routledge, pp. 267–284

Konagaya, Y. (2009) Safuin bunka no keisei to kukan to iu media [Development of surfing culture and media space]. In K. Kanda (Ed.), *Rejya no kukan: shosou to approach* [Leisure space: Aspects and approaches]. Kyoto: Nakanishiya, pp. 59–67.

Laderman, S. (2014) *Empire in Waves: A Political History of Surfing*. Berkeley: University of California Press.

Lanagan, D. (2002) Surfing in the third millennium: Commodifying the visual argot. *The Australian Journal of Anthropology*, 13(3), 283–291.

Lazarow, N., Miller, M. L. and Blackwell, B. (2008) The value of recreational surfing to society. *Tourism in Marine Environments*, 5(2–1), 145–158.

Mainichi Newspaper (2016).*Oshandomu, kyokara kaitaikoji* [Ocean dome demolition starts today]. Available at: http://mainichi.jp/articles/20160804/ddl/k45/020/601000c

Manzenreiter, W. (2013) *Sport and Body Politics in Japan*. London: Routledge.

Manzenreiter, W. and Horne, J. (2006) Leisure and consumer culture in Japan. *Leisure Studies*, 25(4), 411–415.

Martin, S. A. and Assenov, I. (2012) The genesis of a new body of sport tourist literature: A systematic review of surf tourism research (1997–2011). *Journal of Sport and Tourism*, 17(4), 257–287.

McGregor, T. and Wills, S. (2016) *Natural Assets: Surfing a Wave of Economic Growth*. Economics Working Papers Series 2016-06: University of Sydney.

MCTA (1997). *Miyazaki no kanko 1947–1997: Miyazaki shi kanko kyokai 50 juunen kinenshi* [Miyazaki tourism 1947–1997: Miyazaki city 50th Anniversary Memorial Magazine]. Miyazaki: Miyazaki City Tourism Association.

Miyagin Economic Research Institute. (2016) *Miyazakiken jutaku jigyo 'Keizai kouka tou chosa gyomu' Miyazakishi Kisakihama, Hyugashi Okuragahama ni okeru safuin ni kansuru anketo kekka hokokusho* [Miyazaki prefectural government contracted research project on economic impacts: An impact study of surfing at Kisakihama in Miyazaki city and Okuragahama in Hyuga city]. Miyazaki: Author.

Miyahara, H. (2016) *Aoshima Beach Style (vol.2). Free Magazine*. Aoshima: Aoshima Beach Park.

Miyanichi e-Press. (2015) *Miyazaki kengai safua keizai kouka 17 okuen amari Ken chosa* [Economic effects from incoming surfers to Miyazaki estimated to be over 1.7 billion yen]. Available at: http://www.themiyanichi.co.jp/kennai/_16270.html

Miyazaki Kotsu (1997). *Miyazaki Kotsu 70 nenshi* [70 years of Miyazaki transportation]. Miyazaki: Author.

Mizuno, E. (2002) Supōtsu to kaibunka ni tsuite no ichikōsatsu: X sāfu shoppu ni mirareru dansei bun [A study on sports and subculture: 'Male culture' at surf shop X]. *Kyoto Journal of Sociology, 10*, 35–60.

Mizuno, E. (2007) Supōtsu to kanyō-sei: sāfin kyōdōtai ni okeru jendā to rōkarizumu [Sports and tolerance: gender and localism in surfing communities]. In S. Ashina (Ed.), *Tagenteki sekai ni okeru kanyō to kōkyō-sei: Higashiajia no shiten kara*. Kyoto: Koyo Shobō, pp. 198–214.

Mizuno, E. (2015) Nihon ni okeru sāfin o suru josei no 50-nen (1): 1990-nen ikō no sāfin bunka to jendā kōhei [50 years of women surfing in Japan (1): Surfing culture and gender equity since the 1990's]. *Journal of the University of Marketing and Distribution Sciences: Humanities, Arts and Sciences*, *28*(1), 53–76.

Murata, S. (2017) *Kūkan funsō to shite no jizoku-teki supōtsutsūrizumu: Jizoku-teki kaihatsu ga kataranai chiiki no seikatsu-shi* [Sustainable sports tourism as a spatial conflict: Documenting the local life that sustainable development does not talk about]. Tokyo: Shinyosha.

O'Brien, D. and Ponting, J. (2013).Sustainable surf tourism: A community centered approach in Papua New Guinea. *Journal of Sport Management*, 27(2), 158–172.

Ohkawa, N. (n.d.). *Traditional Surfing in Japan: An Unknown History*. Available at: www.nobbywoodsurfboards.com/webpages/cn74/itago.html

Pollack, A. (1993) To surf and ski, the Japanese are heading indoors. *New York Times*. Available at: www.nytimes.com/1993/06/15/world/to-surf-and-ski-the-japanese-are-heading-indoors.html

Ponting, J. and McDonald, M. G. (2013).Performance, agency and change in surfing tourist space. *Annals of Tourism Research*, 43, 415–434.

Preston-Whyte, R. (2002) Constructions of surfing space at Durban, South Africa. *Tourism Geographies*, 4(3), 307–328.

Reis, A. C., de Sousa-Mast, F. R. and Gurgel, L. A. (2014) Rio 2016 and the sport participation legacies. *Leisure Studies*, 33(5), 437–453.

Sakaguchi, K. (Narrator), Yoneda, K. and Sato, T. (Producers) (2010). *Nihon safuin densetsu: Nihon no safuinshi wo tadoru* [The legend of Japanese surfing] [DVD]. Japan: WOWOW.

Sato, K. (1997) Miyazaki no ijin [Great people of Miyazaki]. Miyazaki: Konyakusha.

Spielvogel, L. (2003) *Working Out in Japan: Shaping the Female Body in Tokyo Fitness Clubs.* Durham: Duke University Press.

Statistics Bureau, Japan. (2016) *Statistical Handbook of Japan 2016*. Tokyo: Author.

Stranger, M. (2010) Surface and substructure: Beneath surfing's commodified surface. *Sport in Society*, 13(7–8), 1117–1134.

Surf City Miyazaki. (n.d.). Retrieved from http://surfcity-miyazaki.jp

SURFING Magazine (2015). The Ocean Dome in Japan was the world's best wave pool. Available at: www.surfingmagazine.com/video/the-ocean-dome-in-japan-was-the-worlds-best-wave-pool/#UeCY1sTOMyLghcVK.97

Surfn' Life. (1992) Huge Surf to 1992: Japan Made. *Surfn' Life* (1). Tokyo: Marine Planning.

Tada, O. (2015) Constructing Okinawa as Japan's Hawai'i: From honeymoon boom to resort paradise. *Japanese Studies*, 35(3), 287–302.

The Butterfly Effect (2012).About the Butterfly Effect. Available at: http://betheeffect.com/about-the-butterfly-effect/

The Economist (2012) Surf economics: Beach rush. *The Economist*, 17 March. Available at: www.economist.com/international/2012/03/17/beach-rush

Thompson, G. (2011) Reimagining surf city: Surfing and the making of the post-apartheid beach in South Africa. *The International Journal of the History of Sport*, 28(15), 2115–2129.

Thorpe, H. (2014) *Transnational Mobilities in Action Sport Cultures*. London: Palgrave Macmillan.

Thorpe, H. and Wheaton, B. (2011) 'Generation X games', action sports and the Olympic movement: Understanding the cultural politics of incorporation. *Sociology*, 45(5), 830–847.

Tien-Ming, C. and Chiang-Chuan, L. (2015) The causal relationships among recreational involvement, flow experience, and well-being for surfing activities. *Asia Pacific Journal of Tourism Research*, 20 (Suppl. 1), 1486–1504.

Westcott, M. (Producer/Director) (2013) *Good morning Miyazaki* [Motion Picture]. Canada: DBFilms.

Wheaton, B. (2010) Introducing the consumption and representation of lifestyle sports. *Sport in Society*, 13(7–8), 1057–1081.

Wheaton, B. (2013) *The Cultural Politics of Lifestyle Sports*. London: Routledge.

Yang, E. C. L., Khoo-Lattimore, C. and Arcodia, C. (2017) A narrative review of Asian female travellers: Looking into the future through the past. *Current Issues in Tourism*, 20(10), 1008–1027.

7 Japan's mountain tourism at a crossroads

Insights from the North Japan Alps[1]

Abhik Chakraborty

Mountains as tourism destinations and the case of the North Japan Alps: an overview

The trajectory of mountain tourism development is unique in its temporal and spatial contexts (Richins, Johnsen & Hull, 2016). This is owing to the fact that global mountain systems are highly diverse, with different types of biophysical environments and sociocultural dynamics (Hoorn, Perrigo & Antonelli, 2018; Price et al., 2013). Richins, Johnsen and Hull (2016) point out that the development of mountains as tourist destinations typically commences with a phase of discovery or exploration followed by one of regional development. In many cases, the phase of destination development is closely associated with the opening up of mountains to modern modes of travel and the related development of infrastructure. The interplay of biophysical and sociocultural factors results in complex challenges for managing tourism in mountain destinations, with repercussions for the resident communities of plants and animals and, ultimately, for local residents (Godde, Price & Zimmermann, 2000). Indeed, while many mountain societies possess important traditional knowledge, knowhow and high resilience to challenges, (Byers, Price & Price, 2013), they are typically marginalized during the exploration and developmental phases of mountain designations (Godde, Price & Zimmermann, 2000; Parish, 2002).

Globally, there has been a notable increase of tourism in mountain destinations in recent years (Nepal, 2002; Nepal & Jamal, 2011; Richins et al., 2016), with Snowdon, Slee and Farr (2000) observing that tourism, specifically in the highlands of Europe, has replaced economies of resource extraction. At the same time, Markovic and Petrovic (2013) and Pfister (2000) point out that the visitors to mountain destinations, who are often from urban areas, variously seek thrill, spiritual sustenance and encounters with the unknown, thus transforming mountain destinations into *idealized* spaces. It is estimated by Richins et al. (2016) that currently some 15–20 percent of global tourist visitations occur in mountain regions and, perhaps unsurprisingly, the rising popularity of mountains as tourist destinations and the threats to their biophysical and social environments was recognized in the commemoration of the 'International Year of Mountains' in 2002 by the United Nations Environmental Programme. However, despite several laudable efforts such as this to both raise awareness of the challenges faced by mountain regions and to conserve and develop them in a sustainable manner, mountain destinations around the world are increasingly threatened by visitor pressure, the depletion of resources and global environmental change (CBD, no date; Geneletti & Dawa, 2009; Halofsky & Peterson, 2017; Hoorn et al., 2018; Reece, 2007; Wester et al., 2019).

In Japan, mountain landscapes comprise some of the leading outdoor recreation venues. Geographically, this is self-evident as nearly 80 percent of the Japanese Islands is topographically defined as 'mountains' or 'highlands' (Karan, 2005). Most of these mountains are of low to moderate elevation that allows for easy access, whilst active volcanism throughout the Japanese Islands offers additional tourist resources, such as hot springs. The mountains of the Japanese Islands also have a high ecological value: they frequently feature perpendicular distribution of ecosystems, many of which are not found elsewhere in Japan (Kudo, 2016). Active volcanism and the legacies of Quaternary[2] ice ages have also left lasting imprints on mountain relief and ecosystems of Japanese Islands (Iwata, 2016; Momohara, 2018) and vigorous tectonic and denudation processes are key to the heterogeneity of relief and small-scale ecosystems (Kariya, Takaoka & Sato, 2013). However, these processes and the fine-scale heterogeneity they bring to mountain environments are typically poorly understood in conservation and land management – and even less in tourism planning – resulting in the overall trajectory of the taming of their dynamic properties and the resulting trivialization of the place.

As noted above, Japan's mountain regions comprise some of the most heavily visited destinations in the country. The major peaks and their closely nestled valleys with abundant hot springs collectively represent a significant tourism resource, albeit one that is increasingly popular and at the same time fragile. Longitudinal data on how these sensitive environments are changing over time are scarce, and this situation is reflected in the dearth of research literature on the impact of tourism in mountain destinations. While there are several key literature sources on travel in the highlands – with Fukada Kyuka's account of 100 major mountain destinations perhaps being the most popular and inspiring work among Japanese hikers (Fukada 2003; reprint 2015) – there is comparatively little analysis of how tourism has impacted these landscapes. It is recognized that global environmental change, especially a rapidly changing climate, will leave lasting imprint on these mountains but, so far, little systematic study has been done to understand how mountain destinations will respond to such climate- and environment-related stressors. This lack of reliable ecological as well as visitor monitoring data poses a major challenge for understanding whether these destinations can be managed in a sustainable manner with all their important biophysical properties into the foreseeable future; but it is also worth noting that many of these sites witnessed pioneering nature conservation initiatives, early civilian protest movements against grand development schemes and ongoing conservation and restoration efforts by dedicated groups or individuals, and are therefore endowed with considerable natural and sociocultural resources that can be fruitfully engaged with through various forms of nature-based tourism in order to deepen our understanding of such highland environments.

Below, I provide two instructive case studies of the Kamikochi valley and the Hakuba highland village from the North Japan Alps region in order to illustrate the point that today, these destinations are at a crossroads where anthropogenic drivers increasingly threaten their fragmented and at times fragile resources, while at the same time, some sections of the local society and even some visitors are increasingly aware of this condition. In the face of rapid environmental change, this awareness, though not a fully developed counterweight to exploitation at this point, forms an important social capital that can perhaps be utilized to effect a shift from resource and visitor intensive tourism to a smaller scale and lower impact form of visitation. The insights presented here are based on a 2-year-long ongoing project in the region. The findings are informed by repeated interviews with local inhabitants and visitors, observations and documentation of key landform–landscape–ecosystem features, as well as my subjective reflections out of a growing association with these landscapes that currently increasingly require me to spend most of my summers between their jagged ridgelines and rolling vales.

A brief note on data collection and analysis

The data for this research were collected during a 2-year-long research project in the North Japan Alps between April 2017 and March 2019 (see Figure 7.1). During this period, I visited both sites several times each year (mostly during the May to late October hiking season when all peaks and valleys are safely accessible), and conducted in-depth interviews amongst mountain hut owners, local guides, tour operators, tourists and local researchers. I also traversed the entire Shiroumadake ridgeline and criss-crossed the Kamikochi valley (Mt. Yari and Mt. Hotaka ridges) area a number of times, hiking all the major routes, climbing prominent peaks as well as their smaller neighbours in order to observe the landscape. This narrative is also informed by key literature sources on the concerned areas. Content analysis was undertaken from sources to understand the transition of the environment, whilst empirical data were analysed by coding and by comparing key findings following the grounded theory approach (Charmaz, 2014; Corbin & Strauss 2014).

Outline of the study area: North Japan Alps: Kamikochi and Hakuba

The Japan Alps is a name given to a geologically heterogeneous mountainous region comprising three distinct mountain belts on Honshu Island: the Hida Mountain Belt, the Kiso Mountain Belt and the Akaishi Mountain Belt.

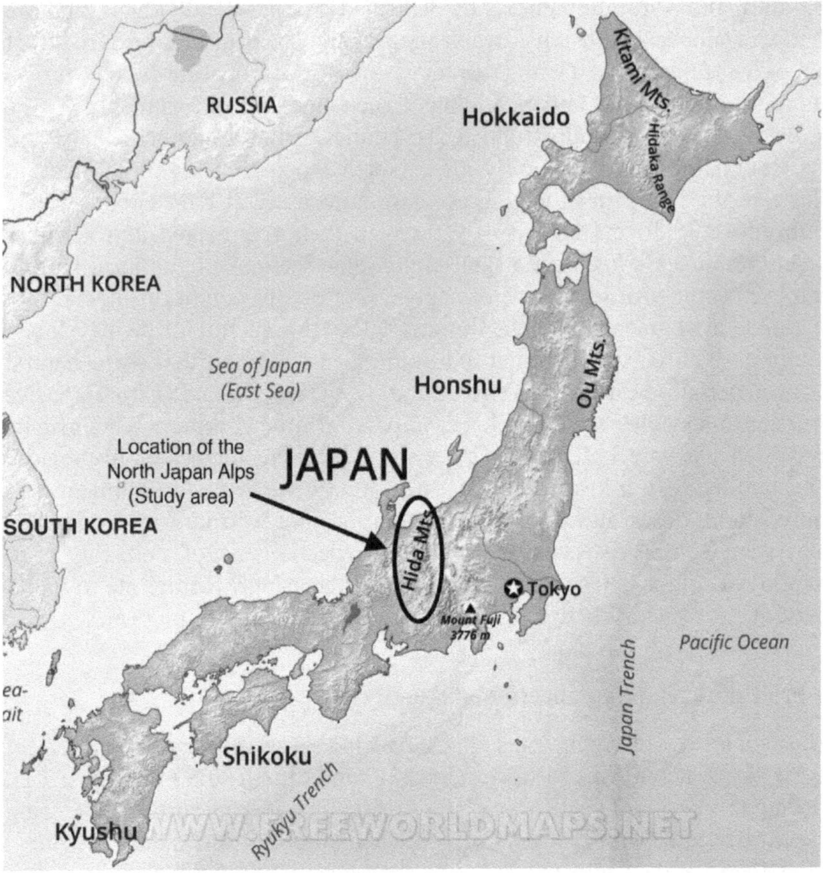

Figure 7.1 Location of the North Japan Alps.

Source: Original map from Free World Maps: www.freeworldmaps.net/asia/japan/, modified by A. Chakraborty.

These 2000 m plus mountains may not be as imposing as their continental counterparts, but as a part of the Circum-Pacific Orogeny they comprise parts of a major global mountain system (Owen, 2013). While they are formed on older bases, the crustal deformation that led to their jagged peak formation is primarily a Quaternary land-formation event, making these mountains young in geological terms and associated with the latest phase of active orogeny in the Japan island arc. The Hida Mountain Belt is also popularly called the North Japan Alps and mean-elevation wise is the highest chain of mountains belonging to a particular group in Japan. Some of the famous peaks in this range are: the eponymous Mt. Yari (3180 m)[3] which is sometimes called 'Japan's Matterhorn'; Mt. Oku-Hotaka (3190 m) that forms the pinnacle of a rocky ridgeline called the 'Hotaka range'; Mt. Tateyama (3015 m), one of the three 'sacred peaks' of Japan); Mt. Tsurugi (2999 m); and Mt. Shirouma (2932 m), located at the centre of a knot of jagged mountains that rises from the picturesque Hakuba valley. Crustal deformation and uplift are the major overall mechanisms behind the formation of the present-day ridgeline, with the intra-arc collision between Northeast Japan and Southwest Japan contributing to vigorous compressional uplift since 500,000 years ago (Isozaki et al., 2010). These mountains also frequently feature older marine and more recent terrestrial volcanic landforms, and have been sculpted by the legacies of Quaternary ice ages and the current Asian Monsoon dominated climate (Nakano et al., 2002).

Initial efforts to map and explore these mountains began during Japan's modernization phase in the late 19th century. The naming of these mountains as the 'Japanese Alps' or 'Japan Alps' is usually credited to William Gowland, a British mining engineer who surveyed the North Japan Alps area for minerals in the 1870s (Fukada, 1964; 2003; Fujioka, 2002). The mountains were later made famous in the accounts of the British missionary Walter Weston, who climbed many major peaks including those surrounding the Kamikochi (the Hotaka range and Mt. Yari) and Hakuba (Mt. Shirouma) areas (Weston, 1896, Reprint 2013). The forays of these men influenced early Japanese alpinists such as Kojima Usui, who famously said that he wanted to climb Mt. Yari due to its formidable appearance (Fujioka, 2002).

Kamikochi is a large valley formed by the Azusa River, located about 50 km from the historic city of Matsumoto (Figure 7.2). The Azusa River is formed by numerous streams issuing from Mt. Yari and the Hotaka range, and eventually becomes Japan's longest river, Shinano. Although the basic feature of the North Japan Alps is crustal deformation and uplift, the highest points of Mt. Hotaka and Mt. Yari are remnants of a large Quaternary caldera volcano which is evidenced by the presence of rocks belonging to a volcano–plutonic complex (Harayama, 2014; 2015). An active stratovolcano, Mt. Yake (2455 m), stands at the other side of the valley; this mountain has erupted numerous times in the recent past and one of those eruptions dammed the Azusa River to create the scenic Taisho-Ike pond (Simkin & Siebert, 1994). Glacial scouring and erosion opened up large valleys both adjacent to the Yari and Hotaka areas during the Quaternary, and the most recent glaciation phases in the Japanese Islands are named after glacial valleys found in these regions (Yoko-o and Karasawa stages, respectively) (Iwata, 2016). The Karasawa cirque (bowl-shaped valley formed by glacial erosion), the Yarisawa valley and the ridges connecting the peaks of Mt. Yari and Mt. Hotaka as well as their lower elevation counterparts that straddle adjacent peaks are popular hiking destinations from summer to autumn. The Azusa River itself is an excellent example of a dynamic mountain stream that exhibits active riverbed formation due to aggressive denudation from the mountains, and vigorous transport and deposition regimes of the river (Iwata, 2016). However, the

Figure 7.2 The Kamikochi valley with Azusa River and Hotaka Range in autumn. Photo: A. Chakroborty.

same dynamic characteristics of this river have resulted in efforts to rein it in a single channel, with the construction of artificial levees and fortification of the embankments, all of which have accompanied the expansion of tourism in the river valley (Iwata & Yamamoto, 2016).

In the early days, the Kamikochi valley was only accessible via a 44 km long road that passed the steep Tokugo Pass whilst during the feudal Edo Period (1603–1868), the Matsumoto fief had logged off a large section of the natural forest (Murakushi, 2005). During the initial years of the modern Meiji state (1868–1912), however, the natural forests recovered as the centralized government restricted logging activities (Murakushi, 2005: 266). Nevertheless, even as logging stopped, roads and hydroelectric dam construction began in the valley, bringing workers and tourists alike. In 1909, 1300 tourists visited Kamikochi as the first large tourist party; most of them went there to visit the hot springs at the foothills of Mt. Yake (Murakushi, 2005: 267). Subsequent efforts to connect Kamikochi to the urban centres culminated in the opening of a bus service from Matsumoto to Kamikochi via the Kama Tunnel in 1933 (Yokoyama, 1971). This continues to be the main mode of tourist transport into the valley today.

Mt. Shirouma (2932 m) is the central peak of a rugged knot of mountains in the Hakuba Village region, which became internationally famous as one of the venues of the 1998 Winter Olympics in Nagano and nowadays is one

Figure 7.3 The Hakuba ridgeline with the three main peaks of Mt. Shirouma-Yari (left), Mt. Shakushi (centre) and Mt. Shirouma (right).

Photo: A. Chakroborty.

of the most popular wintersports destinations in Japan. Mt. Shirouma, along with Mt. Shirouma-Yari (2903 m) and Mt. Shakushi (2812 m), forms the Hakuba-range (Figure 7.3).[4] The area is popular among summer hikers for its steeply rising mountains, denuded landforms, glacier scoured valleys and an abundance of alpine wildflowers. The glacial and denudation landforms of this area have been extensively studied (Iwata, 2014; Kariya, Sato & Komori, 2011; Koizumi, 1993; Sawagaki et al., 2004). Many of the alpine wildflowers found in this area are endemic to these mountains, and the area also features alpine wetland formations in comparatively lower elevations (Ozeki & Matsuda, 1999).

Geologically, the areas adjacent to Mt. Shirouma yield some of the oldest rocks found in the North Alps area; the sections belonging to the Hida-Gaien formation includes Paleozoic rock formations (Nakano et al., 2002). Newer formations, such as the Fossa Magna deposition, bear testimony to the age when the continental plates forming central Japan moved in towards each other. Quaternary volcanism and glaciation also constitute major land formation drivers in this region; the jagged ridgeline formation and landside prone slopes are borne out of various periglacial processes in the Holocene[5] as well as due to erosive agents such as snowmelt, wind and rainfall (Chakraborty, 2018; Nakano et al., 2002).

Regarding the interaction between people and these mountains, there are two main opposing views. On the one hand, Murakushi (2005) contends that owing to the absence of major local mountain belief systems, these mountains were rarely visited by villagers. On the other hand, however, Kikuchi (2005) argues that local people frequently entered these mountains to collect natural resources like sulphur and to enjoy hot springs at lower elevations. Kikuchi (2005) has also compiled an extensive history of the development of tourism in the area that started with a young man called Matsuzawa Teiitsu bravely embarking on a project to construct a hut near the summit of Mt. Shirouma in 1906. Today, a large relief of Matsuzawa greets hikers atop the mountain, and the humble mountain hut has grown into one of the largest accommodations on the North Japan Alps with a capacity of 800 lodgers. In 1950, the Hakuba Oike Ski Ground was opened, and large-scale tourism infrastructure, such as cable cars and gondola lifts, appeared on the scene (Chakraborty, 2018; Kikuchi, 2005). Finally, in 1998 a part of the Nagano Olympics was staged in the Hakuba Village, contributing to its international popularity.

Beginning with the popularization of the mountains in the latter half of the 20th century, a steady deterioration of the alpine plant colonies on the ridgeline was reported. The area adjacent to the three peaks is home to around 400 alpine plant varieties (Chakraborty, 2018), and there were reports of soil compaction and resultant habitat loss of several ridgeline species in the 1970s and 1980s (Tsuchida, 1999). While cordoning off sections of the trail resulted in partial subsequent recovery of the plant species, the issue came under renewed spotlight with the designing of the men's slalom course in the 1998 Olympics across a sensitive plant habitat, and remains an issue of concern today due to heavy summer visitation.

Transition of mountain landscapes, trivialization of place and tourism impacts

As briefly noted above, the North Japan Alps (Hida Mountains) are a major geological feature that represents a part of the long process of the formation of the Japanese Islands. While they are collectively referred to as North Japan Alps owing to the proximity of the major peaks, geologically these mountains are heterogeneous with different stages of Paleozoic landforms overlain by more recent Neogene and Quaternary formations which, in turn, were sculpted by past glaciation events, as well as being modified by ongoing tectonic and erosional processes. These complex drivers form a dynamic environment that is not only remarkable for its physical beauty, but also for its value as the habitat of several alpine and montane ecosystems that are rare in other parts of Japan. However, despite their geological heterogeneity and diversity of relief and biota, these mountain destinations were subjected to a remarkably similar phase of conquest and taming of wild places and tourism development during the 20th century. Today, even as these destinations cope with the lasting legacy of that development, they are confronted by newer challenges such as those emanating from global level environmental change.

Before Walter Weston undertook his journeys through the Japan Alps, mountains in Japan were seen as sacred places where divine spirits dwelled. These areas therefore were out of bounds for the commoners; on the rare occasion when a major peak was climbed, it was undertaken as a pilgrimage. Mt. Yari was first climbed in 1828 by a Buddhist monk called Banryu who nearly perished during the attempt (Fukada, 1964). Hence, a distinct difference between mountain tourism destinations in the European Alps and those in Japan is that the mountains of Japan only began to be climbed in the end of the 19th century. However, as the country modernized, its sacred mountains rapidly became objects of conquest and, later on, became powerful symbols attracting large numbers of urban visitors to the valleys. During the 20th century, wild places all over Japan were 'tamed' by connecting them to cities through roads and tunnels (Totman, 2014), and all major rivers except a handful were dammed to produce energy and to control flooding (Karan, 2005). As tourists started to flock to the valleys, more and more people climbed up the mountains, and these secluded places were 'opened up' with the help of newly developed road networks. A steady trivialization of the place was the outcome, as exemplified by Murakushi (2005) who notes the increase of menus suited to the urban palate in mountain huts and the occasional chorus songs sung by schoolchildren from the warmth of hotels in Kamikochi during the latter half of the 20th century. Jones (2009) also reports of the increasingly casual attire of the hikers in the Kamikochi area of today.

Yet tourism, or rather the perceived benefits of tourism, played a key part in the formulation of the Japan Alps National Park plan, an effort led by the leading voices behind the National Park movement, Tamura Tsuyoshi and Honda Seiroku (Jones, 2016; Murakushi, 2005). In order to stall a plan that would have turned the entire Kamikochi valley into a large dam, Tamura stressed that tourism was the only viable alternative for developing Kamikochi, as it would provide incentives to protect the natural environment into the future (Jones, 2016). The Chubu Sangaku National Park finally came into existence in 1934, with the Hakuba and Tateyama-Kurobe mountainous areas joining Kamikochi as its main components. While the efforts by Tamura, Honda and other like-minded conservationists saved the valley from inundation beneath an artificial reservoir, it also paved the way for touristic exploitation of Kamikochi, notably after the Kama Tunnel was opened in 1933. Subsequently, the area captured popular imagination as a result of its portrayal in Inoue Yasushi's novel 'Hyoheki', and a rapid increase in visitors ensued in the mid-1950s (Shimazu, 1999). Better road access led to an episode of air pollution from vehicular exhaust and intense traffic congestion during the 1960s–1970s, eventually leading to a restriction on private vehicles in 1975 (Shimazu, 1999). Today, a network of supplementary tunnels and roads has spread across the mountains, facilitating speedy access, and in spite of restrictions on access for private cars, intense visitor pressure on the

Kamikochi bus terminal and surrounding areas can be seen during the hiking season.

At Hakuba, the major peaks of Mt. Shirouma, Mt. Shirouma-Yari and Mt. Shakushi rise straight from the narrow valley where the Hakuba Village (connected by the Hakuba Station of Japan Railways Line) is located; the linear distance from the station to the trailhead is less than 10 km. Nevertheless, the mountains were not easily approached before the Meiji era modernization of Japan, as snowfields and widespread alpine moors guarded major trailheads. However, the situation rapidly changed during the Meiji Period, and Hakuba became a favoured destination for early Japanese alpinists. Some local villagers, such as the young Matsuzawa Teiitsu, saw a business opportunity in the rise of popularity of the mountains among alpinists, reflected in his bold decision to set up a mountain hut in an abandoned army checkpost near the summit in 1906.

Detailed descriptions of tourism development at Hakuba can be found in the accounts of Kikuchi (2005) and Murakushi (2005) though, more briefly, the area gained traction as a tourist destination because of the relatively short distance between the valley and the peaks. Walter Weston's trip to Mt. Shirouma in 1893 became the catalyst for tourism development in the valley and in 1903, the Chuo Line from Tokyo was extended to Matsumoto, easing access. Murakushi (2005) points out that unlike Kamikochi, mass tourism in Hakuba took a longer time to develop owing to the steepness of the mountains. Early efforts to protect the region's wildflower population from tourism's impacts can be identified through the Forestry Agency's declaration of the mountain forests as a 'research asset' in 1922 (Murakushi, 2005: 293). Around the same time (in 1919), the 'Hakuba Yama Annai Kumiai' (Hakuba Alpine Guide Agency) was established, representing a pioneering example of raising awareness of hikers about the natural environment of the North Japan Alps. The Hakuba area eventually became a part of the Chubu Sangaku National Park, with Tamura Tsuyoshi reportedly remarking that, in his view, the value of this landscape was second only to Kamikochi (Murakushi, 2005). Murakushi concludes that Hakuba was a rare case where the inscription into the National Park roster was smooth and the process did not involve protest movements, as there was no major development plan for the valley.

The 1950s proved to be the turning point in this case too. Major tourism-related developments commenced in the 1950s with the establishment of the Hakuba Oike Ski-ground and the resulting ease of access to the marshy Tsugaike trailhead, which facilitated entry into the long but technically easier Ko-renge ridge (Chakraborty, 2018; Kikuchi, 2005). The subsequent popularity of the region as a ski resort spread beyond Japan, and the area enjoyed a remarkable growth as a ski-tourism destination between the 1970s and 1990s. The 1998 Nagano Winter Olympics played their part and, at the turn of the century, nearly 30,000 overnight visitors were recorded during the peak skiing season (Kikuchi, 2005). A slow decline in the number of tourists ensued thereafter, possibly indicating a point of saturation for the destination (Chakraborty, 2018).

As these two cases show, while they have considerable fine-scale diversity and somewhat different sociocultural histories, the mountain landscapes of Japan underwent a rapid transition beginning with the Meiji modernization of Japan and subsequently witnessing a remarkable period of intensive change from the 1950s. During the initial phase of modernization, the centralized government consolidated top-down control of land and formulated management plans that favoured economic and industrial development. These development plans drew protests from contemporary conservationists, and the national park movement was borne out of this situation. The second phase of transition of these landscapes during the middle of the 20th century coincided with a nationwide push for economic and infrastructure development, and large-scale construction agendas were prioritized. Tourism played a dual role in this scenario; even as the construction of roadways, dams and hotels accelerated, landscapes that were virtually unknown came under the spotlight. In some cases, the traffic congestion and pollution raised alarm and served as catalysts for restrictions and cleanup, resulting in a partial recovery of some of these landscapes.

Contemporary challenges for sustainable mountain tourism: unfinished business in an era of environmental uncertainty

Today, the Japan Alps have become an easily accessible destination, drawing millions of tourists every year. Although most of the tourism is limited to the lower valleys and hot spring facilities, a total of 1.8 million footfalls were recorded in the North Japan Alps area via the Matsumoto-Azumi gateway (Nagano Prefecture, 2014), whilst the Hakuba village statistics reveal a total of 2.2 million visits in 2015 (Hakuba Village Statistics, 2016). The popularity of Hakuba as a ski-resort rose among domestic and international tourists after the 1998 Winter Olympics, and ski-tourists currently outnumber mountaineers/hikers by an approximate factor of 25 (around 38,000 hikers/mountaineers versus nearly 1 million skiers) (Hakuba Village Statistics, 2016). Hakuba has become popular for international tourists as well, though the total number of international visitors (77,724 in 2014) is small compared to domestic tourist numbers (Hakuba Village Statistics, 2016). These numbers indicate a continuous pressure on the carrying capacity of the mountain ecosystems.

In the case of the Kamikochi valley, the major current challenge is the need to accommodate the pressure from intense visitation during spring to autumn with the dynamism of the natural environment. Increasingly, local planners appear to favour policies that would constrain natural processes such as erosion and riverflow, and that would facilitate further ease of access and comfort of lodging. Unfortunately, this does not augur well for the natural inhabitants of the area as well as for the key geomorphological mechanisms that maintain and diversify habitats. The Kamikochi valley and the Azusa River floodplain are noted for a rich and diverse ecosystem that is shaped by an active river (and its tributary streams). The Asiatic Black Bear (*Ursus thibetanus*), the Serow *(Capricornis crispus)* and the Japanese macaque (*Macaca fuscata*) are notable large mammals of the valley. Equally, the Char (*Salvelinus*) is the most well-known native fish species in the river, whilst the forests are home to a large number of insects and flowering plants (Kamikochi Official Website, no date; Kamikochi Visitor Centre, no date). A mix of broadleaf and coniferous plants forms the crown cover, some notable varieties being the Siebold's Beech (*Fagus crenata*), Japanese Fir (*Abies firma*) and the Japanese Larch (*Larix kaempferi*) (Kamikochi Official Website no date; Kamikochi Visitor Center, no date). Of particular importance is the riparian colony of the Chosenia (*Salix arbutifolia*); the only large natural colony of this riparian plant in Honshu is along the river corridors of Azusa and this colony is now threatened due to river engineering and development plans.

As described earlier, the Kamikochi area was included in the Chubu Sangaku National Park as a 'strictly protected' zone. Tourists cannot collect or remove plants, flowers or loose rocks from this landscape. Yet, paradoxically, the anthropogenic impact in Kamikochi has intensified over the years. Tourism in Kamikochi has steadily shifted from mountaineering to sightseeing over the years with tourists dressed in casual attire increasing (Jones, 2009). This has led to a demand for the provision of 'safety' for tourists in an environment that is inherently dynamic. Even on the higher peaks, fortification of routes with metal ladders has taken place in order to ensure visitor safety. The Azusa River carries a large volume of bedload and has a high rate of in-channel deposition in the area where tourism amenities are concentrated. During peak discharge events, these amenities are threatened by mud and gravel flow, and this propensity of the river to flooding has prompted the construction of weirs and high embankments. To make matters worse, many check weirs were constructed during the 1960s and 1970s on the mountain streams that feed the Azusa River, which made the natural 'silt traps' in tributary stream-mouths redundant. As a result, large boulders and gravel are washed into the main channel during large stormflow events, raising the riverbed (Iwata & Yamamoto, 2016). Iwata and Yamamoto (2016) also mention that some hotels and mountain huts are located inside the floodplain of the Azusa River, and efforts are ongoing at places to restrain the naturally braided river course to a single channel by extensive engineering. Iwata and Yamamoto (2016) point out that the naturally braiding pattern of the river and repeated disturbances, such as channel shifting, sidebar deposition and large flow events, mark the natural characteristics of the landscape, and that restraining the river could wipe out the only natural colony of *Salix arbutifloia* that is adapted to a riparian environment with a high degree of natural perturbation. As Shimazu (2000) points out, construction, at least part of which can be related to the 'hazards discourse,' and the demand for tourist safety have already severely disrupted natural ecosystems of the area.

Iwata and Yamamoto (2016) also note with alarm that visitor walkways have become broader, even in the deep interior of the valley; they object that mountain huts seeking to provide luxury amenities are 'out of sync' with the place. Interestingly, I personally experienced the renovated mountain huts where the managers spoke of catering to sightseers as a business priority. The path to the gateway of the Yari-Hotaka trails at Yoko-o was broadened to allow occasional access by car,[6] and a temporary trail is regularly opened each spring to avoid a landslide prone section of the original trail.

While the general feeling among mountain hut owners in the region is that increased visitation is good (as it brings economic returns), some are concerned that too many tourists are negatively affecting the quality of the visitor experience. A hut operator observed that there is a recent increase of solo campers in the Karasawa cirque, notably in autumn to capture the beautiful foliage around the landscape sculpted by Quaternary glaciation. As the campsite has limited space, many campers erect their tents on areas where damaged vegetation is on its way to partial recovery. In addition, some staff in the mountain huts expressed their concerns over the recent erratic pattern of autumn foliage across the valley, and wondered whether climate change was playing a part in this. The autumn foliage is a major tourist draw, but one that is also affected by the sensitivity of the vegetation fine-tuned to the alpine environment. Enquiries regarding climate change effects elicited two types of responses; while those who have lived or worked in the area for around three decades identified a trend of less snowfall, earlier snowmelt and erratic patterns in autumnal colouration of vegetation, in contrast those who have been associated with the area for shorter periods of time expressed lack of certainty over distinguishable effects. This pattern is expected as climate is a multidecadal phenomenon. However, some of the younger staff were aware of other types of damage in the environment, such as the alteration of forests and rivers, and some of them showed a detectable preference for more adventure-based and perhaps a less intensive form of visitation.

For example, a young hut owner lamented that while hikers are essential for his business, he feels sad that they come and go without noticing much of the environment around them and remain preoccupied with standing on top of just one or two prominent peaks. This focus on conquering prominent peaks arguably results in a myopic view of the total environment. Even an apparent interest in 'nature' can come with its problems. During a number of my own hikes, I observed a number of hikers (notably middle-aged women) who were ecstatic when encountering pretty alpine flower colonies but were equally averse to the insects that provided the crucial function of pollination for these colonies. Such partial and preferential views of the natural environment unfortunately go unchecked as there is no mandatory information session for visitors that could make them aware of the dynamism and vulnerability of the natural environment. Indeed, although the National Park office in Kamikochi has its rangers, they rarely interact with visitors, and nor does the National Park office possess adequate English language material to explain the natural characteristics, conservation challenges and recent change in the area to international visitors. In this knowledge vacuum, tourism is allowed a significant leverage to alter the place by imposing the preferences of visitors over local knowledge and sensitivities.

Though not as much extensively altered as Kamikochi, the landscape of Hakuba has also witnessed an expansion of tourism's footprint over the years. The area is one of the few remaining winter-habitats of the Rock Ptarmigan (*Lagopus muta japonica*) as well as being home to many wildflower varieties. In 1952, the area around the Shirouma range was recognized as a 'Natural Monument' owing to the richness of the resident flora. Tsuchida (1999) undertook an extensive survey of the area in the 1970s and 1980s and found that the rapid increase of hikers from the 1970s directly contributed to damage of wildflower populations, such as *Dicentra peregrine* Makino, concluding that owing to the continuous impact on topsoil, damaged habitats would take decades to recover (Tsuchida, 1999). Tsuchida (1999) also reports that multiple sections of trails were hollowed out to a depth of more than 10 m, a serious condition for alpine flora. During my field trips, I could see that the approach taken by Tsuchida and subsequent specialists to protect plants – of cordoning the walkway with ropes all the way to the summit shoulder – is having a positive effect as vegetation appears to be making a recovery on the slopes. However, I was surprised to find that the mountain hut staff had little or no knowledge for monitoring these sensitive plant colonies, which means damaged niches are vulnerable to expansion of exotic species or other kinds of ecosystem shocks without being reported. A local guide was worried that as much as 25 percent of all wildflowers may become endangered due to climate change and invasion of alien species (his own subjective estimate, not verified with statistical data), indicating the possible impending depletion of a vital tourism resource.

A very different concern was brought up by one of the local guides who expressed concerns that the capacity of the Hakuba village community to provide tourism-related services could be compromised in the near future owing to global level environmental change. Less snowfall on the mountains eventually results in less water in the valley to irrigate crops and, if the situation worsens, local agriculture and the communities that depend upon it might be severely affected.

In addition, changing snowfall patterns are already affecting winter tourism in the village. While a decade ago, steady snowfall starting in November would make skiing possible from late November or early December, in recent years the ski slopes have only opened on a consistent basis from January. One local respondent noted that for the last 10 years, snowfall in the month of December has diminished considerably or has become far too erratic to allow good skiing conditions. Another respondent worried that based on current snowfall trends, the lower ski slopes would have to be used as parking spaces in another two or three decades' time. Moreover, in addition to less or erratic snowfall, thawing is seen to be occurring earlier as well. According to older residents of the village, the spring thaw now occurs approximately a month earlier compared to a decade ago, and this has serious implications for the availability of water during the latter part of summer and autumn. Huts closer to the ridgeline reported thinning of snow patches and difficulty of water provision during the peak of the hiking season. At the same time, change in patterns of freezing and thawing conditions are also affecting highland flower varieties; for example, the Skunk-cabbage (*Lysichiton camtschatcensis* Schott) colonies that are a favourite among early summer tourists in Tsugaike Nature Park are possibly negatively affected by the early onset of the spring thaw.

While the local residents showed a heightened concern for climatic shifts, snowfall change and potential water shortage, awareness of anthropogenic fragmentation of key geomorphic processes was lower, even among locals. The once bothersome Tsugaike wetland has over the years been transformed into a wetland park for casual hikers, but there is currently a perceived problem of siltation of the key channel (Kusukawa) that flows through the wetland. Local staff worried that increased siltation could be bad for business and could damage the extensive wooden walkway that allows visitors to access all sections of the wetland. The predictable response has been channel modification to control silt flow, without taking the natural regime of erosion and transport of mountain streams into consideration. This is somewhat ironic as the wetlands in lower elevations of the Kusukawa basin were systematically landfilled for housing development, leaving only the Tsugaike wetland as the remaining significant wetland in the area (Chakraborty, 2018). Thus, while the attractiveness of the vegetation colonies of Tsugaike make them a significant tourism resource, the complex mechanisms of highland moor formation, drainage and transport patterns and the inherent dynamism of this environment is either weakly perceived or deliberately compromised for sustaining economic returns.

As is clear from this narrative, these two iconic destinations of the North Japan Alps face a number of uncertainties at the moment which locate them at a developmental crossroads, with multiple threats as well as multiple opportunities so far as mountain tourism is concerned. Moreover, these two cases are probably indicative of the overall trends faced by highland destinations in the Japanese Islands in general. From the natural environment perspective, these landscapes have been transformed from wild and dynamic environments shaped by a long history of tectonic, volcanic and erosional agents into tamed spaces for leisure and appreciation by urban visitors. This does not imply that there are no natural mountain ecosystems left; however, the scales of the fundamental geomorphic processes, such as erosion and transport by mountain streams, have been significantly altered as a result of human engineering, and the repercussions for the key biota are only slowly becoming apparent. From the demographic perspective, these destinations are probably somewhat past their prime as overall tourist numbers are declining slightly, but the tourist footprint continues to be large owing to ongoing engineering, the construction of roads and accommodation, and the availability of speedier modes of travel. These trends are added to by the uncertainty of global level environmental change that threatens the sustenance of some of the key natural attractions, such as the possible loss of snowfall amount owing to climate change and its effects at the entire ecosystem level.

Nevertheless, while mountain tourism in the North Japan Alps has steadily shifted towards better access for larger numbers of tourists, and the mountain landscapes as 'place' have been trivialized in general, it is worth remembering that the relationship between tourism and mountain landscapes is not unidirectional. Tourism-related drivers transform landscapes at the destination level but, at the same time, human-induced environmental change affects tourism and raises concerns about ecosystems. It was discussed earlier, for example, how development plans during the early modernization phase united conservationists to protect the natural landscape. Although site-based scientific data on how environmental change is affecting snowfall pattern and ecosystems are not available, most respondents I talked to agreed that they have experienced change in either snowfall intensity or timing, or both, during their lifetime. In Hakuba, there is a palpable sense of uncertainty after the 2015–2016 winter season brought very little snowfall to the ski slopes; mountain hut owners in both Hakuba and Kamikochi reported an overall small decline in *serious mountaineers*, a trend that is consistent with mass access to these areas. In contrast, the information vacuum regarding the formation and vulnerability of the natural environment implies there is a significant potential for smaller scale, less intensive and more ecologically informed modes of tourism.

While these destinations seem to face an uncertain future, at least in the short-term, it is worth referring to Parish's (2002) insightful analogy of the rope for addressing the condition of mountain destinations. Parish (2002) described the condition of mountainous areas as akin to a rope that is both frayed in some places and robust in other sections; while the frayed parts symbolize the overburdening of resources and loss of resilience, the robust sections offer the hope that with appropriate foresight, some of the damage can be mediated. Indeed, there were some indications from this research that visitor concerns about the condition of the environment is changing, along with a change in the *type* of visitors. Guides in Hakuba reported an increase in the number of tourists – particularly women – who seek to understand the environment, its flora, fauna, sights and sounds, rather than solely aiming for the peaks. Specifically, one local guide in Hakuba reported that about 90 percent (his estimate) of women who are over 50 years of age show clear preference to learn about the resident flora whilst most tourists, when cautioned about the fragility of wildflowers, take care not to veer away from the trail. However, more generally, the current head of the Hakuba Alpine Guide Agency was worried that the sudden increase of mass tourism has negatively affected the area. He said that he prefers a village that is in harmony with the mountain landscape and expressed the hope that like Zermatt in the Swiss Alps, Hakuba could strive to develop an identity as a sustainable mountaineering tourism destination based on restrictions on access and control of access modes, which will help elevate the 'sense of place' amongst both residents and visitors. This sentiment was partially echoed by the young hut owner in Kamikochi area who mentioned that he would like to provide more information to visitors on the forest, the mountain streams and their inherent unpredictability, and the dynamic nature of the mountains.

Similarly, another hut owner stressed that he would welcome slower modes of travel that allow visitors to get a fuller flavour of the destination. Such awareness and ideas are indicators that these destinations retain an amount of social capital that can perhaps be fruitfully tapped into to make them more sustainable. A key challenge for sustainable tourism in these destinations will therefore be in the form of whether the local planners would eventually adapt to the changing visitor demographics and natural conditions and show a willingness to gradually phase out intensive visitation. Downscaling of infrastructure and engineering schemes, and focusing on longer staying packages as well as a commitment to develop more adventure-based travel that allow visitors to gain a better sense of the place – notably its unique geological, ecological and aesthetic values – might help these destinations to cope with the legacies of past anthropogenic change and ongoing environmental transformation. But in order to do that, a significant amount of research, and knowledge sharing between locals, National Park administrators and visitors will be required.

Conclusion

This chapter has presented a historical overview of the transition in mountain tourism landscapes of Japan and highlighted some contemporary challenges through analyses of two iconic destinations in the North Japan Alps. Mountains in Japan are vital for the sustenance of ecosystems and human communities and today they attract millions of visitors owing to their close proximity from urban centres. Mountains were seen as sacred domains in preindustrial Japan and, consequently, they were rarely explored prior to the late 19th century. During the late 19th to late 20th centuries, Japan's modernization ushered in an unprecedented wave of development and human modification of these landscapes which fundamentally changed them within a mere hundred years. Whilst the destinations between them offer considerable fine-scale diversity of the natural environment and varied sceneries, the overall trajectory of change – which became conspicuous in the 1950s – has been towards better access and comfort for urban visitors, resulting in a trivialization of their natural, social and historical characteristics. At the same time, these landscapes helped to form pioneering conservation efforts such as the national park movement. Tourism development in these landscapes thus thrust them into a dual trajectory; that of resource exploitation, anthropogenic modification of nature and trivialization of place on the one hand, and preservation through state-funded and locally engaged efforts that helped heighten the sense of place on the other hand. Today, the mountain landscapes face an entirely new set of challenges in the form of global level environmental change along with continued pressure from tourism. However, although a general sense of uncertainty hangs over these landscapes, there is a perceptible awareness among some local stakeholders about the fragility of mountain environments. In the era of global change, there are indications that ecological consciousness is on the rise among some local stakeholders which in the future might transform into the practice of a more place-conscious and sustainable form of tourism in these destinations.

Notes

1 The findings of this research were earlier reported in two journal articles by the author: 'Japan's mountain tourism at a crossroads', *Tourism Planning and Development* 15 (2018); and 'Challenges for environmental sustainability in a mountain destination', *Geo Journal* 84 (2018). This chapter builds on and expands on those works.
2 The current geological period that began nearly 2.58 million years ago.
3 'Yari' means 'spear' in Japanese.
4 Hakuba sanzan or three peaks of Hakuba, in Japanese.
5 The last epoch in the Quaternary era that began 11,700 years ago; although a newer epoch tentatively termed as the Anthropocene has recently been proposed to denote a more recent departure of the earth system from Holocene boundary conditions.
6 Tourist vehicles are not allowed along this path.

References

Byers, A., Price, L. and Price, M. (2013) Introduction to mountains. In M. Price, A. Byers, D. Friend, T. Kohler and L. Price (Eds), *Mountain Geography: Physical and Human Dimensions*. Berkeley: University of California Press, pp. 1–10.

CBD (no date) Mountain biodiversity. Available at: www.cbd.int/mountain/problem.shtml (Accessed 22 June 2019).

Chakraborty, A. (2018) Challenges for environmental sustainability in a mountain destination: Insights from the Shiroumadake District of North Japan Alps'. *Geo Journal*, 84(2), 415–435.

Charmaz, K. (2014) *Constructing Grounded Theory*. London: Sage.

Corbin, J. and Strauss, A. (2014) *Basics of Qualitative Research: Techniques and Procedures for Developing Grounded Theory*. London: Sage.

Free World Maps. (no date) Japan physical map. Available at: www.freeworldmaps.net/asia/japan/ (Accessed 12 July 2019).

Fujioka, N. (2002) Vision or creation? Kojima Usui and the literary landscape of the Japanese Alps. *Comparative Literature Studies*, 39(4), 282–292.

Fukada, K. (1964) *Nihon hyaku meizan* (*One Hundred Mountains of Japan*). Tokyo: Shinchosha [In Japanese].

Fukada, K. (2003) *Nihon hyaku meizan (One Hundred Mountains of Japan)*. Tokyo: Shinchosha (Reprint 2015) [In Japanese].

Geneletti, D. and Dawa, D. (2009) Environmental impact assessment of mountain tourism in developing regions: A study in Ladakh, Indian Himalaya. *Environmental Impact Assessment Review*, 29(4), 229–242.

Godde, P., Price, M. and Zimmermann, F. (2000) Tourism and development in mountain regions: Moving forward into the New Millennium. In P. Godde, M. Price and F. Zimmermann (Eds), *Tourism and Development in Mountain Regions*. Wallingford: CABI, pp. 1–25.

Hakuba Village, (2016) Village statistics. Available at: www.vill.hakuba.lg.jp/somu/statistics/sightseeing.html (Accessed 20 June 2019).

Halofsky, J. and Peterson, D. (Eds) (2017) *Climate Change and Rocky Mountain Ecosystems.* New York: Springer.

Harayama, S. (2014) Kita arupusu wo tsukutta daifunka: Yari-Hotaka karudera to wa (The superexplosion behind the Formation of North Japan Alps: Exploring the Yari Hotaka caldera). *Kagaku* (*Science*), 121, 69–73 [In Japanese].

Harayama, S. (2015) Kamikochi bonchi no chikei keisei-shi daiyonki Yari-Hotaka – Takidani kakosenryokugan kompurekusu (Geomorphic Development of the Kamikochi Basin and Quaternary Yari-Hotaka Caldera and Takidani Granodiorite Complex). *Chishitsugaku Zasshi* (*Journal of Geological Society of Japan), ,* 121(10), 373–389 [In Japanese].

Hoorn, C., Perrigo, A. and Antonelli, A. (Eds) (2018) *Mountains, Climate and Biodiversity*. Oxford: Wiley Blackwell.

Isozaki, Y., Aoki, K., Nakama, T. and Yanai, S. (2010) New insight into a subduction related orogen: A reappraisal of the geotectonic framework and evolution of the Japanese Islands. *Gondwana Research*, 18, 82–105.

Iwata, S. (2014) Tenkoten ni tatsu Nihon Arupusu no hyoga chikei kenkyu: Imamura Gakuro, Iozawa Tomoya to kongo no kadai (The tuning point of glacial geomrphology research in Japanese Alps: Research of Gakuro Imamura, Tomoya Iozawa and beyond). *Dai Yonki Kenkyu (The Quaternary Research*), 53(6), 275–296.

Iwata, S. (2016) Kamikochi no chikei no naritachi (Characteristics of the landforms of Kamikochi). In Kamikochi Shizen Shi Kenkyukai (Kamikochi Nature Research Association) and N. Wakamatsu, N. (Eds), *Kamikochi no shizen shi* (*A Natural History of Kamikochi*). Tokyo: Tokai Daigaku Shuppanbu, pp. 2–18.

Iwata, S. and Yamamoto, N. (2016) Hakai sareru kamikochi no shizen (Destruction of Kamikochi's natural landscape). In Kamikochi Shizen Shi Kenkyukai (Kamikochi Nature Research Association) and N. Wakamatsu (Eds), Kamikochi no shizen shi (*A Natural History of Kamikochi*). Tokyo: Tokai Daigaku Shuppanbu, pp. 146–165.

Jones, T. (2009) Nature-based tourism in a Japanese National Park: A case study of Kamikochi. *Bulletin of University of Tokyo*,121, 87–116.

Jones, T. (2016) The most beautiful valley in Japan: Kamikochi, the Japan Alps and National Parks in Japan. *Arcadia,* 8. Available at: www.environmentandsociety.org/arcadia/most-beautiful-valley-japan-kamikochi-japan-alps-and-national-parks-japan (Accessed 20 June 2019).

Kamikochi Official Website. (no date) Available at: www.kamikochi.or.jp/about/flora-and-animals/ (Accessed 20 June 2019).

Kamikochi Visitor Center. (no date) (Nature Section). Available at: www.kamikochi-vc.or.jp/learn/zukan/plants/outline.html (Accessed 20 June 2019).

Karan, P. (2005) *Japan in the 21st Century: Environment, Economy and Society*. Lexington: The University Press of Kentucky.

Kariya, Y., Sato, G. and Komori, J. (2011) Landslide-induced terminal moraine-like landforms on the east side of Mount Shiroumadake, Northern Japanese Alps. *Geomorphology*, 127(3–4), 156–165.

Kariya, Y., Takaoka, S. and Sato, G. (2013) Kita Arupusu no jisuberi to sangaku no shokusei (Landslides and alpine subalpine vegetation in the Northern Japanese Alps). *Chigaku Zasshi* (*Journal of Geography*), 122(4), 768–790 [In Japanese].

Kikuchi, T. (2005) *Hakuba no hyakunen: kindai tozan hassho no chi to saisho no yamagoya* (*One Hundred Years of Hakuba: The Birth of Modern Alpine Climbing and the First Mountain Hut*). Tokyo: Yama to Keikoku Sha [In Japanese].

Koizumi, T. (1993) *Nihon no yama naze utsukushii: yama no shizengaku he no shotai* (*Why Japan's Mountains are Beautiful: An Introduction to Mountainous Landscapes*). Tokyo: Kokon Shoin [In Japanese].

Kudo, G. (Ed.) (2016) *Structure and Function of Mountain Ecosystems in Japan: Biodiversity and Vulnerability to Climate Change*. Tokyo: Springer.

Markovic, J. and Petrovic, M. (2013) Sport and recreation influence upon mountain area and sustainable tourism development. *Journal of Environmental and Tourism Analyses* 1(1), 80–89.

Momohara, A. (2018) Influence of mountain formation on floral diversification in Japan, based on macrofossil evidence. In C. Hoorn, A. Perrigo and A. Antonelli (Eds), *Mountains, Climate and Biodiversity.* Oxford: Wiley Blackwell, pp. 459–474.

Murakushi, N. (2005) *Kokuritsu koen seiritsu shi no kenkyu* (*Research on the Establishment of National Parks of Japan*). Tokyo: Hosei Daigaku Shuppan [In Japanese].

Nagano Prefecture (2014) Tourism statistics. Available at: www.pref.nagano.lg.jp/kankoki/sangyo/kanko/toukei/documents/26kankouchi.pdf (Accessed 20 June 2019).

Nakano, S., Takeuchi, M., Nagamori, H., Kariya, Y., Okumura, K. and Taguchi, Y. (2002). *Geology of the Shiroumadake District* [In Japanese with English abstract]. Tsukuba: AIST.

Nepal, S. (2002) Mountain ecotourism and sustainable development: Ecology, economics and ethics. *Mountain Research and Development*, 22(2), 104–109.

Nepal, S. and Jamal, T. (2011) Resort-induced changes in small mountain communities in British Columbia, Canada. *Mountain Research and Development*, 31(2), 89–101.

Ozeki, M. and Matsuda, Y. (1999) Tengubara shitsugen ni okeru 19 nenkan no shokusei henka (Vegetation change record of 19 years at the Tengubara Wetland). Nagano Shizen Hogo Kenkyujo Kiyo (*Bulletin of Nagano Nature Conservation Research Institute*), 2, 13–21 [In Japanese].

Owen, L. (2013) Cenozoic evolution of global mountain systems. In P. Owens and O. Slaymaker (Eds), *Mountain Geomorphology*. London: Routledge, pp. 33–58.

Parish, R. (2002) *Mountain Environments.* Harlow: Prentice Hall.

Pfister, R. (2000) Mountain culture as a tourism resource: aboriginal views on the privileges of storytelling. In P. Godde, M. Price and F. Zimmermann (Eds), *Tourism and Development in Mountain Regions*. Wallingford: CABI, pp. 115– 136.

Price, M., Byers, A., Friend, D., Kohler, T. and Price, L. (Eds) (2013) *Mountain Geography: Physical and Human Dimensions.* Berkeley: University of California Press.

Reece, E. (2007) *Lost Mountain: A Year in the Vanishing Wilderness: Radical Strip Mining and the Devastation of Appalachia*. New York: Riverhead Books.

Richins, H., Johnsen, S. and Hull, J.S. (2016) Overview of mountain tourism: Substantive nature, historical context, areas of focus. In H. Richins, H. and J. S. Hull, J.S.

(Eds). *Mountain Tourism: Experiences, Communities, Environments and Sustainable Futures.* Wallingford: CABI, pp. 1–12.

Sawagaki, T., Aoki, T., Hasegawa, H., Iwasaki, S., Iwata, S., and Hirakawa, K. (2004) Later Quaternary glaciations in Japan. *Developments in Quaternary Sciences*, (2), 217–225.

Shimazu, H. (1999) Kankokyaku no kodo kara mita shizen koen riyo no genjo to mondaiten (Current situation and problems of nature park usage from the perspective of tourist mobility). *Bulletin of Rissho University*, 29–45 [In Japanese]. Available at: http://repository.ris.ac.jp/dspace/bitstream/11266/1930/1/KJ00000161074.pdf (Accessed 23 June 2019).

Shimazu, H. (2000) 'Azusagawa joryu Kamikochi Tokusawa Myojin kan no kado ni okeru nen-nen no chikei henka to kankyo tayosei no keisei (Interannual landform changes and formation of environmental diversity in the riverbed at the upper reaches of the Azusa River, Central Japan)'. Chikyu Kankyo Kenkyu (*Global Environmental Research*): 2. 78–91 [In Japanese].

Simkin, T. and Siebert, L. (1994) *Volcanoes of the World.* Tucson, AZ: Geoscience Press.

Snowdon, P., Slee, B. and Farr, H. (2000) The economic impacts of different types of tourism in upland and mountain areas of Europe. In P. Godde, M. Price and F. Zimmermann (Eds), *Tourism and Development in Mountain Regions.* Wallingford: CABI, pp. 137–155.

Totman, C. (2014) *Japan: An Environmental History.* London: I.B. Tauris.

Tsuchida, K. (1999) Kita arupusu shiroumadake ni okeru kozan shokusei no fukugen (Restoration of alpine vegetation at Mt. Shiroumadake, North Japan Alps). *Nihon Seitai Gakkai Shi* (*Japanese Journal of Ecology*), 49, 321–328.

Wester, P., Mishra, A., Mukherji, A. and Shrestha, A.B. (Eds.) (2019) *The Hindu Kush Himalaya Assessment: Mountains, Climate Change, Sustainability and People.* Dordrecht: Springer.

Weston, W. (1896) *Mountaineering and Exploration in the Japanese Alps*. London: John Murray.

Weston, W. (2013) *Mountaineering and Exploration in the Japanese Alps*. Sligo: Hardpress Publishing.

Yokoyama, A. (1971) *Kamikochi kaihatsu shi* (*History of Development at Kamikochi*). Tokyo: Yamato Keikoku Sha [In Japanese].

8 International exchange in tea tourism

Reconceptualizing Japanese green tourism for sustainable farming communities

Amnaj Khaokhrueamuang

Introduction

Rural tourism generally refers to the practice of visiting places outside major metropolitan areas and includes participation in farm-based holidays, ecotourism, adventure, sport and health tourism, educational travel, arts, heritage and historical recreation, festivals and events and food and beverage tourism (Robinson, 2012). Conversely, visiting or taking vacations in some destinations or attractions in the countryside, such as theme parks or inland resorts, may be excluded from rural tourism given that such activities are more 'urban' in character (George, Mair & Reid, 2009). Nevertheless, as a wide-ranging form of tourism that both occurs in rural (non-urban) areas and is essentially rural in nature (Lane, 1994), rural tourism has been developed over the years to become a complex, multifaceted activity that is widely considered as a means of addressing many of the challenges that face contemporary rural communities (Robinson, 2012; Sharpley & Sharpley, 1997). This is certainly the case in Japan, where the concept of rural tourism or, as it is more generally referred to in that country, green tourism, is considered a potential solution to the problems of depopulation and an ageing society in rural areas (Ohe, 2016; Wijaya, 2013).

Nowadays, the term 'green tourism' is frequently heard in Japan. Originally, the concept of green tourism evolved in the late 1980s in Europe as an alternative, sustainable approach to tourism development in general (Font & Tribe, 2001; Jones, 1987) – that is, low impact, environmentally considerate tourism – and as ecotourism in particular. The term 'green tourism' also came to be associated with vacations enjoyed in rural areas with an emphasis on rest and recreation, soaking up the rural idyll and appreciating rural values (Jones, 1987). It was introduced in Japan by the Ministry of Agriculture, Forestry and Fisheries in 1992, although the concept is interpreted there differently compared to in Europe. That is, green tourism in Japan focuses specifically on the touristic experience of urban–rural interaction rather than simply staying and relaxing in the rural environment (Hisegawa, 2005). Hence, Ohe (2003, cited in Wijaya, 2013: 64) defines Japanese rural tourism as green tourism plus activities run by the non-farm sector or rural inhabitants.

It is also important to note that when considering the concept of green tourism in Japan, this interaction between urban and rural people is primarily of relevance to domestic tourism markets. Although international tourism is extremely important to the Japanese economy both nationally and locally (see Chapter 1), there has been limited success with regards to international tourism in the context of green tourism development. Specifically, green tourism destinations in Japan that emphasize exchange between international visitors and local farmers remain small in number owing to many obstacles, not least the language barrier. Hence, although there are a few examples of green tourism attractions that target foreign tourists (Kazunobu & Tomoyoshi, 2010), promoting exchange between international tourists and rural people through green tourism as a means of supporting Japanese agricultural products and communities remains a challenge.

One such agricultural product for which Japan is renowned is tea; indeed, numerous tea production areas exist throughout the country, most notably Shizuoka Prefecture which accounts for about 40 percent of the country's commercial tea production, as well as Kagoshima and Mie Prefectures (ITO EN, n.d). Moreover, Japan is currently ranked eighth in the world, though accounting for just 2 percent of global tea production (Chang, 2015), primarily for domestic consumption. Although increasing, the proportion of tea production that is exported remains relatively low; the export volume of Japanese green tea amounted to 4,000 tonnes in 2015, with the US accounting for about 40 percent, making it a priority export market. However, reflecting the challenges facing the Japanese farming community more generally, productivity in the tea sector has fallen in recent years; having peaked in the late 1970s, tea production has declined since the mid-2000s. In part, this declining productivity is the result of ageing farming communities and, hence, fewer people engaging in tea growing. At the same time, however, another, factor is declining domestic demand. Not only does tea face competition from other beverages such as coffee and soft-drinks, but drinking tea is seen by the younger generation in Japan as an old-fashioned social practice. This phenomenon is reflected in the declining trend of leaf tea consumption but increased demand for PET-bottled (recyclable plastic bottles) green tea beverages in Japan (MAFF, 2017).

Despite these challenges, however, only limited attention has been paid to the potential of supporting the tea growing communities in rural areas in Japan through the development of tea tourism whilst, in particular, there is extremely limited participation in tea tourism by international visitors. Indeed, inbound tourists to Japan are assumed to be seeking only urban tourism experiences because they mostly stay in the cities during their visits. However, there is some evidence to suggest that international repeat visitors seek unique experiences in rural Japan and, as a consequence, increasing international interest in Japanese tea may offer a potential resolution to the decline of Japanese tea industry (Zavadckyte, 2017) and the challenges facing rural tea-producing communities. In other words, marketing Japan's tea farming villages to international tourists and thereby encouraging international exchanges within the Japanese interpretation of green tourism may go some way to enhancing the sustainability of rural communities.

How this inbound tourism might occur and be developed can be demonstrated through the example of Thai tea farmers visiting Tawara, a declining rural community in Nara Prefecture. This exchange project between Thai and Japanese tea farmers was conducted in 2017 in order to encourage tea farmers from Huey Nam Guen village in Thailand's northern Chiang Rai province to share ideas on developing tea production and tea tourism as well as promoting wider awareness on the potential of international tourism to Japanese tea farmers in Tawara. In this chapter, the lessons learned from the project are discussed within the conceptual framework of urban–rural interaction (as implicit it in Japanese understandings of green tourism) based on the elements of tea cultural commodification (see Figure 8.1 below). In so doing, it seeks to suggest ways of sustaining tea communities and to present perspectives on the future development of green tourism through international exchange activities.

However, first it is important to note that when considering international farmer exchanges in tea production and tea tourism, as in the case of the project discussed in this chapter, Japanese interpretations of green tourism which focus on urban–rural interaction (i.e., between urban-based visitors and rural communities) need to be extended to include rural–rural linkages as another strategy of targeting inbound tourists to farming communities implementing green tourism. In other words, the Thai tea farmers from Chiang Rai's Huey Nam Guen village were, in essence, linking rural Thailand with rural Japan because the purpose of their visit was to share knowledge and experiences in tea production and tea tourism development. Hence, even though they were, in a sense, urban tourists similar to international tourists more generally, not least because they spent time in the Japanese cities for purposes such as sightseeing and utilized urban-based sectors, such as accommodation and transportation companies, to facilitate their trip to the farming village, they are fundamentally rural tourists because they are farmers in rural Thailand. Thus, Japanese green tourism based upon the concept of urban–rural linkages can, in this case, be more precisely conceptualized as 'rural–urban–rural' international interaction with productivity and community development as the core travel purpose. The UN's Food and Agriculture Organization refers to this kind of exchange as 'farmer-to-farmer' (FAO, 2017), which aims at sharing knowledge and experience of good practice among rural people from different countries involved in the same productive activities in order to improve organizational capacities and the management of resources, and to provide small producers with access to markets.

It should also be noted that although this chapter is primarily concerned with the outcomes of the tea farmer exchange project in Tawara, it also seeks to broaden the discussion to explore its implications for the development of green tourism in Japan for other upon types of rural products and communities, such as rice or fruit producers. The purpose of this chapter, therefore, is not only to consider the opportunities for and barriers to developing international exchanges through green tourism as a means of sustaining farming communities encountering declines in farm productivity based upon the specific case study of Tawara, but also to reconceptualize Japanese green tourism more broadly as the international exchange of agritourists.

The Japanese concept of green tourism

Rural tourism in Japan comprises a variety of tourist activities undertaken in rural areas and, since the end of the 20th century, has been increasingly considered as an effective tool for revitalizing rural communities. According to Ohe (2003 cited in Wijaya, 2013: 64–65), Japanese rural tourism encompasses a wide range of technical terms related to agriculture, as follows:

- Farm tourism: agricultural activities [for tourists] run by farm households or farm enterprises.
- Agritourism: tourism in farms plus activities run by farmers' cooperatives or other farmers' organizations.
- Green tourism: agritourism plus activities run by municipalities or the third sector.
- Rural tourism: green tourism plus activities run by non-farm sector or rural inhabitants in general.

As mentioned earlier, the concept of rural tourism in Japan is referred to as 'green tourism', which represents a unique conceptualization of tourism in rural areas. Specifically, the term 'green tourism' (*gurīn tsūrizumu*), officially introduced in 1992 by the Ministry of Agriculture, Forestry and Fisheries (MAFF), refers to the concept of rural community revitalization through the promotion of human interaction between urban dwellers and communities located outside the metropolitan area (Funck & Cooper, 2015; Hisegawa, 2005) involving rural-stay leisure activities in mountainous areas or coastal villages (Chen, Qiu, Usio & Nakamura, 2018), but excluding national parks or nature-designed areas (Hashimoto & Telfer, 2010). Hence, green tourism in Japan offers the potential to contribute to sustaining agriculture, forestry and fisheries as well as the communities that are dependent on those activities (Keidanren, 2000). By the end of the 1990s, green tourism had become an integral element of agricultural and regional development policies, and was also referred in the National Development Plan of 1998 as well as in the Basic Law on Food, Agriculture and Villages in 1999. To promote the ideal concept of green tourism, the Organization for Urban–Rural Interchange Revitalization was established in 2000 (Funck & Cooper, 2015).

MAFF categorizes green tourism experiences under five headings: (i) farmers' restaurants for a local food tasting experiences; (ii) farm shops for purchasing local products; (iii) farmhouse accommodation; (iv) experiences related to agriculture, forestry and fisheries; and (v), citizen farms for farming (Wu et al., 2019). MAFF (2013) has attempted to accelerate the development of 'International Green Tourism' around Japan by producing guidebooks in English, Chinese and Korean that provide information on green tourism destinations in many prefectures, such as the green tourism villages of Noto Town of Ishikawa and Tanabe City in Wakayama. The primary purposes of promoting green tourism are: (i) rural revitalization by enhancing the interaction between rural and urban sectors; (ii) conserving the natural environment; and (iii) developing sustainable rural communities (Bixia & Zhenmian, 2013). In this chapter, this concept of green tourism as a nexus of rural–urban interaction for rural revitalization is, as noted above, extended to the notion of rural–urban-rural interaction manifested in the case of an international tea farmer exchange projects as a strategy for targeting the inbound tourist market.

A conceptual framework of Japanese green tourism through the lens of tea tourism

As emphasized above, the development of green tourism in Japan involves two principal processes; the first focuses on agricultural policy to support farmers, whilst the second promotes tourism at in rural and coastal villages through the interaction between rural and urban communities (Arlt, 2006; UNWTO, 2017). Addressing the second, tourism, as one arena of urban–rural interaction (Gon, 2017), involves not only the building of relationships between urban residents and rural communities, specifically farmers, but also strengthens networks of education and business through transferring people between urban and rural markets. Moreover, because of their interest in the natural environment and cultural authenticity, urban-based tourists are also attracted to tourism products in rural areas. Thus, rural destinations are marketed to their urban counterparts for mutual benefit (Nkemngu, 2017). Tea tourism destinations are but one example of the symbiotic linkages between urban and rural markets.

Jolliffe (2007: 9–10) defines tea tourism as tourism that is motivated by an interest in many aspects of tea, including its history, growth, production, processing, blending, traditions and consumption. It may be seen as part of the overall tourism phenomenon in general, as a form of both cultural tourism and culinary tourism, and also as a small, specialist niche product for dedicated tourists. Interestingly, tea tourism offers activities that embrace both urban and rural forms of tourism. For example, buying tea or drinking tea in a café may be thought of as urban tourist activities related to tea culture, whereas learning about tea growing and processing at a tea farm or garden falls within the scope of rural tourism.

Figure 8.1 illustrates how the tea market at both the local and national level benefits from the interaction between urban dwellers, including city tourists, and local people in rural communities. In other words, prices for agricultural commodities in general are dependent on urban markets, whilst prices

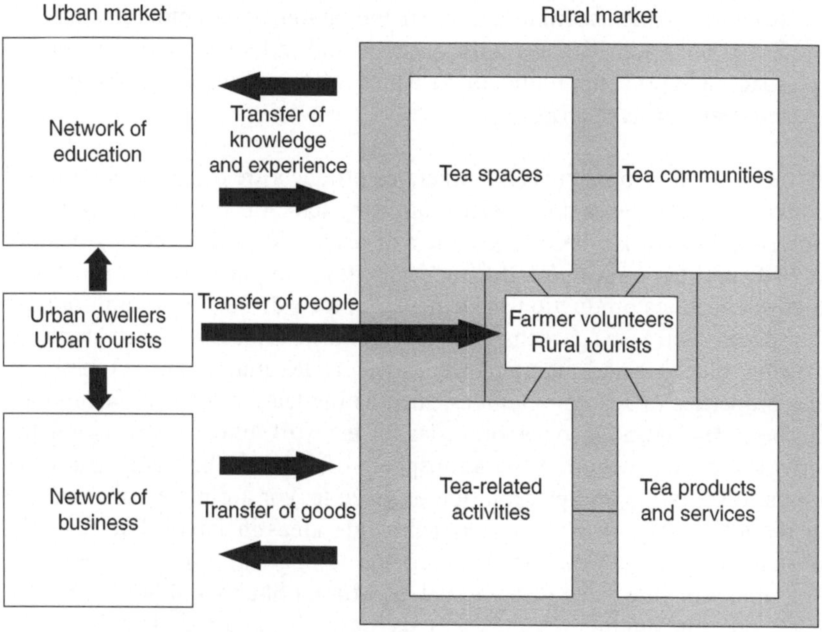

Figure 8.1 A conceptual framework of green tourism based on the four elements of tea cultural commodification in tea tourism.

for green tourism are based on rural markets (Ohe, 2014); the prices of tea products in particular are most influenced by the tea industry. However, green tourism offers a variety of distribution channels for tea and increases the sources of income generated from the rural market based on the four elements of tea cultural commodification: (i) tea spaces; (ii) tea communities; (iii) tea products and services; and (iv), tea-related activities. (Khaokhrueamuang & Chuemchaitrakun, 2019: 108).

- Tea spaces directly involve the tea garden areas which can be commoditized as a place for tea-related tourism activities such as tea plucking, tea festivals and tea folk music performances.
- Tea communities refer to groups of tea producers or farmer households residing in the same environment of tea spaces and enjoying ways of life related to tea production. Tea communities include tangible and intangible culture. The tangible culture concerns tea farmers' or producer's real estate, such as the farmhouse or the tea factory, which can be commoditized as a tea lodge, homestay or farm-inn. The intangible culture represents forms of customs, beliefs, jobs and lifestyles such as tea making and processing methods, tea cuisine, tea ceremonies and tea-telling stories.

- Tea products and services are forms of tea souvenirs, and tea farmer restaurants and accommodation, including tour programmes.
- Tea-related activities are both outdoor and indoor activities in the tea spaces and tea communities divided into two forms; working as volunteers and learning as tourists.

In the local and national urban market, tea products are traditionally shipped from the factory as a mass distribution system for urban consumption. However, those urban-based consumers of tea products have the opportunity to visit rural tea growing/producing destinations to play a (temporary) productive role in tea production as farmer volunteers or simply as rural tourists. Tea spaces, such as tea gardens, are commoditized as places to learn about tea growing, plucking and maintenance, as well as learning about tea cultivars. The properties of tea communities, such as the tea factory and farmhouse, can become classrooms or laboratories for tea workshops, whilst tea-related activities can be created in the atmosphere of '*chado*' (the traditional ceremonial preparation and presentation of green tea) or around events focusing on the tea history of different tea generating areas in Japan. For example, tourists can practice '*chanoyu*' (serving and taking tea) along with learning biographies of major tea masters such as Murata Shuko and Sen no Rikyu, or legends surrounding tea in places such as Nara and Uji. Through these activities, tea products and services are purchased and consumed by those who genuinely understand the process of tea production and appreciate the long history and culture of Japanese tea. Moreover, this commodification of tea production and culture reflects Jolliffe's (2007) definition of tea tourism outlined above.

Based upon the tourism system defined by Leiper (2004), the framework of urban–rural interaction can be extended to the international level if there is cooperation between a rural community in the tourist generating region in different countries and a farming community in the tourist destination region of a host country. One such example, now discussed, is the case of Thai tea farmers from Chiang Rai's tea-growing village visiting Japanese tea producers in Nara's Tawara tea-producing community.

International exchange in green tourism: the case of tea farmers' interaction in Tawara

Green tourism in Tawara

One of the earliest historical records of tea in Japan relates to the royal court of Emperor Shomu in AD 729, in the country's ancient capital city of Nara. Subsequently, the great priest Kobo Daishi Kukai brought tea seeds from China to Japan in AD 806, and local people believe that those seeds were planted in Yamato, present-day Nara Prefecture, enhancing the reputation of Nara's tea growing in the area known as 'Yamatocha' (Baier, 2015).

Tawara is a hill district in the eastern part of Yamato that has been a tea growing area since the Nara period, but the region was merged with Nara City on 1 September 1945. The present territory covers 32.7 sq km and comprises 18 villages. As of October 2017, the total population was 1,750 (847 male, 903 female) living in 771 households. Of these, 246 were farm households. The district is close to the urban fringe area of Nara from where tourists can easily access the villages by public bus, the journey taking about 30 min from the city centre.

Green tourism in the Tawara district commenced in March 2006 with 12 farm properties, and is one of four attractions promoted under the project named 'Naramachikado Hakubustukan', the others being the ancient Kei-Kaido road, the mountain village of Yagyu and Naramachi, the traditional merchant quarter of Nara. Overall, the project aims to present parts of Nara City as a living cultural museum, including farm activities. The government of Nara published a green tourism map of Tawara providing cycling routes and tourist information on farm properties; it is updated every year in Japanese (Figure. 8.2).

By 2017, three groups of 16 enterprises were part of the project in Tawara. The first group comprises agritourist farms and agricultural products, including a tea garden and tea factory, an apple garden, a blueberry orchard,

Figure 8.2 The green tourism route map of Tawara targeting domestic tourists from the city.

Source: Division of Tourism Strategy (kankousenryakuka), Department of Tourism Economy (kankoukeizaibu), City of Nara. www.city.nara.lg.jp/www/contents/1151285908424/simple/tawara.pdf

a Shitake mushroom farm, charcoal making and a pickled vegetable shop. The second group is made up of the artisans, including a pottery, wickerwork, cloth weaving, wood carving, rice straw handicraft and stained glass making, whilst the third group comprises tourism-related services including the tea farmer's restaurant, the marketplace, a traditional house, and souvenir and cycling service centre.

Tourists are able to visit all the sites, but reservations are necessary in the case of experiential agricultural activities because of the farming seasons. Tourists who are particularly interested in tea-related learning activity are required to contact tea farmers during the tea harvesting season, which is from early May and to the middle of July. Regarding tea tourism, the program offers five activities:

- i Tea plucking at an organic tea field.
- ii Observing tea processing at the Tawara Shinonome Kyodoseicha Koujyo tea factory.
- iii Learning Japanese history at the Onoyasuma Ronohaka archaeological tea site.
- iv Experiencing 'Senchado' tea brewing; participating in a traditional tea-tasting game; and enjoying fusion tea cuisine provided by a tea master, Takenichi, at the Takenichinoen Yucha-an (tea farmer restaurant).
- v Shopping for tea products at the Tawara Yamasato Ichiba community farm shop.

In addition, tourists can participate the 'Hatage Helper' volunteer program, working in the organic farms, particularly in the tea gardens, vegetable plots and rice fields. Volunteer tourists can take their own grown and harvested agricultural products home with them. This project supports the consumption of 'safe' foods produced on chemical-free farms. Tawara Natural Farms, a rental farmland managed by a group of local farmers and urban people, is one such example. This group rented an abandoned tea garden and started to grow organic seasonal vegetables, thereby helping to promote green tourism in Tawara. Significantly, the tourist information is provided only in Japanese, representing a major barrier to promoting international tourism to the area.

The tea farmers exchange program

The tea farmers exchange program was established alongside the tea tourism activities available in Tawara. Four Thai tea producers from Huey Nam Guen, a village in Chiang Rai province producing a fermented wild tea known as 'miang', participated in a two-day program in October 2017.This was supported by a research grant from the Center for Tourism Research, Wakayama University, Japan with the collaboration of the Tea Institute, Mae Fah Luang University, Thailand. The exchange activities were designed in line with the Japanese green tourism concept of rural–urban–rural interaction, emphasizing the transfer of knowledge and experience in agriculture and food between hosts and guests. The primary purpose of the visit was to encourage the application of Japanese approaches to tea products and tea tourism management to develop tea tourism at Huey Nam Guen village. At the same time, the program introduced a group of international tea farmers to Tawara as a pilot group of such tourists from Thailand.

Before visiting Tawara, pre-program preparations were at Huey Nam Guen village (e.g., meeting with locals, selection of representatives by the village committees and orientation). In Japan, the first-day's program started with leaving the city centre of Nara by bus, which took about 40 minutes to the village. After that, the green tourism activities began at the Shitake mushroom farm 'Genboku Shitake-en Miyanaka' followed by observing the Japanese tea process at the tea factory 'Tawara Shinonome Kyodoseicha Koujyo', and picking fresh tea buds at the organic tea garden of the Tawara Natural Farm. These were subsequently taken to create the lunch menu at the farmer restaurant 'Takenichinoen Yucha-an'. Here, the Thai guests demonstrated to their Japanese hosts how to cook Thai tea cuisine. They prepared two styles of tea salad 'Yam Bai Miang'. The first one was a sour salad made from pickled Thai tea leaves, and another was mixed with the fresh Japanese tea buds and crispy pork skin

At the same time, a Japanese host, Ms. Takenichi offered Japanese fusion tea dishes. Examples were tea-rice steamed with the roasted tea (hojicha) and cereal, fried bean curd with minced shrimp and tea leaves, and roasted tea pudding. The Japanese tea cuisine at this restaurant uses seasonal local ingredients such as Shitake mushrooms and organic vegetables from Tawara Natural Farm. Here, eating lunch takes up to 2 hours without menus; tourists are both surprised and excited as each dish is served one after the other, offering them the opportunity to talk to the farmers and learn about the food.

After lunch, a Japanese host Ms. Fukui, taught the Thai visitors how to brew a delicious Japanese green tea, 'Sencha', produced in Tawara. Then, Ms. Takenichi introduced the traditional Japanese tea tasting game to them. Subsequently, at tea time, the group of Thai farmers also brewed the Huey Nam Guen's tea products 'Cha Ha Si' (five colours of a tea set). After the end of the program, they took a bus back to stay overnight at the hotel in the city of Nara. On the second day, the Thai visitors came to Tawara again and continued sightseeing at other attractions around tea spaces and communities. They also rode bicycles from the Tawara Yamasato Ichiba, the community market place to pick organic apples at the Kubota farm before visiting the Onoyasuma Ronohaka, the archaeological site located on the tea hill.

Lessons learned from the tea farmers exchange

The activities undertaken during the Thai–Japanese tea farmer exchange based on the concept of rural–urban–rural interaction in green tourism was of benefit to both groups. In Tawara, members of the host community involved in green tourism-related enterprises, such as the farmer restaurant, not only received income from the Thai tourists but established foundations for an international partnership in the tea industry and a network of tea cultural education. Equally, feedback on the program reflected the lessons by the visitors from Huey Nam Guen village with regards to both tea production and tea tourism. More specifically, on their return, the representatives of the Thai tea farmers shared their learning experiences from their visit to Tawara with community members in Thailand, with about 100 villagers attending the feedback session. Four proposed development issues on tea and tourism were highlighted:

i The creation of tea cuisine served in farmer restaurants or homestays.
ii The development of a tea trail for learning about tea culture that links agriculture to gastronomy.
iii The renovation of a traditional farmhouse and tea factory to be used for promoting tea culture through, for example, the development of a tea museum.
iv The building of a community farm shop to sell tea products and other agricultural commodities.

Sustaining farming communities through international exchanges in green tourism

Based upon the framework of Japanese green tourism, conceptualized in Figure 8.1, the interaction of tourists and local communities brings benefit to both urban and rural markets resulting from the networks of education and business. As noted above, in the case of the tea farmer exchange in Tawara, four development guidelines for Huey Nam Guen village can be attributed to such networks.

Education networks

The transfer of knowledge and experience through international exchange activities supports and develops the knowledge and culture of farming communities. For example, in Huey Nam Guen village, the creation of a tea cuisine experience in farmers' restaurants or homestays encourages the villagers to apply their knowledge of tea to food and health benefits, whist the development of the tea trail requires the acquisition of expertise and experience in ecotourism management. Similarly, renovating the traditional farmhouse and tea factory as a tea museum requires knowledge and understanding of the principle of interpretation, whilst building a community farm shop for selling tea products and other agricultural commodities will require farmers to think about marketing strategies. As such knowledge and experience is developed, the local community will acquire the capacity to sustain and build its economy and culture. At the same time, developing tea tourism stimulates the urban market, in a sense establishing an agricultural partnership through tourism. More generally, international exchange also encourages local communities to enhance their understanding of global culture, and offers the opportunity to acquire foreign language skills that give them entry into a variety of jobs and economic sectors (LaPan, 2017).

Business networks

In addition to supporting the knowledge and culture of the farming community, international exchange through green tourism also strengthens the local economy through the transfer of agricultural commodities. That is, in terms of business networks, the development of green tourism enables farmers to establish appropriate prices for farm products independent of the influence of urban markets. In other words, the prices they charge are determined by rural markets (Ohe, 2014). Also, green tourism offers a variety of both distribution channels for agricultural commodities and sources of non-farm income generated from the rural market that is based on the four elements of rurality: rural spaces, rural communities, rural products and services and rural activities (Khaokhrueamuang, 2014), as seen in tea tourism. For the tea industry, the element of rurality has been modified by the process of tea cultural commodification. Thus, the tea-related revenue is attributed to tea spaces, tea communities, tea products and services and tea-related activities in the rural market which can be seen through the linkages with community farm shops. Farm shops are not only places located in the communities and farming spaces, but they also promote local food development and innovation which, in turn, supports rural economies and provides knowledge exchange in the form of information on urban customer expectations (Slocum & Curtis, 2017).

Furthermore, the rural spaces and rural communities can be developed and leveraged to become tourist shopping villages (Murphy et al., 2011), providing tourists with the opportunity to purchase farming products, handicrafts and farm services such as farm stays and farm-related tourism tour program or activities. For example, the tea-growing community of Tawara could develop as the tea shopping village, recreating its community farm shops as multifunction outlets providing not only shopping opportunities but also the exchange of the tea culture and knowledge.

In addition, the development of green tourism can encourage partnerships between urban and rural businesses. This contribution conforms with the three primary stakeholders in agritourism systems identified by Sonnino (2004), namely, agritourism providers, destination marketing organizations (DMO) and agritourists. Agritourism providers in the city work directly with rural tourism businesses, whilst the DMO plays an important role in marketing and connecting international agritourists with agritourism providers and rural communities. In the case of green tourism and tea villages, the DMO as a reservation centre can facilitate foreign tourists by coordinating with a travel agent offered an exchange tour program with tea farmers. This is considered as an example of a 'tourism partnership' between urban and rural economies or 'collaboration' between urban and rural contexts (Boluk, 2017).

Green tourism and the SDGs

The interaction inherent in green tourism based on the four elements of tea cultural commodification represents a means of revitalizing and sustaining the farming community. As such, it potentially contributes to the achievement of the Sustainable Development Goals (SDGs) of the 2030 Agenda for Sustainable Development. Over the next decade, countries will mobilize efforts to end poverty and hunger with strategies to build economic prosperity. At the same time, they will address a range of social needs including education, health, job creation and social and environmental protection while combatting inequality and promoting peace. The concept of green tourism, therefore, offers a specific approach to the design of tourism such that it can contribute to the SDGs. In advocating sustainable development, the SDGs refer to the five 'Ps': people, peace, partnership, planet and prosperity, to which green tourism also contributes:

- **People:** green tourism fulfils the wishes of guests to support rural communities without ignoring the needs or interests of the locals, such as through farm volunteering.
- **Peace:** green tourism avoids conflict and social tension; rather, it contributes to solutions through the participation of the locals in the exchange of knowledge and experience.
- **Partnership:** green tourism contributes to a network of education and business partnerships, providing tourists with space for promoting intercultural learning.
- **Planet:** green tourism reduces negative impacts on the environment and contributes to the conservation of biodiversity through the promotion of sustainable agriculture such as organic farming.
- **Prosperity:** green tourism increases the social and economic benefits for the local people from the transfer of goods, people and wisdom.

Reconceptualizing Japanese green tourism: opportunities and challenges

Although the development of urban–rural interaction in general is advocated in national and global policy, attempts to link rural and urban tourism in particular has garnered limited academic attention (Gon, 2017). Hence, this chapter, through its discussion of green tourism Japan, begins to address this gap in the literature. Beyond this, however, raising the awareness of the importance of international exchanges in green tourism may also contribute to the sustainable development of Japan's rural communities. As Van Leeuwen

(2015) states, flows of visitors affect urban–rural interactions in terms of both the transfer of tourism income and of knowledge and cultural exchange, and international tourists in particular have a role to play in this process. However, in Japan, targeting international tourists as potential participants in green tourism has received limited attention, as manifested in the strategy of promoting *furusato* (hometown) tourism. Consequently, tourism in rural Japan is, with some notable exceptions, primarily domestic, with international tourism spatially limited to a distinctive gaijin-trail (foreigner trail) (Arlt, 2006).

Nevertheless, tourism is expected to become a primary economic sector in Japan in the coming decades, largely as a result of an increase in international arrivals. As discussed in Chapter 1, such growth was stimulated by 'Visit Japan' campaign in 2003, with the goal of achieving 10 million overseas visitors by 2010 (Japan Spotlight, 2016; Soshiroda, 2005) According to the Japan National Tourism Organization, the number of international tourists visiting Japan in 2018 reached 31.20 million, the major markets being China, South Korea and Taiwan (JTB, 2019).

As a result of this growth, partly accounted for by the increasing number of repeat visitors and in response to tourism promotion by local communities, more international tourists are visiting regional cities. Yet although this represents an opportunity for rural or, more specifically, green tourism destinations, these remain primarily designed for Japanese domestic tourism, with only a small number targeting foreign tourists (Kazunobu & Tomoyoshi, 2010). Some of these are destinations based on tea tourism, examples being the Green Tea Tourism Homestay in Nagasaki and the Kyoto Obubu Tea Farms in Kyoto. The Green Tea Tourism Homestay is organized by the Higashisonogi Green Tea Tourism Association, which connects the tea-growing communities in Higashisonogi Town with urban tourists, offers the experience of staying with tea farmers to learn about Japanese green tea. In contrast, Kyoto Obubu Tea Farms,a social enterprise with the mission to bring Japanese tea to the world, arrange various tea-related green tourism activities in the tea fields of Wazuka Town such as seasonal tea picking and rolling, and a tea master course. Moreover, they also offer an international internship program which focuses not only on tea in particular but also on intercultural exchange and social interaction more generally (Jolliffe & Nakashima, 2020).

The foreign tourists participating in these tea tourism programs are assumed to be urban tourists; that is, they primarily visit urban areas during their stay in Japan, However, they become temporary rural tourists when visiting the rural destination, and are indicative of mainly repeat visitors who seek unique experiences in rural Japan. As such, the opportunity exists to attract tourists who are specifically interested in Japanese tea culture. At the same time, however, green tourism is an important option for targeting international tourists to visit Japanese farming communities, even if it does not have universal appeal. For example, for those not explicitly interested in learning about agriculture, green tourism can nevertheless encourage interaction between urban tourists and farmers in terms of economic exchange, such as through shopping for farm products.

In addition, although green tourism is likely to remain a niche area that is embraced by a relatively small proportion of international tourists to Japan, it plays a fundamental role in education, particularly in the transfer of knowledge from farmer to farmer through associations such as the international farmer exchange in tea tourism. Interestingly, based upon the Leiper's model of the tourism system (Leiper, 2004), rural–rural linkages through international farmer exchanges requires an urban organization to play a coordination role. For example, the FAO farmer-to-farmer exchange program empowers rural people to access markets, technologies, information and services while achieving the sustainable management of natural resources. The FAO promotes the exchange visits to industrial wood processing plants, coffee, cocoa and cardamom producer organizations and tea and crafts cooperatives between Guatemala and Nicaragua. In Asia, a mutually beneficial exchange was implemented in Vietnam and Nepal to enhance farmers' knowledge of integrated forest and farm management, cooperatives and forest-based enterprises timber processing and marketing. In Africa, small farmers from Zambia met with local producers of Kenya to improve their capacities in the charcoal and tree nursery value chains (FAO, 2017).

Focusing on Japan, promoting the diversity of food processing through the international farmer exchange based upon the scope of rural–urban–rural linkages may attract more foreigners who involve themselves in farming and food products to gain Japanese know-how and knowledge in many fields of agriculture such as fruit and vegetable technology, rice and sake processing or even tea production. Such a process will create a revised framework of inbound green tourism based not only on urban–rural interaction between international urban tourists but also on groups of farmers as rural foreign tourists. Through the coordination of urban organizations, such as travel agencies or educational institutions, the new green tourism model can move forward dynamically. This movement is in line with the global trend towards tourists seeking to learn about agriculture and to experience different rural lives through programs such as Worldwide Farmers Exchange (WFE) and World Wide Opportunities on Organic Farms.

However, one challenge facing the development of green tourism in Japan remains depopulation and ageing societies in rural areas. Over the past decade, countryside municipalities have implemented so-called U-turn and I-turn policies as a means of encouraging internal migration from the cities to rural areas (Chandran, 2019; Hani, 2001). The U-turn, or return migration of urban city dwellers to their rural hometown, was introduced into Japanese migration research in the late 1960s (Wiltshire, 1979). Meanwhile, the I-turn refers to city residents relocating to rural communities other than their hometown. Besides U-turn and I-turn, J-turn is another concept describing urban dwellers moving to settle in the countryside near their hometown (Lee & Sugiura, 2018). Policies regarding this migration movement relate to the development of green tourism that help generate jobs to those migrants, a process that is expected to support agrarian pathways for the next generation of Japanese farmers in which rural places are more than just sites of agricultural production (Kobayashi &Tanaka, 2018). Hence, the successful promotion of green tourism in Japan is dependent on not only the effective targeting of new international markets but also on appropriate migration/repopulation policies.

In other words, as Ohe (2008) observes, although the number of farmers managing rural tourism in Japan has grown gradually, many issues remain problematic such as a lack of management skills and the ageing of the rural population. Furthermore, green tourism has not substantially impacted or led to any significant change in rural tourism owing to the decline of rural communities and uncompetitiveness in tourist markets (Chen et al., 2018). Hence, the concept of urban–rural interaction through green tourism takes into account its potential to enhance the capacities of local communities and to encourage competition in the inbound tourism market

Figure 8.3 reconceptualizes the interaction between rural hosts and guests at the international level, highlighting the international exchange between a rural community of the tourist generating region and a rural village of the tourist destination region. The exchange may be organized by urban associations such as NGOs, government organizations, educational institutes or private companies which provide support for farming communities in both regions. One example of this in practice is Thai–Japanese tea farmer exchange in Tawara described in the case study in this chapter. As we have seen, Huey Nam Guen is a rural farming village in the tourism generating region (Thailand) whilst Tawara is a community in the tourism destination regions (Japan); the urban associations are Wakayama University and Mae Fah Luang University which together supported the implementation of the exchange program. Such an international exchange stimulates the viability

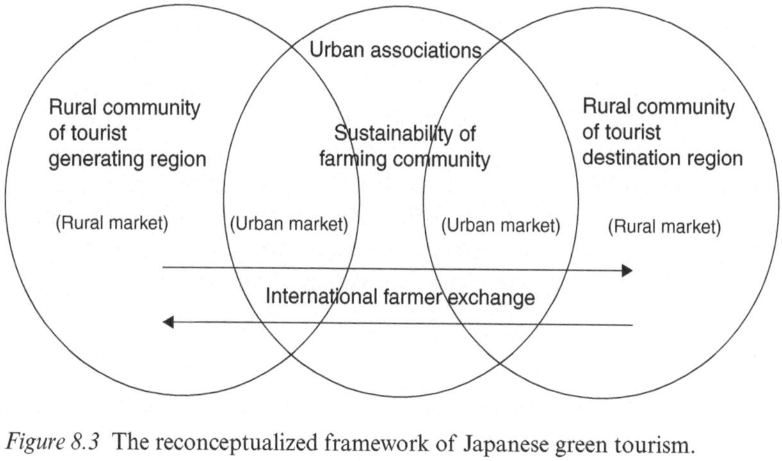

Figure 8.3 The reconceptualized framework of Japanese green tourism.

of the rural and urban markets in both regions through educational travel through supporting a flow of argritourists. Hence, Japan should consider targeting international tourists to suit particular areas. For example, the local government of Shizuoka might target American tourists who are interested in learning about green tea production in Shizuoka, cooperating with the WFE, a non-profit, non-governmental organization in the United States to sponsor programs for young farmers and agricultural students from the USA to Shizuoka. Targeting the American market through tea tourism is possible because of the Japanese green tea boom in the USA (MAFF, 2017) and the increase of American tourists visiting Japan.

The implication of this concept is that Japan can target rural communities in both developing and developed countries. To target agritourists or farmers from developing countries such as Thailand and Asian countries that comprise the primary inbound market, programs that focus on methods of enhancing productivity, such as smart farming, safety food processing and developing agritourism-related businesses should be considered. At the same time, attention can also be focused on targeting visitors from developed countries, particularly to participate in internship or volunteer programs which also promote cultural exchange. Furthermore, this movement should also be advocated by U-turners, I-turners and J-turners who seek jobs in rural tourism and have the ability to cooperate with an international organization and foreign visitors.

Conclusion

Promoting international exchange through the concept of green tourism is a strategy to target inbound tourists visiting the Japanese countryside. Marketing Japan's tea communities to international tourists within a new conceptual framework of green tourism development is therefore recommended. This occurs through the interaction between a rural community in the tourist generating region and a host community in the destination regions, supported by an urban sector organization at international level. In this chapter, the case of Tawara has demonstrated how the knowledge and experience resulting from the interaction between Japanese and Thai tea farmers contributes to supporting both education and tea-related businesses which in turn, may help to activate the regeneration of the tea industry. With this ideal conceptual framework, promoting the rural–urban–rural interaction of green tourism through building the network of education and business between international agritourists and local farmers is also a challenge for sustainable rural development in other declining Japanese farming communities.

References

Arlt, W.G. (2006) Thinking through tourism in Japan. *Tourism and Hospitality Planning & Development*, 3(3), 199–207.

Baier, B. (2015) Nara's Yamato cha green tea: Appreciation of Yamato tea in Nara City. Available at: https://en.japantravel.com/nara/nara-s-yamato-cha-green-tea/23625 (Accessed 16 January 2019).

Bixia, C. and Zhenmian, Q. (2013) Green tourism in Japan: Opportunities for a GIAHS pilot site. *Journal of Resources and Ecology*, 4(3), 285–292.

Boluk, K. (2017) Corporate sustainability as opportunity for tourism partnerships: A case study on Fairmont hotels & resorts. In S. Slocum and C. Kline (Eds.), *Linking Urban and Rural Tourism Strategies in Sustainability*. Wallingford: CABI, pp. 216–232.

Chandran, R. (2019) Doing the 'I turn': Japan taps tourism to lure city dwellers to emptying villages. *The Japan Times News*. Available at: www.japantimes.co.jp/news/2019/07/03/national/turn-japan-taps-tourism-lure-city-dwellers-emptying-villages/#.XYwrw_ZuJPY (Accessed 22 August 2019).

Chang, K. (2015) *World Tea Production and Trade: Current and Future Development.* Rome: UN Food and Agriculture Organization. Available at: www.fao.org/3/a-i4480e.pdf (Accessed on 9 July 2019).

Chen, B., Qiu, Z., Usio, N. and Nakamura, K. (2018) Tourism's impacts on rural livelihood in the sustainability of an aging community in Japan. *Sustainability*, 10, 2896: https://doi.org/10.3390/su10082896

FAO (2017) Reducing rural poverty through farmer-to-farmer exchange. Available at: www.fao.org/3/a-i5862e.pdf (Accessed 19 August 2019).

Font, X. and Tribe, J. (2001) Promoting green tourism: The future of environmental awards. *International Journal of Tourism Research*, 3(1), 9–21.

Funck, C. and Cooper, M. (2015) *Japanese Tourism: Spaces, Places and Structures.* New York/Oxford: Berghahn Books.

George, E., Mair, H. and Reid, D. (2009) *Rural Tourism Development: Localism and Cultural Change.* Bristol: Channel View Publications.

Gon, M. (2017) Discussing rural–urban tourism: A review of the literature. In S. Slocum and C. Kline (Eds), *Linking Urban and Rural Tourism: Strategies in Sustainability*. Wallingford: CABI, pp. 3–19.

Hani, Y. (2001) Easing the way for U and I. *The Japan Times Community.* Available at: www.japantimes.co.jp/community/2001/10/14/general/easing-the-way-for-u-and-i/#.XYwqQPZuJPY (Accessed 22 August 2019).

Hashimoto, A. and Telfer, D.J. (2010) Developing sustainable partnerships in rural tourism: the case of Oita, Japan. *Journal of Policy Research in Tourism, Leisure & Events,* 2(2), 165–183.

Hisegawa, H. (2005) *Rediscovering Japan: Green Tourism*. JFS Newsletter No.34. Available at: www.japanfs.org/en/news/archives/news_id027801.html (Accessed 20 January 2019).

ITO EN (n.d.) Production of Green Tea in Japan. Available at: www.itoen-global.com/allabout_greentea/production.html (Accessed 2 October 2019).

Japan Spotlight (2016) Tourism as a foreign policy tool and engine for growth. *Japan Spotlight*, January/February, pp. 47–51. Available at: www.jef.or.jp/journal/pdf/205th_Interview_02.pdf

Jolliffe, L. (2007) *Tea and Tourism: Tourists, Traditions and Transformations.* Clevedon: Channel View Publication.

Jolliffe, L. and Nakashima, M. (2020) Responsible rural tourism in Japan's tea villages. In G. Musa, V. Nair and A. Hamzah (Eds), *Responsible Rural Tourism in Asia.* Bristol: Channel View Publications, pp. 61–74.

Jones, A. (1987) Green tourism. *Tourism Management*, 8(4), 354–356.

JTB (2019). *Overseas Residents' Visits to Japan by Country and Region.* Tourism Research & Consulting Co. Available at: www.tourism.jp/en/tourism-database/stats/inbound/#annual (Accessed 25 September 2019).

Kazunobu, T. and Tomoyoshi, S. (2010) Gaikoku-jin kankō kyaku o taishō to shi ta gurīntsūrizumu no kanō-sei to kadai – māketingu bunseki no shiten kara [Perspective of green tourism for foreign tourists in rural Japan: From the viewpoint of marketing research]. *E-journal GEO*, 5(1), 35–49.

Keidanren (2000) *Position Paper Concerning Tourism in Japan in the Twenty-First Century.* Tokyo: Federation of Economic Organizations. Available at: www.keidanren.or.jp/english/policy/2000/051.html (Accessed 2 October 2019).

Khaokhrueamuang, A. (2014) The commodification of rurality in Mae Kam Pong village for Japanese long-stay tourism in Chiang Mai. In P. Porananond and V. King (Eds), *Rethinking Asian Tourism: Culture, Encounters and Local Response.* Newcastle-upon-Tyne: Cambridge Scholars Publishing, pp. 166–196.

Khaokhrueamuang, A. and Chuemchaitrakun, P. (2019) Tea cultural commodification in sustainable tourism: Perspectives from Thai and Japanese farmer exchange [Special Issue]. *Rajabhat Chiang Mai Research Journal*, 104–117.

Kobayashi, M. and Tanaka, K. (2018) Agrarian pathways for the next generation of Japanese farmers. *Canadian Journal of Development Studies,* 40(2), 272–290.

Lane, B. (1994) What is rural tourism? *Journal of Sustainable Tourism*, 2(1–2), 7–21.

LaPan, C. (2017) From centre to periphery: Inequality, indigeneity and domestic tourism in Guatemala. In S. Slocum and C. Kline C. (Eds), *Linking Urban and Rural Tourism: Strategies in Sustainability*. Wallingford: CABI, pp. 128–143.

Lee, J.Y. and Sugiura, H. (2018) Key factors in determining internal migration to rural areas and its promoting measures: A case study of Hirosaki City, Aomori Prefecture. *Policy Research Institute, Ministry of Finance, Japan, Public Policy Review*, 14(1),153–176.

Leiper, N. (2004) *Tourism Management*, 3rd Edn. Sydney: Pearson Education.

MAFF (2013) *Green Tourism: Enjoy Japanese Rural Areas.* JTB Tourism Research and Consulting Company.

MAFF (2017) *Current Outlook of Japanese Tea.* Ministry of Agriculture, Forestry and Fisheries. Available at: www.maff.go.jp/e/policies/agri/attach/pdf/index-1.pdf (Accessed 9 July 2017).

Murphy, L., Benckendorff, P., Moscardo, G. and Pearce, P. (2011) *Tourist Shopping Villages: Forms and Functions*. London: Routledge.

Nkemngu, A. (2017) The urban–rural tourism mix: A partnership of convenience or sustainability imperative. In S. Slocum and K. Kline (Eds), *Linking Urban and Rural Tourism: Strategies in Sustainability*. Wallingford: CABI, pp. 115–127.

Ohe, Y. (2008) Characteristics and issues of rural tourism in Japan. In C.A. Brebbia and F.D. Pineda (Eds), *Sustainable Tourism III*. Southampton: WIT Press, pp. 305–316.

Ohe, Y. (2014) Characterizing rural tourism in Japan: Features and challenges. In P. Diaz and M. Schmitz (Eds), *Cultural Tourism.* Southampton, WIT Press, pp. 63–75.

Ohe, Y. (2016) Community-based rural tourism in super-aging Japan: Challenges and evolution. *Anais Brasileiros de Estudos Turísticos: ABET*, 6(1), 28–41.

Robinson, P. (2012) Rural tourism. In P. Robinson (Ed.), *Tourism: The Key Concepts*. London: Routledge, pp. 173–175.

Sharpley, R. and Sharpley, J. (1997) *Rural Tourism: An Introduction*. London: International Thomson Business Press.

Slocum, S. and Curtis, K. (2017) The urban–rural tourism relationship: A case of suburban farm shops. In S. Slocum and C. Kline (Eds), *Linking Urban and Rural Tourism: Strategies in Sustainability*. Wallingford: CABI, pp. 82–93.

Sonnino, R. (2004) For a 'piece of bread'? Interpreting sustainable development through agritourism in Southern Tuscany. *Sociologia Ruralis*, 44(3), 285–300.

Soshiroda, A. (2005) Inbound tourism policy in Japan from 1859 to 2003. *Annals of Tourism Research,* 32(4), 1100–1120.

UNWTO (2017) *International Rural Tourism Development: An Asia-Pacific Perspective*. Madrid: UN World Tourism Organization.

UNWTO (2018). *UNWTO Tourism Highlights 2018*. Madrid: UN World Tourism Organization.

Van Leeuwen, E. (2015) Urban–rural synergies: An explorative study at the NUTS3 level. *Applied Spatial Analysis and Policy*, 8(3), 273–289.

Wijaya, N. (2013) Contemporary problems in Japan's rural areas and opportunities for developing rural tourism: A case of Yamashiro district in Yamaguchi prefecture. *Journal of East Asian Studies,* 11, 59–72.

Wiltshire, R. (1979) Research on reverse migration in Japan: (I) reverse migration and the concept of 'U-Turn'. *The Science Reports of the Tohoku University, 7th Series, Geography*, 29(1), 63–68.

Wu, Y., Gao, M., Sato, K., Lu, T., Zhang, F., Xing, S., Brandt, K., Uemura, M. and Furuya, K. (2019) Study on green tourism experiences: Cases of popular strawberry farms in Chiba Prefecture. *Japan Geoscience Union Meeting 2019*. Available at: https://confit.atlas.jp/guide/event/jpgu2019/subject/HSC06-P03/detail (Accessed 20 August 2019).

Zavadckyte, S. (2017) *Japanese Tea: A Comprehensive Guide*. Kyoto: Kyoto Obubu Publishing House.

9 Pilgrimage tourism in regional communities

The case of Tanabe City and Kumano Kodo

Ricardo Nicolas Progano and Kumi Kato

Introduction

Travel motivated by spirituality-related reasons, specifically pilgrimage, has long been an important part of tourism (Digance, 2006; Kaelber, 2006). In contemporary society, however, pilgrimage carries a much wider connotation beyond traditional religion, including travel for reasons of health, wellness and self-improvement, reflecting the fact that 'the common ground between tourism and religion has moved beyond what has been termed "religious tourism"' (Suntikul & Butler, 2018: 1). Moreover, visitors, especially those who travel by foot, have specific interests closely related to sustainability and this mode of travel, which may be defined as slow tourism (Dickinson & Lumsdon, 2010; McGrath & Sharpley, 2018), helps shape the kinds of tourism services provided by surrounding communities and direct destination planning and development. This is observed in the case of the World Heritage pilgrimage trail, Kumano Kodo, located in Wakayama, Japan. Referring to the evolving meaning of spirituality and tourism both globally and in Japan, this chapter explores the significance of today's spiritualities in destination management from the perspective of local communities.

Spirituality in contemporary society

Secularization and the 'privatization of religion'

Secularization theories have often anticipated the devaluing or marginalization of the influence of religions on society through the advance of modern scientific knowledge, urbanization, industrialization and rationalization (Hanegraaff, 2000; Heelas, 2006; Houtman & Manscini, 2002; Stark, 1999). While secularization has increased (Houtman & Manscini, 2002; Reader, 2012), particularly noted in the decline of institutional religions' influence (Casanova 2006), spirituality is still present in contemporary society. However, it has undergone changes in how it is understood and practiced (Hanegraaff, 2000; Houtman & Manscini, 2002; Casanova, 2006; Heelas, 2006; Okamoto,

2015). Modern spirituality is characterized by indifference to institutional religious authorities on matters that were traditionally under their guidance, while still holding private beliefs (Ito, 2003; Casanova, 2006; Okamoto, 2015). Because religion's influence has been reduced in public spheres, the interpretation of the world and values has diversified. At the same time, religion has been confined to the private sphere of human life. This situation has been referred to as 'privatization of religion' (Okamoto, 2015), which has led to the 'customization' of religion, where individuals are free to accept or reject certain aspects of their own religion, as well as combine elements from different traditions, even detached from temporal or spatial contexts, in their own private faith (Casanova, 2006). At the same time, modern societies' egalitarianism, awareness of different religions and government neutrality towards religion have played a significant role in the acceptance of faith diversity, and its later individualization, as societies become reluctant to enforce religious conformity (Bruce, 2006; Heelas, 2006).

Contemporary spirituality in Japan

In Japan, the New Age movement is known as Spirit World (*seishin sekai*) and began to appear in the early 1980s. In contrast to its development in the West, its arrival did not involve criticism of contemporary social institutions, such as capitalism and Christianity, but was more consumption-oriented; however, it still lacked unified organizations or teachings. The media played a central role in the expansion of the Spirit World movement, a phenomenon that has also been observed by researchers in other contexts (Pravdova & Radosinska, 2013). For example, even the word 'spiritual' is used instead of the Japanese word *rei* because of its use by spiritual counsellors in television programs, in particular by Hiroyuki Ehara, a self-proclaimed spiritual counsellor (Ito, 2003; Horie, 2009; Kotera, 2011). The relationship between spirituality and the media in Japan can be traced back to the 1910s and its discussion in newspapers and on radio, but was deepened by the introduction of spirituality-related content in television during the 1970s. The 1995 Tokyo subway sarin attack, perpetrated by the cult Aum Shinrikyo, resulted in a reduction of this television content until the 2000s, when spirituality-related themes re-emerged, though presented in a lighter, non-affirmative entertainment format related to life improvement. This was done in order to distance the concept of spirituality from any negative associations related to dangerous cults or the advent of a New Age (Horie, 2009; Kotera, 2011). In this context, the word 'spirituality' gained popularity over the Japanese *rei* as it avoided such negative connotations. Also, because the word 'spirituality' is of foreign origin, it was perceived as more advanced and fashionable (Horie, 2009). These contemporary spirituality-related elements coexist with a secularization process in Japanese society which has continued since Second World War (Reader, 2012). However, secularized East Asian societies, including Japan, still demonstrate high levels of spirituality while not publicly acknowledging any affiliation to specific religions (Reed, 2007). In sum, it can be observed that Japan's spiritual landscape is defined by a complex map that mixes contemporary spirituality with secularization and limited religious affiliation.

Contemporary spirituality and pilgrimage tourism

Tourism has developed at a variety of sacred sites (Shackley, 2001; Stausberg, 2011), with pilgrimage sites in particular not only becoming important tourism destinations but also attracting increasing attention in the tourism research literature since the 1970s (Collins-Kreiner, 2016). Still, pilgrimage as a research concept in tourism studies barely existed before the 1990s (Collins-Kreiner, 2016). Vukonic (1996) argues that the inter-relationship between tourism and religion represents an undeniable fact that deserves scholarly interest and analysis, yet observes that there had been few studies into this inter-relationship prior to the publication of his now widely cited work. Nowadays, the number of academic studies on religion and tourism has increased (Butler & Suntikul, 2018; Raj & Griffin, 2015; Timothy & Olsen, 2006), whilst it also enjoys a dedicated academic journal, the *International Journal of Religious Tourism and Pilgrimage*.

However, research into tourism and religion has not been conducted evenly with regards to geographic locations. In particular, there is still a comparative lack of research undertaken at Asian pilgrimage sites, despite their importance in tourism. Indeed, the Asia-Pacific area has the greatest number of pilgrims and travellers for religious events, for both international and domestic tourism (UNWTO, 2011). Also, according to the UNWTO/GTERC (2017), this region has more generally become the second most visited after Europe, accounting for 308 million arrivals, as well as being the fastest growing, registering 9 percent growth in 2016. This demonstrates the great importance of pilgrimage-related tourism in Asia-Pacific, as well as the immense opportunities for academic research on the subject. However, research on pilgrimage tourism is considered scarce in the Asia-Pacific area (UNWTO, 2011).

Yet, it has been pointed out that tourism can be a medium for religiousness in contemporary society (Vukonic, 1996; Norman, 2004; Stausberg, 2011), and that spirituality has become a major tourism market (Attix, 2015; Timothy & Conover, 2006). The importance of these developments is evident in that the UNWTO has organized both the First UNWTO International Conference on Spiritual Tourism for Sustainable Development (21–22 September 2013, Ninh Binh Province, Vietnam) and the First UNWTO International Congress on Tourism and Pilgrimages (17–20 September 2014, Santiago de Compostela, Spain). There has been a diversification in the purpose for visiting sacred sites, with secular motives and contemporary spirituality replacing or coexisting with traditional faith. Norman (2009) and Okamoto (2015) note that while the number of people undertaking the Camino de Santiago pilgrimage route has been steadily increasing, many of them were motivated not by devotion but, rather, by New Age books, self-discovery or outdoor exercise such as walking. Norman (2004) analyzed different sources of spiritual tourism and identified a wide variety of motivations amongst spiritual tourists, finding that many, even if they visit places of worship, show little interest in religious authority. Tourists also travel to experience alternative practices, as individual and unique experiences are central to contemporary spirituality seen in examples of yoga tourism (Lehto, Brown, Chen & Morrison, 2006).

The pilgrimage tradition in Japan

Characteristics of Japanese pilgrimage

Stoddard has stated that 'converting the concept of pilgrimage into an operational definition that will apply equally well to all cultural settings is difficult. This is partly because of differences in linguistic usage' (Stoddard, 1997: 49). Consequently, for the purposes of this chapter it is necessary to consider the definition of pilgrimage in Japanese culture. Reader and Swanson (1997) note that, in the Japanese language, there are multiple words that can be translated as 'pilgrimage'. Some of these terms overlap and can be used interchangeably, but others are mutually exclusive. The following is a summary of the most utilized terms:

- *Junrei* (巡礼): refers to visiting multiple religious sites that are linked together in a circuit. During the Heian period, it implied ascetic practices and visiting miraculous sites, but the term is no longer limited to this use nowadays. It is widely used when referring to specific pilgrimage routes.
- *Henro* (遍路): while it refers to a circuitous route, this word is used when referring specifically to the Shikoku Henro pilgrimage in Shikoku island. It also refers to the pilgrims themselves.
- *Meguri* (巡り/稼り): literally meaning to 'go around', this refers to a pilgrimage to a number of shrines and temples. It is most commonly used when the sites of a pilgrimage route are not dedicated to the same divinity, but when a group of them that are related in some way.
- *Moude* (詣): denotes a visit to worship at a shrine of temple. It may be added as a suffix to the names of sacred sites or areas to indicate a special journey of worship.
- *Mairi* (参り): signifies the process and practice of making special visits to a sacred place that is not part of one's everyday life. It is also widely used to refer to the pilgrimage to Ise.
- *Sankei* (参詣): indicates the process of visiting sacred sites and making pilgrimages. It is not used to refer to a specific pilgrimage.

When considering these definitions, it is evident that Japanese pilgrimages may be based not only on a single site as is common in the West, but also on multiple sites that are linked together (Kadota, 2013). This form of pilgrimage is not exclusive to Japan, but it represents one the main characteristics of Japanese pilgrimages, namely, a tendency to link multiple sacred sites in a numerical structured route. These multiple-site routes do not have a single 'centre' but, instead, encompass a defined geographical area within which the sites that should be visited are located (Reader & Swanson, 1997). Another characteristic is that there is a greater scope in the performance of their itineraries in the sense that there is not a single route that must be traversed. Rather, pilgrims have the choice of visiting the sites in an order that suits them best. Also, pilgrims have the alternative option of completing the pilgrimage in a number of journeys over a period of time, since the focus is on visiting all the sites rather than the journey. This pattern can be seen, for example, in the Shikoku Henro.

Historical outline of Japanese pilgrimage

Spiritual travel in the form of pilgrimage to different sacred sites in the country has a long tradition in Japan, its origin lying in the ritual travels of mountain ascetics along the Kii peninsula and Shikoku Island, as well as the official visits to the Ise Grand Shrine. During the Heian period (794–1185), Emperors and aristocrats would travel from the capital city of Kyoto to the mountain ranges of Kumano, located in the Kii peninsula, which is the largest of Honshu, the main island of Japan. It is characterized by mountainous terrain covered by dense forests. Their travel included ritual purifications and other religious practices, but also leisure activities. In the Kamakura period (1185–1333), the new military class joined the aristocracy in the practice of undertaking pilgrimages whilst the subsequent Muromachi period (1336–1573) witnessed the emergence of mass pilgrimages, with large numbers of peasants also undertaking the Kumano pilgrimage. During this period, the pilgrimage to the Ise Grand Shrine (located in present day Mie prefecture) also began to gain popularity among larger sectors of society (Hoshino, 1981). However, it was during the Edo period (1603–1868) that, perhaps, the complementary relationship between spirituality and tourism became more clearly evident. During this era, the Tokugawa government imposed travel restrictions that prevented the population from freely traveling but, nevertheless, the authorities would allow travel under special conditions such as pilgrimages (Funck & Cooper, 2013; Guichard-Anguis, 2009). During this period, conditions for travel improved as well. Later, advances in agriculture improved the life of the common folk (Hoshino, 1981), resulting in the popularization of travel among the population and mass tourism as a combination of pilgrimage and sightseeing. The most visited pilgrimage destination became the Ise Grand Shrine, which replaced Kumano in popularity (Funck & Cooper, 2013; Reader, 2014) yet, while Ise peregrinations became more a form of leisure travel, other pilgrimages, such as the Saikoku pilgrimage (the 33 Kannon temples in the West) and the Shikoku Henro (the 88 temples on

Shikoku Island), were undertaken by travelers primarily motivated by religion (Funck & Cooper, 2013).

In the period immediately following the end of the Second World War, the difficult socioeconomic situation led to a decline in the overall number of pilgrims in Japan (Hoshino, 1981; Reader, 2014). Moreover, during the post-war period, the Saikoku pilgrimage gradually fell in popularity and whilst the Shikoku Henro pilgrimage started to attract a growing number of visitors, particularly from the 1960s onwards (Reader, 2014). The increasing popularity of the Shikoku Henro reflected both transformations in public perceptions of it and also developments in transport and infrastructure since the 1950s (Reader, 2014). In particular, different stakeholders, including the mass media, transport companies and tourism offices, have played a key role in transforming perceptions of Shikoku Henro as a distant, dangerous place into an enjoyable pilgrimage tourism destination where visitors can take on journeys of self-discovery in a setting where 'traditional' Japanese culture lives on (Mori, 2005; Reader, 2014). Because of its new image as a repository of traditional Japanese culture which is believed to have been lost in Westernized urban areas, Shikoku Henro effectively replaced the Saikoku pilgrimage and became the 'national' pilgrimage (Reader, 2014). In this sense, the post-war 'pilgrimage boom' has been originated, in part, as a search for identity by the Japanese people (Hoshino, 1981).

Pilgrimage in contemporary Japanese tourism

Pilgrimage sites as tourism destinations

Pilgrimage is an important element in contemporary tourism in Japan (Mori, 2005; Okamoto, 2015). Local authorities have realized the importance of their pilgrimage routes and sites, and actively promote them to help revitalize their regional economies. This has also prompted the revival of pilgrimage routes such as Kumano Kodo. Other authorities, noting the tourism potential of pilgrimage, have created routes that link different sites, such as in Nagasaki's Churches and Christian Sites. In addition, as Reader (2014) observes, religious organizations have also realized the potential of pilgrimage for attracting visitors and have created modern pilgrimage routes by linking different locations. For example, in 1987, monks from the Fukuoka prefecture developed a new Kannon pilgrimage in northern Kyushu Island consisting of 33 sites, similar to the Saikoku pilgrimage. To help promote this new pilgrimage, the monks collaborated with transport companies and tourist offices. Similarly, monks from Suo Oshima Island, located in Yamaguchi prefecture, have also developed an 88 stage pilgrimage based on the figure of Kodo Daishi. The original pilgrimage, which had been abandoned, was restored in 1988 as a means of increasing visitor numbers and revitalizing the local economy through tourism development (Reader, 2014).

Tourism, however, has not always been developed smoothly and both environmental and social conflicts have arisen (McGuire, 2013). Tourism development has also come into conflict with religious traditions and practices associated with pilgrimage. For example, Jimura (2016b) and Amada (2012) explain how tourism development facilitated access and signalling in dangerous mountain passes that were previously only accessible by being guided by Shugendo practitioners, who also provided interpretations of the place based on their traditional beliefs. However, easier access brought by tourism development conflicted with the Shugendo practitioners' role: not only could places be easily reached without the assistance of guides, but they were interpreted only as landmarks with no emphasis their religious significance. Similar observations are made by McGuire (2013), who observed that the increased accessibility brought by tourism infrastructure transformed the mountains' physical environment and reduced the danger of ascending them, an important aspect in Shugendo practices.

The importance of the UNESCO World Heritage designations

UNESCO World Heritage Sites (WHS) are also closely related with tourism development in Japanese pilgrimage sites. Since its belated signing of the World Heritage Convention in 1992, Japan was rapidly awarded a number of World Heritage designations; by 1993, the country had four WHS, a number that increased to 20 by the end of the 20th Century. Subsequently, since 2000, Japan has experienced a nationwide 'World Heritage boom', as local communities have realized the potential of tourism development for economic revitalization through a UNESCO designation (McGuire, 2013; Matsui, 2013; Sawamura, 2016). Not only have local authorities and organizations been working towards obtaining a designation, but also the national government, through the Agency for Cultural Affairs, has supported this movement. Indeed, during the period between 2006 and 2007, the Agency actively encouraged prefectural authorities to make applications to the UNESCO Tentative List (Sawamura, 2016). Yet, it has been noted that obtaining a WHS registration does not necessarily mean an increase in tourism; specifically, already well-established tourism destinations did not gain a significant increase in tourist numbers post-designation. However, for those destinations that were not well-established, WHS designation did have a positive effect in increasing visitors (Jimura, 2016a).

Inevitably, pilgrimage sites and routes are no exception to the 'World Heritage boom'. The first Japanese pilgrimage route to be designated as a UNESCO WHS was the 'Sacred Sites and Pilgrimage Routes in the Kii Mountain Range' in 2004, encompassing a large system of pilgrimage routes and temples located in the Kii Peninsula. This designation had an influence on other pilgrimage routes such as Shikoku Henro, whose related stakeholders have started to work towards achieving UNESCO designation (Mori, 2005; Okamoto, 2015). Christian churches located in the Nagasaki prefecture were also inscribed in the tentative World Heritage list in 2004, under the support of different stakeholders but principally of the Sekai Isan no Kai, a group that included media, government officials, local companies and church members (Matsui, 2013). They were finally designated as a WHS of cultural significance in 2018. Although the sites are not connected through pilgrimage routes, different routes have been created to organically connect the churches and to commodify them for tourism consumption. For example, in 2005, a guidebook was published presenting the churches as sacred sites of a pilgrimage referred to as the Nagasaki pilgrimage. Local cultural and natural sites were presented as well (Matsui, 2013; Okamoto, 2015).

UNESCO WHS designation and subsequent tourism development has also come into conflict with local traditions. McGuire (2013), for example, points out that UNESCO's assumptions about shared, universal human values conflicts with Mount Omine's traditional prohibition of women. Specifically, women have traditionally been banned from climbing Mount Omine, one element of the 'Sacred Sites and Pilgrimage Routes in the Kii Mountain Range'. However, UNESCO's designation has challenged this restriction, which in turn has prompted a group of Shugendo practitioners to question whether UNESCO has the right to overturn a 1300 years old tradition. They have also voiced concerns that, if the ban is lifted and women and, subsequently, day hikers are allowed, the sacredness of Mount Omine will be lost. In the process, it will become a profane mountain similar to Mount Fuji which, since the post-war period, has been climbed primarily as a leisure activity (Okamoto, 2015). It is important to point out that prohibiting women from accessing sacred mountains is not a tradition exclusive to Mount Omine; the practice has also existed in other in other parts of Japan (Nomoto, 2006). In addition, Jimura (2016b) has noted that, owing to the designation of the 'Sacred Sites and Pilgrimage Routes in the Kii Mountain Range' site, the religious practices of Shugendo have been hindered as a result of the installing of signposts. In Shugendo, senior practitioners, called *sendatsu*, traditionally act as religious guides and pathfinders along the mountainous trails. However, new signposts have undermined the role of the *sendatsu* because they show the names and the correct paths to follow.

Contemporary spirituality and its influence on tourism

While travelling for religious purposes is a deep-rooted institution in Japan, modern spirituality has reinterpreted traditional sites and given new meaning to them. Two of the most influential phenomena are *iyashi* and powerspot. The concept of *iyashi* has become uniquitous in Japan since 1988 (Matsui, 2013), being associated with different products and services, including travelling. *Iyashi* can be understood as physical and mental restoration, or a form of holistic healing. This '*iyashi* boom' has also influenced tourism promotion and activities at pilgrimage sites. For example, the four prefectures of Shikoku Island partnered with the railway operator JR Shikoku to promote

Shikoku Henro as a place for healing (Matsui, 2013). Since the late 1990s, Shikoku Henro has increasingly been presented in the mass media, including both magazines and television shows, as a place of *iyashi* (Mori, 2005), with this process accelerating since the early 2000s (Kadota, 2013).

Similarly, the concept of 'powerspot' first emerged in 1986, although the popularity of the phenomenon started to increase exponentially since 2006 (Suga, 2010) with media coverage of different powerspots on television and in magazines, as well as the trend of celebrities visiting them, being important factors in its expansion (Kotera, 2011; Okamoto & Kawasaki, 2012). The term generally refers to specific places where it is believed by some that 'energy' or 'power' gathers which can confer a wide array of worldly benefits, such as good fortune, healing or luck in romance. Powerspots are akin to power places, a concept that emerged in the West during the 1960s, which include sites such as Stonehenge, the Pyramids and Mount Everest (Stausberg, 2011). This 'powerspot boom' is particularly popular amongst women, the segment most commonly targeted by media outlets (Okamoto, 2015). Publications tend to present the powerspots with beautiful photographs, along with data on access and opening hours; in addition, information on accommodation, local specialties and food may also be provided. As such, they are similar to tourist guides (Okamoto, 2015; Tsukada & Omi, 2011). Unsurprisingly, perhaps, local administrations and tourism organizations also recognized the potential of the 'powerspot boom' and began to present their own destinations as powerspots (Okamoto, 2015). In some cases, however, conflicts may arise because visitors' motivations and behaviour may not be in line with what the site administrators consider suitable (Okamoto, 2015; Okamoto & Kawasaki, 2012). Japanese powerspots are broadly divided into two types: natural sites such as Yakushima Island, and Shinto-related sites such as the Kumano Hongu Taisha, the Ise Grand Shrine or Izumo Shrine. In contrast, Buddhist temples are scarce (Tsukada & Omi, 2011).

Case study: Kumano Kodo and the community of Tanabe city

General outline of Tanabe city

As mentioned earlier, the tourism promotion of pilgrimage sites in Japan is connected to regional revitalization. Since the early 2000s, the Japanese government has prioritized tourism development, focusing in particular on the inbound tourism market, as a means of revitalizing regional economies suffering of low economic activity and ageing populations. More specifically, tourism has also been developed to assist in the recovery of areas affected by the 2011 Tohoku earthquake and tsunami disaster (Prime Minister of Japan and Cabinet, 2012). Different initiatives have been undertaken in recent years to promote tourism development, such as the Visit Japan Campaign in 2003, the Tourism Nation Promotion Basic Law in 2007, the Tourism Nation

Promotion Basic Plan in 2007 (renewed in 2012) and the establishment of the Japan Tourism Agency in 2008.

Tanabe city, located in Wakayama prefecture, is an interesting example of regional revitalization through tourism utilizing pilgrimage-related resources. The city is located on the west-central part of the Kii peninsula of Honshu Island, the main land mass of Japan. The current city was formed in 2005 through the amalgamation of four nearby towns: Honguu, Ryuujin, Oto and Nakahechi. This merger transformed Tanabe into the largest city in the Kansai region in terms of surface area, covering a total of 1,026.91 km^2. Geographically, it comprises a coastline situated on its western border, while 88.9 percent of the city's area is covered with mountain forests. As of 2016, Tanabe's population was 74,770, a decline from 79,119 in 2010. Along with this demographic change, Tanabe's economy has undergone significant transformation in recent years. In particular, traditional industries have experienced a decline in their workforce, with the number employed in farming falling from 3,369 in 2010 to 2,918 in 2015 and from 1,828 in 2010 to 1,508 in forestry (Tanabe City, 2016). In this situation, tourism is expected to play a central role in community revitalization (Tanabe Commerce and Industry Conference, 2009).

Following the amalgamation of the four towns, falling within Tanabe city's borders are a number of important tourist resources, including 60 km of pilgrimage routes comprising the Nakahechi and the Kohechi routes, as well as the Kumano Hongu Shrine and Oyunohara, the former location of Kumano Hongu Shrine. The Nakahechi route, which links the three main shrines in a circuit, is the most popular route for tourists and has its origin in the Heian period, when aristocrats would travel from the imperial capital in Kyoto. Along this route can be found numerous Oji, small shrines dedicated to lesser protective deities where pilgrims used to perform offerings. There are also a number of important Onsen (hot springs) that have been popular amongst visitors for several centuries, the most famous being Yu no Mine, located in the former Hongu town of Tanabe city. Hiking, regional cuisine, handcrafts and local festivals are other tourism assets that can be found.

Establishment of the Kumano Tourism Bureau and development of inbound tourism

The year 2004 was key for Tanabe's tourism development, when different sites of Kumano Kodo were designated as part of the UNESCO World Heritage Site 'Sacred Sites and Pilgrimage Routes in the Kii Mountain Range' under the category of cultural landscape. The designation provided local communities with a new way to strengthen the tourism industry in order to revitalize the regional economy and attract both national and international tourists. It also changed the community's perception of the pilgrimage routes. As noted above, the area benefits from other tourism attractions, with its hot springs being one of the main ones along with local cuisine, health tourism and cultural resources (Tanabe City, 2016; Tanabe Commerce and Industry Conference, 2009). To coordinate tourism policies effectively in the newly formed city, the Kumano Tourism Bureau was established in April 2006 as a local-level destination management organization (DMO). While the local tourism associations of each of the constituent four towns still function, the Bureau plays an important management role by collaborating with them in order to further develop the local tourism sector. The Bureau's organization is established as a joint public–private tourism promotion agency, and includes the four old towns' administrations as well as the present Tanabe city administration. Its working staff not only consists of locals who have worked beforehand in regional tourism but also an international staff member who has played a significant role in the development of inbound tourism.

The Bureau's main functions are divided in two general fields: tourism promotion and grassroots tourism development. Tourism promotion consists of activities such as the development of a multilingual website and information pamphlets, overseas promotions (including joint tourism promotions with Galicia, Spain), support for press and travel agents, and giving lectures and presentations. Grassroot tourism development is concerned with activities related to management and stakeholder coordination, such as stakeholder networking, grassroot hospitality seminars for stakeholders involved in tourism, assistance to local guides and the creation of bilingual material aimed at bus schedules and area maps. Improvements in tourism infrastructure, such as information centres and signs, are also included in grassroot tourism development.

As part of its tourism policies, Tanabe city has also decided to increase its number of inbound Western tourists by promoting its WHS. Such a policy might be considered surprising given that Japan's main source of international visitors are Asian countries. However, the Bureau had noticed that the nearby Buddhist complex of Koyasan, also registered in the Sacred Sites and Pilgrimage Routes in the Kii Mountain Range site, was receiving an influx of French tourists who were interested in Japanese history and culture. Therefore, the Bureau made the decision to focus on Western tourists and small groups, anticipating that, compared with Asian tourists, they would stay for longer and spend more money because they would walk the Kumano pilgrimage routes and enjoy the traditional Japanese scenery and spiritual culture. In particular, independent Western tourists were the main market that the Bureau aimed to reach (Japan Tourism Agency, n.d.; Nagatani, 2017).

In order to provide a better service, the Bureau promoted the dissemination of bilingual maps, bus timetables, audio guides and signs. These efforts were paired with educational seminars for local stakeholders in order to improve hospitality and services aimed at international visitors. Finally, an important initiative was the establishment in 2010 of a travel agency, named Kumano Travel, which includes its own online reservation system where visitors can book lodging facilities, tours and other services. The reservation system is aimed particularly at small groups and individual tourists, and is mainly utilized by foreigners: in 2016, out of a total of 11,442 tourists who used it, 7,744 were international visitors (approximately 67 percent), an increase of 49 percent in comparison to 2015. The adoption of this technological tool has simplified and speeded up the planning and booking process for visitors, in particular international visitors. At the same time, it has benefited the local establishments by processing reservations with no intermediaries and thus ensuring that earnings stay within the region, as well as creating employment. It has also become an important source of funding for the Bureau, amounting to 80 percent of its 2016 annual budget (which amounted to 60,000,000 yen, or roughly US$540,000), while the rest was funded by the Tanabe city local government.

As a result of these efforts, the number of visitors to Tanabe city has been steadily growing. The Heisei 28 Tourist Motivation Research Report indicates that in 2016 Tanabe attracted 3,650,588 visitors. However, only 407,427 (11.1 percent) of them stayed overnight in Tanabe, hence increasing the number of lodgers being one of the challenges for further tourism development. Regarding international tourists, 30,958 were foreigners and the rest, domestic. However, because this report only counts overnight visitors and the existence of 65,426 unidentified visitors, the number of international tourists may be bigger than reported (Wakayama Tourism Agency, 2017). Nevertheless, as the statistics reveal, Japanese domestic tourists account for the largest proportion of visitors; however, their number has been gradually decreasing, while the number of international tourists is on an upward trend.

Sustainable tourism development and regional community

While visitor numbers have experienced robust growth since World Heritage designation in 2004, tourism has impacted differently on the communities located in the Nakahechi section of Tanabe city whilst not all members of particular communities have experienced the same impacts. Past research has found similar results in other pilgrimage sites (Fernandes, Pimenta, Goncalves & Rachao, 2012; Knight, 1996; Lois-Gonzalez & Santos, 2015; Progano, 2018; Rizzello & Trono, 2013), although the subject remains understudied in the field of pilgrimage tourism. As spirituality-related tourism continues to grow in contemporary society, studies regarding its impacts on host communities are of great importance for both tourism researchers and policymakers.

Research (Progano, 2018) conducted at the Chikatsuyu community, located in the Nakahechi town and a nodal point of the Nakahechi trail, sheds some light on the subject. Results partially confirm what previous studies have revealed about economic impacts being low along trails because of the characteristics of pilgrimage tourism itself (Fernandes et al., 2012; Lois-Gonzalez & Santos 2015). Positive economic results were perceived to be mostly restricted to occupations related directly to tourism, such as tourism agents and luggage transporters. Occupations indirectly involved with tourism, such as cafeterias and grocery shops, also reported economic benefits, but to a lower degree. The rest of the community reported neither positive nor negative economic impacts. However, the community was satisfied overall with the scale of tourism development carried out in Chikatsuyu. They perceived that, considering the role of Chikatsuyu as a middle point in the Nakahechi route, the scale of tourism development was acceptable. The Bureau has taken a similar stance in their tourism planning, prioritizing sustainable tourism development based on small groups and individual visitors (Tanabe City Kumano Tourism Bureau, n.d.).

While economic impacts were seen as restricted, social impacts were positively perceived across the whole community. Local events were regarded to have had a positive impact by almost all participants, who mentioned the creation of new festivals or an increase of participants. The community's awareness and conservation of their own traditions, such as the lion dance and other activities, was regarded as positive in relation to tourism, although not all participants perceived it. The presence of international tourists was noted to be particularly positive, as the community commented on friendly experiences with them, despite cultural and language differences. The Bureau's efforts in grassroots tourism development aimed at hosting international visitors could be another factor affecting the overall positive image of international visitors as they helped the host community to be prepared for their arrival and efficiently interact with them (Tanabe Kumano Tourism Bureau, n.d.).

Regional revitalization through positive economic impacts of tourism is, however, not the only parameter to measure the benefits from tourism development. Local communities surrounding the pilgrimage routes are also looking for a sustainable lifestyle. Chikatsuyu, for example, is now gaining an influx of younger settlers who have moved to the area seeking a more sustainable, safe and healthy environment for their families. Younger cohorts relocating from urban to rural environments may be called as I-turns (those moving to a new place), U-turns (those going back to their hometown) or J-turns (those relocating to new areas from their hometown). They make up a post-development phenomenon of people that value intangible qualities of life (i.e., family time, healthy environment, safe food, community unity) more than materialistic values, even if their income may be reduced or life may become less convenient. This phenomenon increased particularly after the East Japan Earthquake and Tsunami on 11 March 2011 that appeared to challenge the current material-dependent and energy-intensive lifestyle. The lifestyle of those choosing to live in rural areas tends to be sustainability-oriented, interested with a focus on locally produced organic food, natural clothing materials, chemical-free products, hand-made products, alternative energy and spiritual wellbeing. Businesses arising from these interests include shops, restaurants, accommodation or guiding, instructing and teaching (local knowledge, food, exercise, spiritual guidance) and collectively create tourism opportunities that promote lower-impact lifestyle and sustainability.

Tourism can also be channelled to provide a means for caring for the natural and cultural heritage of the Nakahechi trail, with the Conservation Walks being an example of it. Many parts of the pilgrimage trails are subject to erosion owing to steep gradients and high precipitation (the average rainfall is around 2730 mm annually). Consequently, these become inaccessible to vehicles and supporting soil supplements need to be carried on foot by walkers. This has turned out to be an attractive volunteer tourism activity for walkers who are willing to contribute to the conservation of the trails. The Conservation Walk program (*Michi-bushin*) incorporates traditional ideas of community *fushin*, which traditionally refers to community members who contribute labour or funds to build or repair houses and supporting infrastructure. Nowadays, the term *fushin* is used for volunteer work in building or restoring public facilities (this includes road, canals, disaster recovery). Today, the Conservation Walks attract over 2,000 participants annually, including corporate groups using opportunities for CSR and staff training, school and community groups. Individual participation may comprise a package tour of transport to and from the restoration site and overnight lodging provided.

Taking into consideration these observations, it could be argued that a slower tourism growth could be an alternative for both the host community and the visitors, in particular because the host community did not mention any wish for an accelerated tourism development and emphasized their appreciation for close interactions with tourists, which are more likely to happen in small-scale tourism, as well as a slower, sustainable lifestyle. This paradigm shift towards a balanced tourism growth that prioritizes quality of life over fast economic development has also been observed in previous studies (Rivera, Croes & Lee, 2016). This approach can be an alternative development strategy (Koens & Thomas, 2015) in contrast to recent cases of overtourism (Hagi, 2017; Seraphin, Sheeran & Pilato, 2018). As the residents seem to prefer an overall small development that favours close interactions with visitors, a balance in development may be of utmost importance for sustainable tourism policies. In this regard, a slow tourism approach (Caffyn, 2012; Kato & Progano, 2017; Lumsdon & McGrath, 2011) could be a valid alternative as slow travel already exists in Kumano Kodo.

Matos (2004) discusses the concept of slow tourism as a potential market development in which tourism becomes the process of enjoying the destination without causing a heavy environmental impact. Two core principles are proposed: taking time, especially in relation to perception of nature as well as understanding local culture; and encouraging attachment to a place with regard to the special characteristics of the locality, its architecture for example, as a contribution to the sense of place. Gardner (2009) proposes several principles for slow tourism, including that tourism is a state of mind; locality is important; the journey is intrinsic to the tourism experience; culture through language and engagement with local people; and tourists should make opportunities by seeking out the unexpected. Further criteria suggested by Caffyn (2009), applying Honore's definition of slow (Honore, 2004) that implies forming meaningful connections with people, places, culture, food, heritage and the environment, applies qualities related to personal wellness and personal growth of travelers: relaxing and refreshing the mind and body; creative and unstructured play for children and learning and skills activity–personal development. In sum, slow travel involves qualities of valuing time, appreciation of process, appreciating locality (nature, landscape, people, culture, food) and attempting to reduce environmental impact.

With the above discussions in mind, pilgrimage as a contemporary spiritual practice is central, as is walking as a mode of tourism, with slowness allowing more conscious engagement with locality (environment, culture and people). Walking tourism along the pilgrimage routes may be seen as a 'co-creation process and development process by walkers, surrounding communities and the environment' leading to sustainable modes of tourism practice. Although a pilgrimage journey may not always be 'on foot', walking is central to its meanings. Walking provides time and rhythm, allowing senses to be awakened and optimizing spatial and temporal engagement in situ (Ingold & Vergunst, 2008; Omar, Patterson & Pegg, 2012; Slavin, 2015). The associated slowness facilitates more locally engaged, mindful and lower impact tourism, and allows for more attention to people and place. Thus, slow tourism has significant implications for tourism development, setting particular parameters by demanding more intentional, locally engaged sustainable tourism.

Conclusions

Pilgrimage tourism has become a valuable part of contemporary Japan's efforts towards the revitalization of its regional economies, whose main industries have experienced a decline and its communities are at risk owing to population decrease. In this regard, spiritual-related heritage, such as pilgrimage sites, play an important role as tourism resources which can attract both domestic and overseas visitors who may have little connections to the traditional faith of the place itself. Pilgrimage-related tourism development, intertwined with the contemporary social background, has embraced spiritual phenomena such as *iyashi* and powerspots, which appeal to a larger demographic than traditional faith nowadays. However, traditions have not disappeared entirely, forming a mosaic of tradition, consumerism, nostalgia and individualized spirituality.

Visitors' interest in the spirituality of the place is commensurate with the community's hopes for a sustainable way of living that emphasizes slow development. The motives of the visitors and service providers serve to co-create a mode of tourism that is slow and sustainable. This kind of tourism can enhance such quality, raising awareness, publicizing more widely its significance and outcomes, making direct and indirect contributions to sustainability. Walking, a central aspect in pilgrimage tourism, can be a quiet yet powerful force for the sustainable management of a place, allowing tourists to be spiritually engaged, while supporting a regional communities' way of life, as they collectively set the pace and rhythm of development. Management of such heritage sites highlights spiritual significance and facilitates a wide range of public participation in educational and informative conservation programs. This creates a model for sustainable stakeholder collaboration in tourism development as a dynamic application of cultural interpretation founded in community strength.

References

Amada, A. (2012) Kumano. In R. Okamoto, E. Hoshino and H. Yamanaka, H. (Eds). *Sacred Pilgrimage Tourism (Seichijunrei tuurizumu).* Japan: Kobundo, pp. 4–97.

Attix, S. A. (2015) New Age-oriented special interest travel: An exploratory study. *Tourism Recreation Research*, 27(2), 51–58.

Bruce, S. (2006) Secularization and the impotence of individualized religion. *The Hedgehog Review,* 8(1–2), 35–45.

Caffyn, A. (2009) The slow route to new markets. *Tourism Insights* (September 2009), n.p.

Caffyn, A. (2012) Advocating and implementing slow tourism. *Tourism Recreation Research*, 37(1), 77–80.

Casanova, J. (2006) Rethinking secularization: A global comparative perspective. *The Hedgehog Review*, 8(1–2), 7–22.

Collins-Kreiner, N. (2016) The lifecycle of concepts: The case of 'pilgrimage tourism'. *Tourism Geographies*, 18(3), 322–334.

Dickinson, J. and Lumsdon, L. (2010) *Slow Travel and Tourism.* London: Earthscan.

Digance, J. (2006) Religious and secular pilgrimage: Journeys redolent with meaning. In D. Timothy & D. Olsen (Eds.), *Tourism, Religion and Spiritual Journeys.* London: Routledge, pp. 36–48.

Fernandes, C., Pimenta, E., Goncalves, F. and Rachao, S. (2012) A new research approach for religious tourism: The case study of the Portuguese route to Santiago. *International Journal of Tourism Policy*, 4(2), 83–94.

Funck, C. and Cooper, M. (2013) *Japanese Tourism: Spaces, Places and Structures.* New York and Oxford: Berghahn Books.

Gardner, N. (2009) A manifesto for slow travel. *Hidden Europe Magazine*, 25, 10–14.

Guichard-Anguis, S. (2009) Introduction: The culture of travel (tabi no bunka) and Japanese tourism. In S. Guichard-Anguis and O. Moon (Eds). *Japanese Tourism and Travel Culture*. London and New York: Routledge, pp. 1–18.

Hagi, K. (2017) Kyoto residents now complain about hordes of overseas tourists. *Asahi Shinbun*. Available at: www.asahi.com

Hanegraaff, W. J. (2000) New Age religion and secularization. *Numen,* 47(3), 288–312.

Heelas, P. (2006) Challenging secularization theory: The growth of 'New Age' spiritualities of life. *The Hedgehog Review,* 8(1–2), 46–58.

Honore, C. (2004) *In Praise of Slow*. London: Orion.

Horie, N. (2009) Spirituality and the spiritual in Japan: Translation and transformation. *Journal of Alternative Spiritualities and New Age Studies,* 5, n.p.

Hoshino, E. (1981) *Pilgrimage: Phenomenology of the Sacred and Profane* (*Junrei: Sei to soku no genshougaku*). Tokyo: Kodansha.

Houtman, D. and Manscini, P. (2002) Why do churches become empty, while New Age grows? Secularization and religious change in the Netherlands. *Journal for Scientific Study of Religion*, 41(3), 455–473.

Ingold, T. and Vergunst, J. (Eds) (2008) *Ways of Walking: Ethnography and Practice on Foot*. Aldershot: Ashgate.

Ito, M. (2003) *Contemporary Society and Spirituality* (*Gendai shakai to Supirichuaritei*). Tokyo: Keisuisha.

Japan Tourism Agency (n.d.). *Japanese-Style DMO Formation and Establishment Plan (Nihonban DMO keisei kakuritsu keikaku)*. Retrieved from the Japanese Ministry of Land, Infrastructure and Transport website: www.mlit.go.jp/common/001154820.pdf.

Jimura, T. (2016a) The relationship between World Heritage designation and local identity. In L. Bourdeau, M. Gravari-Barbas and M. Robinson. (Eds), *World Heritage, Tourism and Identity: Inscription and Co-Production*. London: Routledge, pp. 81–91.

Jimura, T. (2016b) World Heritage Site management: A case study of sacred sites and pilgrimage routes in the Kii mountain range, Japan. *Journal of Heritage Tourism,* 11(4), 1–13.

Kadota, H. (2013). *Ethnography of Pilgrimage Tourism – The Consumed Religious Experience* (*Junrei tsuurizumu minzokushi – Shouhi sareru shuukyou keiken*). Japan: Shinwasha.

Kaelber, L. (2006) Paradigms of travel: From medieval pilgrimage to the postmodern virtual tour. In D. Timothy and D. Olsen (Eds.), *Tourism, Religion and Spiritual Journeys*. London: Routledge, pp. 49–63.

Kato, K. and Progano, R. N. (2017) Spiritual (walking) tourism as a foundation for sustainable destination development: Kumano-kodo pilgrimage, Wakayama, Japan. *Tourism Management Perspectives*, 24, 243–251.

Knight, J. (1996) Competing hospitalities in Japanese rural tourism. *Annals of Tourism Research*, 23(1), 165–180.

Koens, K. and Thomas, R. (2015) Is small beautiful? Understanding the contribution of small businesses in township tourism to economic development. *Development Southern Africa*, 32(3), 320–332.

Kotera, A. (2011) What is Powerspot? Preliminary research on the study of its social background and its reception (Pawasupotto toha nanika – Shakaiteki haikei no kentou to sono juyou nitsuite no yobiteki chousa). *Touyou Eiwa Jogakuin Daigaku Jinbun Shakaigakka Ronshu,* 29, 87–110.

Lehto, X., Brown, S., Chen, Y. and Morrison, A. (2006) Yoga tourism as a niche within the wellness tourism market. *Tourism Recreation Research*, 31(1), 25–35.

Lois-Gonzalez, R. and Santos, X. (2015). Tourists and pilgrims on their way to Santiago. Motives, Caminos and final destinations. *Journal of Tourism and Cultural Change,* 13(2), 149–164.

Lumsdon, L. and McGrath, P. (2011) Developing a conceptual framework for slow travel: A grounded theory approach. *Journal of Sustainable Tourism*, 19(3), 265–279.

Matos, R. (2004) Can slow tourism bring new life to Alpine regions? In K. Weiermair and C. Mathies (Eds), The Tourism and Leisure Industry: Shaping the Future. New York: Haworth Hospitality Press, pp. 93–103.

Matsui, T. (2013) *Language and Marketing: A Social History of the 'Healing Boom' in Japan* (*Kotoba to maaketeingu: Iyashi buummu no shouhishakaisi*). Tokyo: Sekigakusha.

McGrath, P. and Sharpley, R. (2018) Slow travel and tourism: New concept or new label? In M. Clancy (Ed.), *Slow Tourism, Food and Cities: Pace and the Search for the 'Good Life'*. London: Routledge, pp. 49–62.

McGuire, M. (2013. What's at stake in designating Japan's sacred mountains as UNESCO World Heritage Sites? Shugendo practices in the Kii Peninsula. *Japanese Journal of Religious Studies*, 40(2), 323–354.

Mori, M. (2005) *The Modernization of Shikoku Henro: From a 'Modern Pilgrimage' to a 'Healing Journey'* (*Shikoku Henro no kindaika: Modan Henro kara iyashi no tabi made*). Tokyo: Sogensha.

Nagatani, K. (2017, February) (Economy + regional onset) Inviting international visitors to Kumano – Local human resources and a Canadian advising. ((Keizai + Chiikihatsu) Kumano he, kaigaikyaku izanau jimoto ni jinzai, kanadajin ga jogen). *Asahi Shinbun.* Retrieved from: database.asahi.com/library2/main/top.php

Nomoto, K (2006) *Landscape Study on Gods and Nature* (*Kami to shizen no keikanron*). Tokyo: Kodansha Gakujutsu Bunko.

Norman, A. (2004) *Spiritual Tourism: Religion and Spirituality in Contemporary Travel* (BA Honors Thesis). Australia: University of Sydney.

Norman, A. (2009) The unexpected real: Negotiating fantasy and reality on the road to Santiago. *Literature and Aesthetics,* 19(2), 50–71.

Okamoto, R. (2015) *Pilgrimage to Sacred Sites: From World Heritage to Anime Setting* (*Seichi junrei: Sekai isan kara anime butai made*). Tokyo: Chuokoron Shinsha.

Okamoto, R. and Kawasaki, N. (2012) Meiji Jingu – Kiyomasa no Ido. In R. Okamoto, E. Hoshino & H. Yamanaka (Eds), *Sacred Pilgrimage Tourism (Seichijunrei tsuurizumu)*. Tokyo: Kobundo, pp. 142–145.

Omar, W., Patterson, I. and Pegg, S. (2012) A green pathway for future tourism success: Walking trails in Kuala Lumpur. *Tourism Planning & Development*, 9(1), 57–76.

Pravdova, H. and Radosinska, J. (2013) Media culture in the context of transformation of contemporary spiritual values and spirituality. *European Journal of Science and Theology*, 9(6), 169–178.

Prime Minister of Japan and Cabinet (2012) *Tourism Nation Basic Promotion* Plan (*Kankou Kokuritsu Suishin Honkeikaku*). Retrieved from Japan Tourism Agency website: www.mlit.go.jp/common/000208713.pdf

Progano, R. N. (2018) Residents' perceptions of socio-economic impacts on pilgrimage trails: How does the community perceive pilgrimage tourism? *Asian Journal of Tourism Research*, 3(2), 148–178.

Raj, R. and Griffin, K. (2015) *Religious Tourism and Pilgrimage Management: An International Perspective* (2nd ed.). Wallingford: CABI.

Reader, I. (2012) Secularisation, R.I.P.? Nonsense! The 'rush hour away from the gods' and the decline of religion in contemporary Japan. *Journal of Religion in Japan*, 1, 7–36.

Reader, I. (2014). *Pilgrimage in the Marketplace*. London: Routledge.

Reader I. and Swanson, P. (1997) Pilgrimage in the Japanese religious tradition. *Japanese Journal of Religious Studies,* 24(3–4), 225–270.

Reed, S. T. (2007) Analyzing secularization and religiosity in Asia. *Japanese Journal of Political Science*, 8(3), 327–339.

Rivera, M., Croes, R. and Lee, S. (2016) Tourism development and happiness: A residents' perspective. *Journal of Destination Marketing & Management*, 5(1), 5–15.

Rizzello, K. and Trono, A. (2013) The pilgrimage to the San Nicola Shrine in Bari and its impacts. *International Journal of Religious Tourism and Pilgrimage*, 1(1), 24–40.

Sawamura, A. (2016). World Heritage registration and tourism trends (revised edition): Analyzing 10 cases in Japan (Sekai isan touroku to kankou doukou (shuuseika hikkou): Nihon no juu jirei kara). *The Journal of Economics of Niigata University, 100*, 117–128.

Seraphin, H., Sheeran, P. and Pilatos, M. (2018) Over-tourism and the fall of Venice as a destination. *Journal of Destination Marketing & Management,* 9, 374–376.

Shackley, M. (2001) *Managing Sacred Sites*. London: Continuum.

Sharpley, R. (2009) Tourism, religion and spirituality. In T. Jamal and M. Robinson (Eds), *The Sage Handbook of Tourism Studies*. London: Sage Publications, pp. 237–252.

Slavin, S. (2015) Walking as spiritual practice: The pilgrimage to Santiago de Compostela. In A. Norman and C. Cusack (Eds.), *Religion, Pilgrimage and Tourism*. New York: Routledge, pp. 283–299.

Stark, R. (1999) Secularization, R.I.P. *Sociology of Religion*, 60(3), 249–273.

Stausberg, M. (2011) *Religion and Tourism: Crossroads, Destinations and Encounters*. London and New York: Routledge.

Stoddard, R. (1997) Defining and classifying pilgrimages. *Geography Faculty Publications,* 2, 41–60.

Suga, N. (2010) Shinto shrines as power spots (Pawaasupotto toshite no jinja). In K. Ishii (Ed.), *Where is Shinto Going? (Shinto ha doko he ikuka)*. Tokyo: Perikansha, pp. 232–252.

Suntikul, W. and Butler, R (2018) Tourism and religion: Origins, interactions and issues. In R. Butler and W. Suntikul (Eds.), *Tourism and Religion: Issues and Implications*. Bristol: Channel View Publications, pp. 1–9.

Tanabe City (2016) *Tanabe City Statistics Book, Heisei 28 November Edition* (*Tanabeshi tokeisho Heisei 28 11 gatsu hakkou*). Tanabe: Tanabe city.

Tanabe Commerce and Industry Conference (2009) *Examination on Kumano* (*Kumano Kentei*). Tanabe: Tanabe Commerce and Industry Conference.

Tanabe Kumano Tourism Bureau (n.d.). *Towards a Sustainable Destination Opened to the World* (*Sekai ni akareta jizokukanouna kankouchi wo mezashite*). Retrieved from the Tanabe Kumano Tourism Bureau website: www.tb-kumano.jp/download/.

Timothy, D. and Conover, P. (2006) Nature religion, self-spirituality and New Age tourism. In D. Timothy and D. Olsen (Eds.), *Tourism, Religion and Spiritual Journeys*. London and New York: Routledge, pp. 139–155.

Tsukada, H. and Omi, T. (2011) A deluge of 'religious' information on new religions, power spots, funeral services and Buddhist statues. *Nanzan Institute for Religion & Culture Bulletin,* 35, 24–47.

UNWTO (2011) *Religious Tourism in Asia and the Pacific*. Madrid: UN World Tourism.

UNWTO & GTERC (2017) *Asia Tourism Trends 2017 Edition*. Madrid: UN World Tourism Organization & Global Tourism Economy Research Center.

Vukonic, B. (1996) *Tourism and Religion*. Oxford: Pergamon.

Wakayama Tourism Agency (2017) *Heisei 28 Tourist Motivation Research Report* (*Kankoukyaku Douki Chousa Hokusho Heisei 28 Nen*). Wakayama: Wakayama Tourism Agency.

10 Confronting difficult pasts

The case of 'kamikaze' tourism

Richard Sharpley and Kumi Kato

Introduction

From a pragmatic perspective, 'tourism is far more dependent on peace than peace is on tourism' (Hall, Timothy & Duval, 2004: 3). However, it has long been suggested that tourism itself might be an effective vehicle of peace and mutual understanding. In the years following the First World War, for example, travel was promoted as a means of encouraging peaceful relations between former adversarial nations whilst, half a century later, the United Nation's International Tourism Year in 1967 was endowed with the slogan 'Tourism : Passport to Peace' (Wohlmuther & Wintersteiner, 2014: 17). Subsequently, the World Tourism Organization (1980) claimed that international tourism could be a vital force for peace, a theme that was adopted and built upon by D'Amore (1988) who founded the International Institute for Peace through Tourism in 1986. Since then, increasing attention has been paid to the relationship between tourism and peace (Blanchard & Higgins-Desboilles, 2013; Moufakkir & Kelly, 2010), the belief being that social and cultural connections through tourism 'spur dialogue and exchange, break down cultural barriers and promote the values of tolerance, mutual understanding and respect' (Rifai, 2013: 11).

The extent to which this occurs in practice is, inevitably, the subject of debate (Litvin, 1998). Such is the scale and diversity of tourism, the enormous variety of contexts in which it occurs and, indeed, the multitude of ways in which it may (or may not) promote peace, understanding or reconciliation that generalizations cannot be made. Moreover, with increasing evidence of so-called 'overtourism' (see Chapter 14 in this volume), the outcome of tourism development may in fact be resentment and conflict. Nevertheless, in certain circumstances, particularly where tourism is developed and promoted at sites related to difficult pasts – or what might be referred to broadly as 'dark tourism' (Sharpley & Stone, 2009; Stone et al., 2018) – the opportunity exists, through appropriate commemoration and interpretation, to encourage reconciliation and understanding (Batten, 2009). More specifically, dark heritage sites of or related to conflict between peoples and nations, such as battlefield sites, museums dedicated to violent pasts or sites of genocide or memorials to those who have lost their lives in acts of terror or violence, represent a legitimized space where, in principal at least, visitors with direct, indirect or even no connection with the event, its victims and indeed perpetrators may congress to both remember those who suffered in a violent past and also to reconcile differences in order to establish a more peaceful present and future (Gurler & Ozer, 2013), to ensure that 'never again' can such events occur (Lollis, 2014).

The purpose of this chapter is to consider how Japan not only confronts but interprets one such particular difficult, dark past – the kamikaze phenomenon – through tourism. As Nelson (2003: 445) observes more generally, 'since the end of World War II, diverse interpretations over how to represent, acknowledge and atone for Japan's aggressive exploits throughout Asia and the Pacific have occasioned as much controversy and conflict as they have closure'. Amongst these exploits, the suicide missions of kamikaze pilots, more formally referred to as *tokkō*, or Special Attack Force (Sheftall, 2008: 155), arguably remain the most controversial. It is generally accepted that between October 1944, when the first official kamikaze mission took off from Mabalacat in the Philippines (Axell & Kase, 2002: 40), and the formal end of the Pacific War on 2 Setember 1945, approximately 3,000 Japanese airmen died in kamikaze attacks, although some suggest the figure could be as high as 5,800 (Sheftall, 2008). Not only are they commemorated at a variety of museums and shrines in Japan but also, in recent years, increasing academic attention has been paid to the varying nature of this commemoration (Allen & Sakamoto, 2013; Axell & Kase, 2002; Inuzuka, 2016; Nelson, 2003; Sakamoto, 2015; Sheftall, 2008; Yoshida, 2004). However, these studies are typically concerned with competing discourses of the kamikaze phenomenon within Japan. In contrast, the extent to which these commemorative sites as (dark) tourism attractions make a contribution to wider understanding and reconciliation through tourism, specifically to an international audience, has not been considered. This is, arguably, a significant omission, not least given Japan's policy of increasing international arrivals to 40 million by 2020, and 60 million by 2030 and the formal objective of the 'enhancement of mutual international understanding' (MLIT, 2012: 4) through international tourism (see Chapter 1).

Therefore, focusing primarly on the Chiran Peace Museum which, located in Kagoshima Prefecture, is dedicated to the lives and deaths of 1,036 kamikaze pilots, but also making reference to the lesser-known Uzurano airfield in Kasai, Hyogo Prefecture, where kamikze pilots were trained and where the original runway and other infrastructure is still in existence, this chapter seeks to address this gap in the literature. In order to do so, it is necessary to explore briefly the historical context in which the kamikaze attacks were conceived, authorized and undertaken. The first task, however, is to consider the potential of dark tourism more generally to contribute to peace, understanding and reconciliation as a conceptual framework for the subsequent discussion.

Dark tourism, peace and reconciliation

The concept of dark tourism was first proposed more than two decades ago (Foley & Lennon, 1996). Since then, it has attracted ever-increasing academic attention within both tourism studies and other related disciplines, such as heritage and museum studies, as well as becoming a popular theme within the media. It remains, however, a contentious subject, not least with regards to definition. Not only has the term 'dark tourism' been applied to such a broad diversity of sites and attractions that it has become 'increasingly diluted and fuzzy' (Sharpley, 2009a: 6) – indeed, it has been suggested that the concept should be abandoned altogether (Bowman & Pezzullo, 2009) – but there also remains a lack of consensus over whether it is primarily a supply based ('dark' places) or experiential ('dark' tourists) phenomenon (Ashworth & Isaac, 2015).

A detailed discussion of these issues is beyond the scope of this chapter (see Light, 2017 for a comprehensive review). Fundamental to understanding what dark tourism 'is', however, is the relationship bewteen a site or attraction and the event it represents. In other words, dark tourism is concerned with tourism at places of or associated with death, disaster or suffering or, more precisely, places of or associated with dark events. And the significance of dark tourism lies in the manner in which dark sites or attractions mediate between the visitor and the events with which they are associated.

Of course, the nature of such events varies enormously, from, for example, the natural death of one person to multiple violent deaths resulting from atrocity or conflict (Ashworth & Hartmann, 2005). Equally, dark tourism sites also vary enormously in type, scale and significance (Stone, 2006), with graveyards, battlefields, museums, commemoration sites, disaster sites, former prisons and even 'houses of horror' falling under the umbrella of dark tourism. Amongst this diversity of sites, however, post-conflict sites and destinations, such as memorials or museums commemorating difficult pasts, arguably offer the greatest potential to not only provide visitors with the opportunity for commemoration, reflection or learning, but also to facilitate the post-conflict recovery of societies more generally (Friedrich, Stone & Rukesha, 2018). In other words, as places where all stakeholders, or those with both a direct and indirect interest in the conflict, importantly including both victims and perpetrators, may come together, post-conflict sites may foster accountability, justice, debate and reconciliation, or 'transforming relations of hostility and resentment to friendly and harmonious ones' (Bar-Siman-Tov, 2004: 4), as a necessary foundation for peace-building (Buckley-Zistel & Schaefer, 2014).

The extent to which such peace-building may actually occur is, inevitably, subject to a number of forces and influences. On the one hand, post-conflict sites offer the opportunity for education or telling 'the truth', in so doing encouraging all stakeholders to engage critically with the event, to communicate and to overcome past differences in a spirit of tolerance and forgiveness.

For this to take place, however, not only should the site be accessible and welcoming to all who wish to visit, but it should also be managed and presented in a manner that fosters reconciliation, including the appropriate or accurate representation of all stakeholders' heritages or stories (Poria, 2007; Sharpley, 2009b). Moreover, there should ideally exist both acknowledgement of the need for and a wider culture of reconciliation; in short, for dark tourism act as a catalyst of peace and understanding, all stakeholders should from the outset be committed to achieving such peace and understanding.

On the other hand, the very nature of dark events or difficult pasts almost inevitably means that a number of challenges face their effective or appropriate representation and commemoration. First and foremost, and as argued elsewhere (Sharpley, 2009b), atrocities, tragedies and other dark events not only have the potential to be exploited for commercial gain through tourism, with quite evident ethical implications (Kelman & Dodds, 2009); they are also highly susceptible to political influence. In other words, the development and interpretation of dark sites may be undertaken to convey particular political messages, reflecting what Light (2007: 747) refers to more generally as the 'cultural politics of tourism development'. It has long been recognized that, owing to its significance and visibility as a social and economic phenomenon, tourism may be exploited for political or ideological purposes (Richter, 1983), most usually to affirm or stengthen cultural identity. As Cano and Mysyk, 2004: 880) observe, 'the state may assume the role of marketer of cultural meanings, in which it attempts to make a statement about national identity by promoting [through tourism] selected aspects of a country's cultural patrimony'. Of particular significance to this chapter, such cultural mediation is often intended for an internal, domestic audience, the purpose being to construct and maintain a national identity (Palmer, 1999). Equally, the message may also be intended for an international audience. For example, the Kigali Genocide Memorial, which commemorates the victims of the Rwandan Genocide of 1994, not only promotes peace and reconciliation amongst Rwandans themselves but also seeks to convey a politically loaded message, not least legitimizing the contemporary government of the country, to international visitors (Friedrich, Stone & Rukesha, 2018; Sharpley & Freidrich, 2016)

The outcome of political intervention in commemoration or memorialization at dark sites is but one potential source of what is referred to as dissonance or dissonant heritage. According to Tunbridge and Ashworth (1996: 6), heritage is 'a product of the present, purposefully developed in response to the current needs or demands for it and shaped by those requirements' or, as Seaton (2001: 125) puts it, 'as much a product of present perspectives as past events'. In other words, history is presented as heritage in such a manner as to satisfy contemporary political and social needs. At the same time, however, '…all heritage is someone's heritage and therefore logically not someone else's, and the original meaning of an inheritance implies the existence of disinheritance' (Tunbridge & Ashworth 1996: 21).

This 'disinheritance' occurs when there exists a 'lack of congruence at a particular time or place between people and the heritage with which they identify' (Ashworth & Hartmann, (2005: 253) or, alternatively stated, dissonant heritage is manifested when the past is represented or interpreted in such a way that, for particular people or groups, 'their' past is distorted or displaced (see also Smith, 2006).

All heritage is susceptible to dissonance, not only because multiple groups are likely to have an interest in a particular heritage, implying multiple stories and multiple (and potentially competing) interpretations, but also because, over time, people's relationship with and understanding of past events may change. However, that dissonance may be enhanced when the representation or interpretation of past events is purposefully designed to create or enhance a group's or national identity, to 'align with a particular ... group, to (re)interpret past events to meet contemporary political agendas, to erase or deny and particular past [or] to celebrate victory' (Sharpley, 2009, b: 150–151). And significantly, Ashworth and Hartmann (2005: 254) argue that not only does 'human tragedy imbue [dissonance] with a capacity to amplify the effects and thus render more serious what otherwise would be dismissable as marginal or trivial', but also that dissonance is inevitable in dark heritage construction and interpretation. As a consequence, dissonant dark heritage may reinforce cultural stereotypes, enhance rather than diminish the divisions or animosity between different stakeholders and, in the extreme, perhaps even legitimize the event(s) and those involved as perpetrators and act as a catalyst for similar events in the future.

As discussed later in this chapter, the manner in which the kamikaze phenomenon is represented and commemorated in Japan is arguably subject to a relatively significant degree of dissonance with regards to both the interpretation of the role and sacrifice of the pilots themselves and the 'disinheritance' of others. Consequently, it will be suggested in particular that the Chiran Peace Musuem, established to 'expose the tragic loss of [the pilots' lives], so that we all might better understand the need for everlasting peace and attempt to encourage such tragedies to never be repeated' (Chiran, 2017), emphasizes the pilots' heroism but is less successful than it might be in contributing to wider understanding and reconciliation amongst all stakeholders. Before doing so however, the next section considers briefly the historical and cultural circumstances surrounding the kamikaze phenomenon.

Kamikaze: the historical context

According to Axell and Kase (2002: 3), 'For the younger generation, the kamikaze phenomenon is a curiosity from the past; to the older it is a reminder of a cruel epoch'. More than 70 years after the end of the Pacific War, however, the narrative surrounding it remains highly contested within Japan (Jeans, 2005; Sheftall, 2008) and, in all likelihood, misunderstood beyond the country's borders. Moreover, to further complicate matters, the word 'kamikaze' is imbued with a certain historical romanticism. Translating as 'divine wind', it was the name given to a powerful typhoon that, in 1281, destroyed the invading fleet of the Mongolian Emperor Kublai Khan as it approached Japan, leading the Japanese to believe that, if once again facing attack, their country would be similarly saved by a divine wind (Chiran, 2017: 11). Nevertheless, although controversy continues to surround the strategic decision to employ kamikaze tactics, both the generally acknowledged historical events leading up to the official commencement of kamikaze attacks in 1944 as well as the broader cultural context, specifically the influence of so-called *bushido*, or the code of honour, provide a framework for understanding the reasoning behind the attacks and the apparent willingness of the young pilots to undertake the missions, as well as for assessing the manner in which they are commemorated.

The Pacific War commenced on 8 December 1941 when, following the attack on Pearl Harbor, Britain and the USA declared war on Japan. The events leading up to the attack are complex (Costello, 1982). However, it is generally considered to have been the inevitable outcome of, initially, Japan's foreign policy, specifically its 1931 invasion and occupation of Manchuria (north-east China), primarily to exploit the region's natural resources, and its subsequent intention to establish a 'New Order in East Asia' (Axell & Kase, 2002: 21). This resulted in growing opposition amongst Western nations to Japan's international activities, culminating in the USA imposing, amongst other things, an embargo on the export of iron and aviation fuel to Japan. As a country with few natural resources of its own, this was seen as a threat to not only its international ambitions but its very survival, although the commencement of hostilities against the US in 1941 was also seen to be an opportunity to liberate the region from the control of the world's then colonial powers. Hence, as described without a hint of irony in the Chiran Peace Museum's booklet, *The Mind of the Kamikaze* (Chiran, 2017: 8):

> After the attack at Pearl Harbour, Japanese forces advanced into countries in Southeast Asia. Their aim was not only to liberate countries from their colonizers, but also to obtain much needed strategic minerals...of which Japan had virtually none.

In the 6 months following Pearl Harbor, Japan enjoyed notable military success although by mid-1942, her armed forces began to suffer a number of setbacks. In particular, the Battle of Midway in June of that year, which severely depleted the country's naval air power, is seen by many as turning point in the Pacific War. By mid-1944, it had become recognized by Japan's military leaders that the continuing advances of numerically superior American forces could no longer be countered by conventional tactics and, hence, with the Japanese-occupied Philippines facing imminent attack, and in the face of overwhelming odds, the first Special Attack missions were launched against American ships in the October of that year. Subsequently, the kamikaze campaign reached its peak near Okinawa during the following summer, finally ending with the cessation of hostilities in August 1945.

The strategy of sending young pilots on suicide missions was not universally accepted amongst Japan's military leaders, with some questioning the rationale of sending men to certain death but with less certainty with regards to successful outcomes (Axell & Kase, 2002); it is suggested, for example, that as few as 10 percent of pilots actually crashed into their targets (BBC, 2017). Equally, there is evidence to suggest that, for a variety of reasons, many of those who flew on these missions were not in fact willing volunteers (Sheftall, 2005) and, indeed, that some deliberately ditched their aircraft in the ocean rather than crashing into ships. However, 'they … knew there was no alternative. To refuse to fly was to show a lack of duty to their country and their parents' (Chiran, 2017: 19). More specifically, a significant factor in the large number of young men willing to become *tokkō* pilots was the tradition of honourable death that was deeply entrenched in the culture of the Japanese military. This, in turn, reflected the ancient Code of the Samurai or *bushido* which, according to Nitobé (1908), translates literally as 'military knight ways'. Bushido represents a moral code, a set of principles that the ancient warrior class, or Samurai, were required and instructed to observe in all aspects of their lives. Such principles included justice, courage, benevolence, politeness, truthfulness, honour and loyalty (Nitobé, 1908).

Underpinning *bushido* are the twin influences of Buddhism and Shintoism. On the one hand, Buddhism encourages a sense trust in fate, acceptance of the inevitable and stoic composure in the face of adversity. On the other hand, Shintoism promotes loyalty above all to the Emperor, filial piety (respect for parents and elders) and, in particular, reverence for ancestral memory inasmuch as it is believed that the souls of the dead remain behind to watch over the living. Specifically, in Shintoism, those who die tragically or heroically can be worshiped as gods, and it is for this reason that the great majority of Japan's war dead, including kamikaze pilots, are commemorated at the Yasukuni Shrine, a 'high profile and controversial' (Sakamoto, 2015: 160) Shinto shrine built in 1869 to house their spirits. For the young Japanese pilots, then, undertaking the kamikaze missions offered the opportunity to display not only their loyalty to their country and Emperor (who at that time was also considered a diety) by dying honourably in warfare, but also their respect for their families (especially their mothers) as revealed in the final letters they wrote before setting off on their final journey. Moreover, they did so secure in the belief that they would become deities and that their spirits would remain with and look over those they left behind, the living.

In short, 'From today's perspective, Japanese military's large scale and systematic use of young men to launch "suicide attacks" seems nothing short of collective madness or extreme state coercion' (Sakamoto, 2015: 164). However, the kamikaze phenomemon may be understood, though not necessarily justified, on the basis that a proud nation defined by fatalism, stoicism and loyalty to country and Emperor was facing inevitable defeat, and that the armed forces by and large believed in – or perhaps felt obliged to pursue – honour in death, to fight to the end. Hence, the kamikaze strategy was arguably seen as both a practical and honourable way to continue inflicting damage on the enemy, the objective perhaps being to negotiate eventual surrender from a stronger position. However, it is not a straightforward story; indeed, there are multiple stories and interpretations. The question is, therefore: to what extent are these stories presented as a basis for all stakeholders – not only older and younger generations of Japanese but also visitors from other countries – to confront and reconcile with such a difficult past? This question is now addressed through the case study of the Chiran Peace Museum.

Commemorating the kamikaze: Chiran Peace Museum

Chiran Peace Museum is located near the site of the former Chiran air base at the southern end of the island of Kyushu from where, in the final months of the Pacific War in 1945, many kamikaze missions were launched against the American fleet at Okinawa. In fact, 'more kamikaze pilots took off from Chiran than anywhere else' (Chiran, 2017: 5). Unlike at Uzurano, discussed shortly, relatively little evidence of the airfield remains but, in order to commemorate the pilots who flew out on their final missions, the Chiran Resource Centre was initially constructed by the town in 1975 to preserve and display memorablia, such as letters, photographs and others materials, from the kamikaze operations at Chiran. In the mid-1980s, the Centre was redeveloped with state funding and opened in 1987 as the Chiran Peace Museum. Commemorating 1,036 kamikaze pilots (their average age was 21.6 years), almost half of whom flew out of Chiran, the museum now attacts more than 500,000 domestic vistors annually, as well as around 10,000 international visitors.

The dominance of the domestic market is not, perhaps, surprising given the museum's location, it being relatively distant from the typical international tourist 'trail' in Japan. However, its 'message' is also clearly directed towards a domestic as opposed to an international audience. Moreover, that message has long been controversial, falling into the wider debates surrounding the manner in which Japan commemorates and represents its role in the Pacific War. For example, Yoshida (2004: 16) suggests there are three distinctive kinds of museums in Japan: those which glorify war and 'romanticize the sacrifices of deceased soldiers'; those that present strong anti-war messages; and those that tread the middle ground, attempting to 'satisfy both critics and supporters of Japan's wartime past'. As a 'peace museum', it might be assumed that Chiran would fall into the second or third categories but, for Yoshida (2004), it is an example of the first. And this is immediately evident on visiting the musuem, as experienced by the authors of this chapter (see also Inuzuka, 2016 for a detailed discussion of kamkaze commemoration at Chiran).

On approaching the museum building, visitors first encounter a full-sized replica of a a kamikaze plane (constructed for a movie) and, nearby,

Figure 10.1 Replica kamikaze plane, Chiran Peace Museum.
Photo: R. Sharpley.

the statues of both a kamikaze pilot (Figures 10.1 and 10.2) and of Tome Torihama who owned a restaurant in the town where pilots would go, sometimes with their families, prior to departing on their final mission. She became known as *tokko no haha* or the 'mother of special attack pilots' (Inuzuka, 2016: 151) and her restaurant is now a small museum. These installations immediately establish the focus of the museum on the pilots themeselves and, on entering the main building, this is confirmed when visitors are first confronted with a large mural (the 'Chiran Requiem') of a pilot being carried from his burning plane by angels (Figure 10.3). This serves to both represent the Shinto deification of the kamikaze pilots and their honourable death but also, more significantly, sets the overall tone of the museum's perspective on the kamikaze phenomenon. In front of the mural, a sign (one of very few in the museum translated into English) states that '... the Peace Hall was built here in commemoration of the pilots who died heroically in the skies and to impart the historical realities behind their lives, as well as pray for enduring peace'. It is also worth noting that audio tours in English are available and, in the introduction to the tour, visitors are informed that although the term 'kamikaze' is more popularly used, the museum refers to the pilots as *tokkō* because they attacked only military and not civilian targets. In so doing, an attempt is being made to not only to legitimize the kamikaze strategy but also, perhaps, to distinguish it from the more recent Islamic suicide bombings, in

Figure 10.2 Statue of a kamikaze pilot, Chiran Peace Museum.
Photo: R. Sharpley.

particular, the 9/11 attacks, with which commemtators on the kamikaze phenomenon sometimes draw comparisons (Axell, 2002).

Next to the mural, a continual video shows original footage of the kamikaze missions, emphasizing the challenges facing the pilots in reaching their targets. Many planes are seen crashing into the sea, but successful attacks are also shown. Also, in an adjacent room, physical items are on display, including the wreckage of a plane recovered from the seabed off the coast of Kagoshima (Figure 10.4). These, along with other artefacts on diplay elsewhere in the museum, such as clothing and other equipment used by the

Figure 10.3 Mural in Chiran Peace Museum.
Photo: R. Sharpley.

Figure 10.4 Wreckage of a Japanese 'Zero', Chiran Peace Museum.
Photo: R. Sharpley.

pilots, are tangible and uncontroversial exhibits; at the same time, a replica of an A-frame barracks hut in which pilots would have been billetted prior to their missions, located between the museum and a modern shrine dedicated to them, demonstrates the spartan conditions in which they spent their final days.

Most of the museum, however, is given over to photographs of each of the 1,036 pilots commemorated. Accompanying each photograph is a description of the individual's background, age, education, family and military experience. In addition, original copies of their final letters to family and loved ones can be seen in display cases beneath the photographs although, with the exception of a small number of such letters displayed together in a separate section of the museum, these are not accompanied by translations into English. Hence, the experience and understanding of international non-Japanese speaking visitors of these letters is, in a sense, controlled by the (no doubt intentional) selection of exhibits explained and translated in the audio tour, as well as those few translated letters on display. In addition, a selection of translated letters is provided in the museum's official booklet on the kamikaze phenomenon (Chiran, 2017), reproduced under themes such as 'The Mothers', 'Love for Children' and 'Friendships'. Collectively, these letters convey a message of duty, of selfless willing sacrifice, of respect and love for parents and family, and of happiness for dying for Emperor and country. For example, one letter says simply: 'Dear Parents and everyone in my family, at last the long awaited chance has come. It will be my honor to descend into the ocean with my enemy' (Chiran, 2017: 26). Another says

> Dear Mother and Brother, Now I will go. I feel truly happy… I am recalling how every morning when I left to go to elementary school I would say 'See you later'. Well, now I am leaving and I shall not return. I am truly satisfied.
>
> (p. 32)

whilst a father writes to his daughters

> Your father can't be the horse you ride on through life, so the two of you take care of each other. Please know your father is a happy man. He is riding a vehicle that will chase away our enemies. Become as great as your father.
>
> (p. 63)

This theme of willing sacrifice is further emphasized by stories highlighted during the audio tour. One, also related elsewhere (Axell & Kase, 2002), tells of how a wife killed herself and her children so that her husband would not held back from fulfilling his duty; another, based on a photograph of five young pilots aged between 17 and 19 who are smiling and playing with a puppy, suggests that they were happy even though they were allegedly departing on their mission the day after the photograph was taken. Nevertheless, it is also acknowledged that, in contrast, at night time many young pilots could be heard 'weeping bitterly ... because they were afraid of death and felt a deep sorrow for their very short life' (Chiran, 2017: 19).

Overall, then, not only does the museum focus specifically on the young Japanese pilots as apparently willing actors in the kamikaze campaign, but also the narrative surrounding them emphasizes their individuality, youth and humanity, their honour and sacrifice. In other words, there is, as Allen and Sakamoto (2013: 1050) put it, a 'quasi hero worship ... evident throughout the exhibits'. Not only is this focus on willing sacrifice 'consistent with Japanese revisionist approaches to the memories of the Asia-Pacific War' (Inuzuka, 2016: 157) but it is also a clear manifestation of dissonance as considered earlier in this chapter. That is, the narrative presented at the museum can itself be challenged – and indeed, has been through the stories of those pilots and others who survived the war (Sheftall, 2005) and who have, therefore, been disinherited by the museum's interpretation – yet, of equal if not greater significance, other stories are also not told. No attempt is made to locate the kamikaze phenomenon within the context of the war or Japan's role in it; no reference is made to the dissent of some commanders and pilots opposed to the kamikaze strategy or to its quite evident futility given that so few planes hit their targets and that defeat was inevitable. No reference is made to the prevailing culture of honourable death within the Japanese military at that time, a culture which inspired a sense of obligation rather than willingness amongst kamikaze pilots; and nor is it acknowledged that, in essence, those pilots were state-trained suicide bombers (Allen & Sakamoto, 2013). Moreover, also absent is any reference to their targets or victims, to the equally futile deaths of the crews of American ships that suffered direct hits. In short, the museum does not offer the knowledge or stories to enable its visitors, whether domestic or international, to begin to understand how and why, in the context of the last year of the Pacific War, the kamikaze phenomenon occurred and, in so doing, confront a difficult past. Rather, they are presented with a very specific, highly politicized and emotionally laden story that, for some visitors, might elicit feelings of respect and sympathy but for others, bemusement and perhaps even anger.

In turn, this raises the inevitable question of whether the Chiran Peace Museum, as dark tourism destination, might promote peace and reconciliation, whether it in fact succeeds in its stated objective of promoting peace and, indeed, is living up to its name as a 'peace museum'. On the one hand, many comments made by visitors, reproduced both in the museum's book *The Mind of the Kamikaze* and in a volume available for visitors to read in a seating area in the museum, suggest that it does. Typically, reference is made to the futility and waste of war and to the need for peace, although such arguably inevitable generic sentiments are largely expressed through the lens of the sacrifice of the young kamikaze pilots; that is, it is not the (untold) stories or heritages of the Pacific War that encourage visitors to write 'never again', but the senseless loss of thousands of young lives. Moreover, some visitors are of the opinion that the pilots' sacrifice laid the foundations of contemporary peace, a subliminal message that, perhaps, the museum intends to convey. On the other hand, some other comments reveal a more critical perspective; one, for example, states: 'I am from Singapore ... my country men also suffered ... we lived through 3 half [sic] years of Japanese occupation' whilst another visitor (from Norway) simply observes: 'What a strange thing to call this place a peace museum!' (Chiran Visitors Book: 10–11). The latter comment reflects the conclusions of others who critique Chiran's interpretation of the kamikaze phenomenon, there being broad consensus that a museum that not only honours the pilots as heroes but implies that their actions led to peace (whereas in fact they prolonged conflict) cannot be thought of as a peace museum. As Inuzuka (2016: 126) observes, peace 'is not integrated rationally into the displays' messages and pacifism is promoted in a rather ambiguous way'. Thus, whilst it may feed a contemporary national identity amongst some Japanese visitors, the museum's quite evident dissonance, or its disinheritance of many potential tourists from both Japan and overseas, suggests that rather than stimulating understanding, reconciliation and peace in the broader sense, it in fact denies visitors the opportunity to confront collectively and openly a difficult, controversial past. In contrast, however, the potential to offer such an opportunity exists at another site associated with the kamikaze phenomenon, the Uzurano airfield.

Uzurano: a dark tourism site for peace?

Situated in a valley in Kasai, some 80 km or 90 minutes' drive west of Osaka, Uzurano airfield was constructed between 1942 and 1943 as a training base for the Himeji Naval Air Corps. It was also the site of the Kawanishi aircraft factory which built fighter planes. The Air Corps based at Uzurano consisted of approximately 500 young men who trained there, sometimes lodging with local families at weekends; after 30 hours of flight training they were dispersed to units elsewhere. In 1945, however, a Special Attack corps, the *Shrisagi* or White Eagle squadron, was formed and 63 pilots, following training, eventually died in kamikaze operations near Okinawa.

Uniquely, the original runway is still in existence (Figure 10.5), as are other structures dating back to when the airfield was a military base, including air-raid shelters, ammunition stores and gun emplacements. In addition, alongside the runway is a memorial to the 63 kamikaze pilots trained at the airfield (Figure 10.6) whilst, in recent years, a small museum

has been established by local volunteers, displaying artefacts such as a pilot's uniform, parts of aircraft, models of aircraft and copies of letters written by the pilots prior to departing. This attracts about 1500 primarily domestic visitors a year. Following its sale to Kasai City, investments were made in 2018 in improving access to the structures around the airfield, the provision of signage and a cycle route, and the construction of a gatehouse

Figure 10.5 Original runway, Uzurano airfield.
Photo: R. Sharpley.

Figure 10.6 Memorial to 63 kamikaze pilots, Uzurano airfield.
Photo: R. Sharpley.

Figure 10.7 Entrance to an air-raid shelter.

on the site of the original one (Figures 10.7 and 10.8). There are also plans to construct a walkway from the local train station at which young trainee pilots would arrive at Uzurano.

Significantly, the Uzurano Heritage Tourism Promotion Committee has also been created to manage the future development of the site. Comprising, amongst others, representatives from local transport providers, Kasai City's tourism committee, the local residents' council and Uzurano Heritage guide volunteers, it is seeking ways of encouraging peace education through developing tourism at the site, with guides telling the stories surrounding the airfield's role in the kamikaze operations. At the same time, however, it is also exploiting the site for other touristic activities, such as cycling and hot-air ballooning. On the one hand, this may increase the economic benefits of tourism accruing to the area; on the other hand, it may serve to trivialize visits

Figure 10.8 Reconstructed gate house.

to what is arguably one of the more important yet still largely unknown sites related to the kamikaze phenomenon in Japan. That is, the original runway and other structures represent an authentic and, undoubtedly, thought-provoking and emotive setting to not only commemorate the pilots who lost their lives but also to present the multiple stories surrounding the Pacific War in general and the kamikaze in particular. It is, for example, impossible to walk along the runway without imaging the pilots taking off on the first leg of their final journey, whilst the numerous shelters and bunkers may easily become spaces for the honest, wide-ranging and imaginative interpretation and explanation of the Pacific War and the decisions leading up to the authorization of the kamikaze strategy. And this would complement the already vital role played by local volunteer guides, mostly drawn from the local community, who keep alive the stories of the airfield and the pilots from the perspective of the community and who, importantly, are committed to promoting peace through those stories. As noted earlier, however, a commitment to presenting these multiple stories or heritages (not simply commemorating the pilots, for this more than adequately undertaken elsewhere) is required amongst all stakeholders if Uzurano airfield is to fulfil its potential as a dark tourism site to promote peace.

Conclusion

Much controversy continues to surround the manner in which Japan confronts its difficult history relating to the Pacific War. The differing perspectives on approaches to this history, manifested in various museums around the country, is widely debated in the literature although typically, the representation and interpretation of events varies from pro-war to strong anti-war messages (Yoshida, 2004). As discussed in this chapter, however, the kamikaze phenomenon and its commemoration and interpretation at the Chiran Peace Museum, the country's principal site dedicated solely to the kamikaze, undoubtedly follows the former perspective. That is, through its dominant theme of heroic sacrifice, it conveys an explicit message of not peace rather than war but peace through war. In so doing, it fails to confront not only the multiple realities of the kamikaze phenomenon in the past, but also the realities of contemporary conflicts. Consequently, it is missing the opportunity to allow both domestic and potentially increasing numbers of international tourists to reconcile with a difficult past as well as to offer a more contemporary peace message. Alternatively, the potential exists to fulfil this 'peace through tourism' role at Uzurano airfield, where the authentic surroundings provide a context for visitors to engage more objectively with the multiple stories of the Pacific War and the kamikaze. However, it remains questionable whether consensus can be achieved on presenting such stories.

Acknowledgement

The authors would like to acknowledge the valuable contribution to this chapter made by the students in the Dark Tourism Study Group, Faculty of Tourism, Wakayama University, specifically with regards to the Uzurano airfield case study: Katayama Toyohito, Ito Soichiro, Shintani Honoka, Iwayasu Ryosuke and Koike Natsumi.

References

Allen, M. and Sakamoto, R. (2013) War and peace: War memories and museums in Japan. *History Compass* 11(12), 1047–1058.

Ashworth, G. and Hartmann, R. (2005) *Horror and Human Tragedy Revisited: The Management of Sites of Atrocities for Tourism*. New York: Cognizant Books.

Ashworth, G. and Isaac, R. (2015) Have we illuminated the dark? Shifting perspectives on 'dark' tourism. *Tourism Recreation Research*, 40(3), 316–325.

Axell, A. (2002) The kamikaze mindset. *History Today*, 52(9), 3–4.

Axell, A. and Kase, H. (2002) *Kamikaze: Japan's Suicide Gods*. Harlow: Pearson Education

Bar-Siman-Tov, Y. (2004) Introduction: Why reconciliation? In Y. Bar-Siman-Tov (Ed.), *From Conflict Resolution to Reconciliation*. Oxford: Oxford University Press, pp. 3–9.

Batten, B. (2009) The Myall Creek Memorial: History, identity and reconciliation. In W. Logan and K. Reeves (Eds), *Places of Pain and Shame: Dealing with 'Difficult Heritage'*. London: Routledge, pp. 82–96.

BBC (2017) How Japan's youth see the kamikaze pilots of WW2. BBC News. Available at: www.bbc.co.uk/news/world-asia-39351262 (accessed 3 February 2018).

Blanchard, L. and Higgins-Desboilles, F. (2013) *Peace Through Tourism: Promoting Human Security Through International Citizenship*. London: Routledge.

Bowman, M. and Pezzullo, P. (2009) What's so 'dark' about 'dark tourism'? Death, tours, and performance. *Tourist Studies*, 9(3),187–202.

Buckley-Zistel, S and Schaefer, S. (2014) *Memorials in Times of Transition*. Cambridge: Intersentia Ltd.

Cano, L. and Mysyk, A. (2004) Cultural tourism, the state and the Day of the Dead. *Annals of Tourism Research*, 31(4), 879–898.

Chiran (2017) *The Mind of the Kamikaze*. Minamikyushu-shi: Chiran Peace Museum.

Costello, J. (1982) *The Pacific War 1941–1945*. New York: Harper Collins

D'Amore, L. (1988) Tourism: A vital force for peace. *Tourism Management*, 9(2),151–154.

Foley, M. and Lennon, J. (1996) Editorial: Heart of darkness. *International Journal of Heritage Studies*, 2(4), 195–197.

Friedrich, M. Stone, P. and Rukesha, P. (2018) Dark tourism, difficult heritage and memorialisation: A case of the Rwandan Genocide. In P. Stone et al. (Eds), *The Palgrave Handbook of Dark Tourism Studies*. London: Palgrave Macmillan, pp. 261–289.

Gurler, E. and Ozer, B. (2013) The effects of public memorials on social memory and urban identity. *Procedia: Social and Behavioral Sciences*, 82, 858–863.

Hall, C.M., Timothy, D. and Duval, D. (2004) Security and tourism: Towards a new understanding. *Journal of Travel & Tourism Marketing*, 15(2–3), 1–18.

Inuzuka, A. (2016) Memories of the Tokko: An analysis of the Chiran Peace Museum for kamikaze pilots. *Howard Journal of Communications*, 27(2), 145–166.

Jeans, R. (2005) Victims or victimizers? Museums, textbooks, and the war debate in contemporary Japan. *The Journal of Military History*, 69(1), 149–195.

Kelman, I. and Dodds, R. (2009) Developing a code of ethics for disaster tourism. *International Journal of Mass Emergencies and Disasters*, 27(3), 272–296.

Light, D. (2007) Dracula tourism in Romania: Cultural identity and the state. *Annals of Tourism Research*, 34(3), 746–765.

Light, D. (2017) Progress in dark tourism and thanatourism research: An uneasy relationship with heritage tourism. *Tourism Management*, 61, 275–301.

Litvin, S. (1998) Tourism: The world's peace industry? *Journal of Travel Research*, 37(1), 63–66.

Lollis, E. (2014) Peace as a destination: Peace tourism around the world. In C. Wohlmuther and W. Wintersteiner (Eds), *International Handbook on Tourism and Peace*. Klagenfurt: Drava, pp. 294–309.

MLIT (2012) *Tourism Nation Promotion Basic Plan*. Ministry of Land, Infrastructure and Transport. Available at: www.mlit.go.jp/common/000234920.pdf

Moufakkir, O. and Kelly, I. (2010) *Tourism, Progress and Peace*. Wallingford: CABI.

Nelson, J. (2003) Social memory as ritual practice: commemorating spirits of the military dead at Yasukuni Shinto shrine. *Journal of Asian Studies*, 62(2), 443–467.

Nitobé, I. (1908) *Bushido: The Soul of Japan*. Ebook available at: www.gutenberg.org/files/12096/12096-h/12096-h.htm (accessed 2 December 2018).

Palmer, C. (1999) Tourism and the symbols of identity. *Tourism Management*, 20(3), 313–321.

Poria, Y. (2007) Establishing cooperation between Israel and Poland to save Auschwitz Concentration Camp: Globalising the responsibility for the massacre. *International Journal of Tourism Policy,* 1(1), 45–57.

Richter, L. (1983) Tourism, politics and political science: A case of not so benign neglect. *Annals of Tourism Research*, 10(3), 313–335.

Rifai, T. (2013) Foreword. In C. Wohlmuther and W. Wintersteiner (Eds), *International Handbook on Tourism and Peace*. Klagenfurt: Drava, p. 11.

Sakamoto, R. (2015) Mobilizing effect for collective war memory. *Cultural Studies,* 29(2), 158–184.

Seaton, A. (2001) Sources of slavery – Destinations of slavery: The silences and disclosures of slavery heritage in the UK and US. *International Journal of Hospitality and Tourism Administration*, 2(3–4), 107–129.

Sharpley, R. (2009a) Shedding light on dark tourism: An introduction. In R. Sharpley and P. Stone (Eds) *The Darker Side of Travel: The Theory and Practice of Dark Tourism.* Bristol: Channel View Publications, pp. 3–22.

Sharpley, R. (2009b) Dark tourism and political ideology: Towards a governance model. In: R. Sharpley & P. Stone (Eds), *The Darker Side of Travel: The Theory and Practice of Dark Tourism.* Bristol: Channel View Publications, pp. 145–163.

Sharpley, R. and Friedrich, M. (2016) Genocide tourism in Rwanda: Contesting the concept of the 'dark tourist'. In G. Hooper and J. Lennon (Eds) *Dark Tourism: Practice and Interpretation*. London: Routledge, pp. 134–146.

Sharpley, R. and Stone P. (Eds) (2009) *The Darker Side of Travel: The Theory and Practice of Dark Tourism.* Bristol: Channel View Publications.

Sheftall, M. (2005) *Blossoms in the Wind: Human Legacies of the Kamikaze*. New York: NAL Caliber.

Sheftall, M. (2008) Japanese war veterans and kamikaze memorialization: A case study of defeat remembrance as revitalization movement. In J. Macleod (Ed.), *Defeat and Memory: Cultural Histories of Military Defeat in the Modern Era*. Houndmills: Palgrave Macmillan, pp. 154–174.

Smith, L. (2006) *Uses of Heritage*. London: Routledge.

Stone, P. (2006) A dark tourism spectrum: Towards a typology of death and macabre related tourist sites, attractions and exhibitions. *Tourism: An Interdisciplinary International Journal*, 54(2), 145–160.

Stone, P., Hartmann, R., Seaton, A., Sharpley, R. and White, L. (2018) *The Palgrave Handbook of Dark Tourism Studies*. London: Palgrave Macmillan.

Tunbridge, J. and Ashworth, G. (1996) *Dissonant Heritage: The Management of the Past as a Resource in Conflict*. Chichester: John Wiley & Sons.

Wohlmuther, C. and Wintersteiner, W. (2014) Tourism and peace: An emerging field of action and research. In C. Wohlmuther and W. Wintersteiner (Eds), *International Handbook on Tourism and Peace*. Klagenfurt: Drava, pp. 17–28.

WTO (1980) *Manila Declaration on World Tourism*. Madrid: World Tourism Organization.

Yoshida, T. (2004) Whom should we remember? *Journal of Museum Education*, 29, 2–3.

11 Whaling heritage and tourism development – 'sliced, diced and boiled down'

Simon Wearne

Introduction

For an industry which throughout history and for many countries has been both celebrated and of much economic and cultural significance (Ellis, 1999; Hawley, 1958), whaling has become highly controversial. The relationship between humans and whales, while scientifically close, might once have been described as a grand contest, one which pitted humans against whales; nowadays, it is an activity that pits humans against humans across cultural divides, not least because of widespread contemporary support for the protection of whales. Not surprisingly, then, contemporary whaling is the focus of much media interest whilst, in particular, Japan's continuing support for and engagement in commercial whaling not only stirs passionate argument but also creates a flash point in international debates.

In essence, the controversy surrounding Japan's engagement in whaling can be 'boiled down' to entrenched positions amongst key players. On the one hand, there exists an international community of criticism which generally, or through targeted activism, opposes Japan's whaling. On the other hand, within Japan there are whaling communities both past and present, as well as whaling companies, industries and political leaders, who are either respectful and supportive of whaling or openly promote the killing and consumption of whales. Thus, it is perhaps inevitable that both sides in this debate seem resigned to permanent disagreement. For example, key commentators at an International Whaling Panel Discussion in 2015 predicted that there would be no change to this impasse; indeed, in his concluding summary of the proceedings, the panel moderator, Nobuyuki Yagi, Associate Professor at the University of Tokyo Graduate School of Agricultural and Life Sciences, conceded the issue would remain unresolved with all parties continuing to 'agree to disagree'.

This impasse is exemplified by the Sea Shepherd Conservation Society's (SSCS) abandonment in 2017 of both of its signature campaigns, namely, its Southern Ocean anti-whaling campaign and its Taiji Cove protest (code named 'Infinite Patience') against Japan's annual dolphin hunt at Taiji in Wakayama Prefecture. This has not only resulted in what might be referred to 'activist impotence' amongst anti-whaling groups and organizations (see SSCS, 2020; and AFD, 2019) which have been frustrated by their inability to stop whaling and dolphin hunting (Kozaki, 2017) but also, since then, transformations in the form of dialogue creation or shifts in policy have seemed increasingly unlikely. What was once referred to as 'difference of opinion' by the Australian Prime Minister Malcolm Turnbull during his Tokyo visit of December 2015 might now be better described as a difference of culture, with a kind of 'expert impotence' in evidence amongst both national governments and international specialist commentators.

It is against this background that this chapter proposes a means of confronting what might be considered by some as both a difficult past and a difficult present for Japan through the development of tourism based upon the country's whaling heritage. More specifically, it describes an ongoing project being undertaken by the author to assemble and collate both tangible and intangible elements of Japan's whaling heritage, particularly that related Taiji, from the 17th century a whaling town located in Wakayama prefecture, and to garner publicity and regional support for the ultimate objective of developing a tourism experience that offers an accessible understanding and balanced explanation of whaling in Japan past and present. In so doing, the project seeks to address the contemporary impasse in the whaling controversy within a framework of tourism and cultural sustainability.

As part of the project, the history of whaling in Japan and, specifically, in Taiji is explored in depth. This history is summarized in this chapter and, subsequently, the development of the project and its outcomes at the time of writing are discussed. First, however, contemporary positions and debates regarding whaling in Japan are highlighted to provide the background against which the project is being undertaken.

Japanese whaling: entrenched positions

The contemporary impasse in debates over the state of Japanese whaling and attitudes towards it can be considered critically through the following example; the brief history of Japanese whaling outlined later in the chapter adds further context to and understanding of the distinct differences in attitudes that are revealed in this example.

On Friday, 11 December 2015, The Institute for Cetacean Research (ICR) organized a discussion panel in Kanagawa comprising leading international experts on the topic, which was attended by the author of this chapter. Working under the banner of *Revisiting Japan's Whaling policy – An Overseas Perspective*, it had the following agenda:

> It is over 30 years since the International Whaling Commission (IWC) passed the so-called 'commercial whaling moratorium'. Since then, in order to lift the moratorium, Japan has been collecting scientific data through research involving whaling and has supported moves to create

> a Revised Management System (RMS) which contains a procedure for setting conservative catch limits for commercial whaling. Japan also played a leading role in an initiative aimed at normalizing the workings of the IWC, which is widely regarded to be a dysfunctional international institution. What should be done to achieve the 'sustainable use of whale resources based on science' and what is behind opposition to commercial whaling? In this panel discussion, four specialists (and a moderator) in international fishery/environmental policy will discuss the future of Japan's whaling policy from an overseas perspective. (unpublished conference proceedings handout)

The panel members were Gavin Carter, from the UK but based in the US, Carlos Mazal from South America, Eugene Lapointe of France and Glenn Inwood from New Zealand. Collectively, their skillset was impressive and comprehensive, embracing: research and education in marine policy, law of the sea and environmental economics; advisory aspects related to Japan's whale research programmes; governance of public policies and the sustainable use of marine resources; and, communication and media. The moderator was the aforementioned Yagi from the University of Tokyo.

The stated objective 'what is behind opposition to commercial whaling' was covered by keywords 'international ecopolitics'. Throughout the discussion there seemed to be reticence about this key issue. This might have been diplomacy given that the discussion was being hosted by the ICR, 'a unique organization in Japan specializing in the biological and social sciences related to whales' (ICR, n.d.) Arguably, it would have been more useful to engage in open, accurate and critical dialogue, addressing issues of central significance to an audience that included informed observers. Indeed, despite 2.5 hours of expert presentation, the discussion by the panel revealed nothing new. There was no media critique, despite the media being widely acknowledged as a key factor in what is behind opposition to commercial whaling. Opposition to Japan's whaling was identified by the moderator in summary as a key point throughout the discussion; however, as noted in the introduction to this chapter, no agreement was likely to be achieved.

Whilst the ambiguous outcomes of the panel discussions in this example are evidence of the continuing impasse in the whaling debate in general and Japan's whaling in particular, a number of other possible contributory factors can be suggested, as now considered.

Anti-whaling (activism) and anti-anti-whaling (alternative activism)

Current responses and strategies by Japan tend to be passive and defensive, and only become more assertive when faced with external criticism. The country appears to lack the will or confidence to proactively defend itself when challenged about its whaling activities. This may be due to a limited understanding of how the story could be framed positively; equally, it might reflect the recognized widespread lack of knowledge amongst Japanese people of their own whaling history and tradition and how it is entwined with Western whaling history – hence the objective of the project discussed in this chapter. In fact, background research for the project revealed traditional Japanese whaling heritage to be a major asset. Activists, on the other hand, are uncomfortable with the notion of tradition, as became evident from the vitriolic responses in early 2012 to an exhibition of photographs (installed in mid-December by the author) at the Ishigaki Art museum in Taiji with the unambiguous title 'Tradition Taiji'. One aim of the exhibition was to gauge responses while unequivocally stating through creative works that tradition was actual and relevant. An interesting next stage in this interplay of activism and alternative activism was the apparent disengagement by the anti-whaling movement. There was immediate counter claim (Scott West – personal communication, Taiji, 18 December 2011), but no follow-up, perhaps an indication that there is simply nothing more for anti-whaling activists to say, as suggested in an email to the author from an SSCS representative.

Ineffective activism, PR failures and misinformation

Two things are evident in the broader work being undertaken by the author beyond this chapter. First, current activism is not working and second, there have been PR failures, misinformation and a lack of understanding both inside and outside Japan. With regards to the former, dialogue is absent and change is non-existent despite an energetic and expensive international campaign by SSCS called 'Cove Guardians' targeting Taiji, which ran for about 13 years. With the SSCS organization in a state of disarray and near collapse in the US in 2015/2016 and with a court case successfully prosecuted by Japan to exclude them from their annual flagship campaign, the Southern Ocean operation, it is clear that SSCS is facing significant challenges. For example, on the eve of opening day of the 2016/17 dolphin hunting season in Taiji, SSCS leader, Paul Watson, was forced to address speculation that his efforts were failing (Watson 2016). By the opening day of the 2017 hunting season, Sea Shepherd's 'Infinite Patience' campaign had run out: 'Sea Shepherd cannot in good conscience send others (volunteers) into a situation where their freedom and rights are threatened' (Watson, 2017).

Nevertheless, one last act of defiance by an independent activist caused international embarrassment for Japan at the beginning of dolphin hunting season in Taiji. On 1 September 2016, the Japanese coast guard interrogated a protestor who they had apprehended on a jet-ski on the sea in front of the cove. Two officers then held up a banner which read 'Save Dolphins', and photographed the evidence with the cove as the backdrop. At the same time, the activist secretly filmed the encounter from the same angle and promptly distributed it to international media. The coastguard officers had apparently checked everything on the activist's jet-ski and on request, all items other than the video were handed over. The video was subsequently posted on the SSCS

Facebook page (Paul Watson Facebook site, 1 November 2016). The activists considered this day a victory. From an objective viewpoint, however, this incident progressed the cause of neither side; the only winner was the media. Moreover, the annual dolphin hunt in Taiji continues, and the resolve of the fishers and Japanese officials is hardened.

With regards to the latter factors, Japan's public relations in general continues to be found wanting on the whaling issue, which could be handled much more effectively with clear, confident, knowledgeable and comprehensive interpretation. The three main locations for display of Kumano whaling history, Taiji, Kushimoto and Wakayama City, are poor by international standards; in fact, this is admitted by staff at two of the three locations (personal communication: Taiji, January, 2012; Kushimoto February, 2020). Leadership, particularly in Wakayama Prefectural Government and Central Government in Tokyo, continues to be unable to respond adequately, avoids discussion and is resistant to consultation (personal communications with the Governor's Secretariat and officers of the Tourism Promotions Division, Wakayama Prefectural Government).

More specifically, Japan has been widely criticized for labelling Antarctic whaling as research but has not been able to offer a strong defence. This is arguably the country's most significant whaling-related PR failure. Undoubtedly, whaling is at the forefront of scientific understanding of the Southern Ocean and any research by Japan for whales, or indeed any other international research in Antarctica, contributes to closing the gap in our knowledge of ocean science in general and climate science in particular. For that reason, Japanese whale research can be considered by no means pointless. Yet, at a time when it was heavily criticized in the mass media, research collaboration between 20 countries including Japan and Australia in the Antarctic for *The Biogeographic Atlas of the Southern Ocean, 2014* to which Japan was a major contributor (iAtlas, n.d.) goes unreported (personal communication with Michael Stoddart[1]). In addition, it could be argued that 'research hunting' of whales is necessary for Japan because, unlike its greatest critics, the United States and Australia, the country has faced food security issues numerous times in its history and understands the value of whale as a food/protein source. In this context, and as an example of misinformation, the species hunted by Japan are not designated as threatened and are hunted under sustainable quotas (BBC, 2019). The view that Japan hunts all whales is simply untrue.

Negative perceptions and impact on tourism

As a consequence of ineffective PR and misinformation, Japan and Wakayama Prefecture in particular suffers from negative perceptions of whaling. This is potentially manifested in lost business in both tourism and more generally for a region that is highly attractive historically, culturally and environmentally. Indeed, Wakayama's Prefectural Tourism Office actively discriminates against Taiji's tourism market by suppressing the town's history (it is omitted from Wakayama Tourism Brochure for Wakayama Prefecture, 2011), thereby impacting negatively on domestic visitation. Moreover, by not encouraging international media to visit Taiji and stipulating that journalists who seek cooperation from the department in reporting Wakayama's attractions must not mention whaling (personal conversation with visiting European media, 2018), international tourism is also being suppressed. In so doing, not only are Japanese people being denied the opportunity to discover, learn about and celebrate a more complete story of their country's whaling heritage, but also international visitors who might be equally enthusiastic to learn about heritage are being denied the opportunity. Moreover, it is arguably only right that a more balanced picture is provided of whaling in Japan to counter the more familiar negative story presented in the media. Fundamental to that is a fuller understanding of the history of whaling and specifically, Japanese whaling in Taiji. Exploring the history of international whaling from 1812 to 1878, when it was dominated by the Americans and British, provides insight for the purposes of this chapter and a narrative that is not widely known.

Japanese and international whaling: a brief history

Two key themes, emphasized in this brief historical review are firstly, the American demand for whale oil and other whale products and the dominance of America in whaling from the early 19th century. Second, there are issues of resistance to that dominance, philosophically, environmentally and politically from the same time to the present day. It is also important to state from the outset that there exists a constant misconception and inconsistency between what is written about Japanese whaling in contemporary descriptions including media reporting (e.g., Denyer, 2019a,b) and what can be found by more detailed inquiry. Specifically, the assertion that pre-1868 Japanese whaling was 'artisanal' and, therefore, pre-industrial or non-industrial cannot be supported; scholarly work by Nakazawa (2010) and Hawley (1958) reveals that the whale catching and processing practiced during Edo era Japan(1603–1868), beginning in Taiji from 1606 (Hashiura, 1969), was the basis for the island nation's first complex system of manufacture. In other words, a resource, in this case, the whale, was caught, processed and distributed, with various parts of the whale being distributed and further processed for an equally wide variety of purposes. The entire operation was undertaken under strict regulations of Tokugawa Japan, adhering to a deep ceremonial and spiritual formality and with a methodical division of labour in the sequences of the hunt from preparation to return, and the subsequent dissecting, rendering, recording, accounting and distributing the materials derived from the whale resource. This history is supported in the fine text by Hawley (1958) and brought to life in the illustrative woodcut/block prints in his book (see Figures 11.1 and 11.2).

Figure 11.1 Spine of Frank Hawley's *Miscellanea Japonica II, Volume 1, Whales and Whaling in Japan.*

Photo: S. Wearne.

Figure 11.2 'The Harpooners Carouse'.

Photo: S. Wearne, with permission from the trustees of Hawley's estate.

Figure 11.3 Sekobune.
Photo: S. Wearne.

Fundamental to this early whaling was the chaser boat, *sekobune*, claimed to be the fastest human powered wooden boat in the world (see Figure 11.3). This sleek whale chase boat, capable of matching the speed of fast whales which were not previously catchable, was a key piece of equipment in getting the net around the whale and subduing to exact a clean kill. The attributes of the *sekobune* and its crew were unrivalled anywhere in the world and were a key factor in the relationship the Edo whalers had with the whale. While European whalers were 'sitting ducks' as they rowed their heavier boats near the whale, the Japanese chased and moved skilfully at speed with the whale. Hence, the notion of the whale as a foe, while prevalent in European texts, is not the way the Japanese regarded the whale. With impressive performance, superior functionality and specialized visual impact, this craft is a key object in the interpretation of the story of Japan's whaling. To understand the *sekobune* in its traditional role is to grasp the very essence of whaling in Japan and appreciate this technology in a contemporary context. According to Tokugawa traditions based on Shinto and Buddhist beliefs, the departure to catch, the taking of the whale with a solemn communication to the animal and the procedure of bringing it to shore, was marked at every stage with ceremony and respect. In Taiji, this all concluded on the beach in front of the shrine next to a freshwater spring. A whale provided survival for seven villages and gratefulness and respectful sorrow was more common sentiment in the Japanese whale hunt (Wakayama Prefecture, 2016).

Whale war

In tracing relevant history of whaling in the 19th century, a constant theme recurs. The notion of war, a whale war, battle and contest is ever present. For example, some accounts of pre-modern Japanese whaling refer to the military precision and coordination of boatmanship (Hawks, 1856); it is easy to imagine, when reading of the methods employed, that the teamwork and discipline required is analogous to that of a 'sea army' as described in C.W. Nicol's historical fiction 'Isana' (Nicol, 1988). In fact, when contemplating a strategy for defending Japan against foreign intrusion in the 19th century, the feasibility of a floating force based on the skill, speed and manoeuvrability of the *sekobune* crews was considered (Hashiura, 1969; Hawks, 1856).

Words such as battle, conquering, the idea of killing or being killed have long accompanied descriptions of whaling and the analogy of war against the whale has been present for centuries. Humans doing battle with the whale runs through storylines from the tale of Jonah and the Whale in the Old Testament and Moby Dick (Melville, 1950 (1851)) and news media where humans were depicted standing 'victorious' on top of a dead whale. In Kato (2015), we see how reporting bloody scenes of the whale hunt became off-putting to audiences, first with the inception of colour film newsreels in the 1950s and later, with colour TV.

Research reveals seven key events informing both contemporary and historic perceptions of war and whaling for Japan. First was the Matthew Perry Expedition of an American Squadron to Japan to secure re-supply for whaling (see Figure 11.4). This was 'gunboat diplomacy' with the largest

Figure 11.4 The coming of the Black Ships (kurofune).
Photo: S. Wearne.

Figure 11.5 Whale Wars DVD covers. 'The second coming of The Black Ships'. **Photo:** S Wearne.

floating force of firepower ever taken from one country to another in 1852 (Hawks, 1856).

Fifty years later, Japan defeated the Russians in 1904 and secured a ship through the spoils of war which became the first Japanese factory whaling ship (Ellis, 1999). Then the US Great White Fleet made a 14-month tour of the globe from 1907 to 1909 as a demonstration of naval power (McKinley, 2017). The War Time Civil Control Administration (1942) (Taiji History Archive, personal access 2011), and General Douglas McArthur – GHQ Tokyo and Brisbane, representing the victorious United States in the Pacific after Second World War gave Japan special permission to hunt whales with whaling harpoon guns to feed the starving country (Hawley, 1948). Next, the 1960's witnessed a period known as 'The Whaling Olympics', during which a multinational fleet of factory ships took more whales than at any time in history under the IWC 'Olympic System' (Japan Whaling Association, n.d.). Finally, emerging from the war between Greenpeace and SSCS, was Animal Planet's *Whale Wars,* a 6-year run of a factually flawed but financially profitable TV series depicting Sea Shepherd's 'war' with the Japanese whalers in the Southern Ocean (Figure 11.5).

Japanese whaling in the 19th century

The history of whaling in Japan during the 19th century contrasts starkly with the contemporary perceptions of Japan's whaling activities. American domination is a theme and, for the purposes of this chapter, the story begins in 1812 when the Perry name became famous in the United States. The naval achievements of Oliver Hazard Perry, the older brother of Commodore Matthew Calbraith Perry (who was to play a significant role in opening Japan up to the West through the 1854 Treaty of Kanagawa), became legendary during the Battle of Lake Erie in the 1812 War with Britain. An American naval commander from Rhode Island, he was born 23 August 1785 and died on the same date in 1819. His battle flag read, 'Don't Give Up the Ship' and his decisive victory at Lake Erie not only neutralized the British expansion in North America / Canada, but also assisted in defeating a coalition of Native Americans who were trying to defend their land. Victory was sealed in 1813 and so was any prospect of the dominion of Native Americans over their land.

With the collapse of Atlantic whale stocks, the Pacific had become the new target in the late 18th century although it took more than 30 years for the Northwest Pacific to be discovered (Ellis, 1999: 158). By 1816, trans-Atlantic trade was flowing with cheap British imports, creating a challenge for US-made goods. In 1817, the Rush-Bagot Treaty (or Rush-Bagot Disarmament) saw peace and stability restored and in the following year, The Treaty of 1818 between the US and Britain delivered agreement on territory boundaries and fisheries between the US and British North America (Canada). Subsequently, by 1820, Britain and the US were whaling together again. Hawley (1958) cites 'the terser phrasing' from the logbooks of the first two ships to be 'on Japan', in the late Northern Spring and early Summer of 1820 (exactly 200 years before this chapter). The *Maro* of Nantucket took 2,425 barrels of sperm oil and the *Syren* of London took 2,768. East of Tohoku, 36N and 168E, Captain Joseph Allen of the *Maro* 'saw no ships on the coast but plenty of whales'. This marked the beginning of the end for the traditional industrial whaling of Japan. Moreover, it was the strongest indication that swarming around Japan were countries which would, with attitudes of superiority, prevail on and in Japan whether Japan liked it or not. The Japan Ground was about to be 'fished' out between 1820 and 1840, and Japan had no say in it. Oil (from whales) was big business and demand was driven by growth, with fossil fuel/hydrocarbon-based oil not being discovered until 1859.

Exploding Wests and Japan's whales

On 15 March 1820, Maine became the 23rd state of the US and on 24 April and the Land Act of 1820 reduced the price of land in the Northwest Territory and Missouri Territory, establishing favourable conditions for Americans and others to settle in the west. Demand grew for goods such as oil for lamps and lubrication, soap and baleen (from the Baleen whales) which was used in the days before plastics. Combs, umbrellas and, famously, corsetry which was becoming fashionable at the time were also based on whale products. Exponential growth of population and economies followed, a process referred to as 'Exploding Wests' in James Belich's (2009) description of the boom and bust cycle which accompanied the settling of North America. Edo Japan remained closed to the outside world but it didn't stop a growing frustration from whalers.

In 1831, 11 years after whaling began around Japan and 20 years before Matthew Perry was to arrive, Bourne Russell, an English-born Australian mariner arrived in Japan from Sydney (Jones, 2008). Russell's actions to gain cooperation from resistant Japanese on Hokkaido, while lacking diplomacy, were decisive, single-minded and evidence of a right which he regarded as universal and one which must be understood by the Japanese. He claimed to come in friendship but also in great need, and relations turned aggressive when government troops labelled Russell and his party as 'the enemy'. This is hardly surprising when the letter he intended for The Emperor states 'I burnt their villages, and took what I wanted… I hope all my countrymen will do the same, burn and destroy all your towns and villages that refuse that Hospitality which is due to every Man' (Jones, 2008). Russell threatened that all his countrymen were well capable of this and that even 'your City of Jeddo' (Japan's capital of Edo – present day Tokyo) would not be exempt. He then stated that his intended purpose was to procure an order from the Emperor to permit foreign ships to, 'enter any of your Ports, and be supplied with such refreshments as they want, and they will pay for all they get, and should it be mutually advantageous to Trade, you must allow your subjects to do so' (Jones, 2008). In his letter to The Emperor, dated 15 April 1831, Russell offered conditional respect, signing himself as 'Your Obedient Servant an Englishman'; this 'obedience' is consistent with the way Western whalers came to Japan.

The increasingly besieged and isolated Japanese found their supply of whales being diminished by a multinational fleet estimated to number 760 ships (Ellis, 1999) which operated in the Japan Ground in this period. In particular, the grey whales along the Korean and Japanese coasts were decimated (Ellis, 1999: 244) As Russell's account demonstrates, the 'inconvenience' of operating without access to Japan was rewarded with a relatively short but obviously fruitful period for Western Whalers. Japanese mariners were increasingly the meat in the sandwich, which caught between the outside world and Edo Japan policy, was fast becoming a whaleburger with the lot, and diplomatic, economic, industrial and cultural clashes were inevitable.

In his history-based fictional account *Harpoon* set in the 1840s, Nicol (1988: 15) writes that the whalers of Taiji have observed a western whale boat on their coast. 'Are there no whales on their own shores? The Barbarians must never learn these waters'. The Taiji people knew little or nothing of how the whale migrations worked beyond their shores and could not have imagined how widespread the depletion of whale stocks was in other regions (Maury, 1853); and in *Harpoon*, a young man is convinced the western technique with large ships must be superior, but the wisdom of the 'Tuomi', the Taiji whalers' look out and operations observer, retorts:

> No Shinsuke, our way has evolved to be the best and only way for us. Let the foreigners do what they do, their way, but for us it is not only stupid, but dangerous to dream about copying them.
>
> (Nicol, 1988: 57)

By the 1840s, Western pressure was building to open Japan. James Biddle made the first official attempt by the US to negotiate with the Japanese at Edo Bay (Tokyo) in July 1846 but failed (Blumberg,1985: 20). Also by 1840, the Japan Ground was exhausted. America and Europe redirected their whaling efforts to the Sea of Ohthosk north of Japan and back to the Southern Hemisphere. The years 1847 to 1865 witnessed 1391 voyages to the Sea of Okhotsk. 426 whales were taken including 341 Right Whales (the prized species in Taiji). The other species taken were Bowhead whales and these were almost totally eliminated during this period. The total voyages to the Bering Sea at the same time were much higher. The Bering Sea was considered fished out by 1854 with the catch rate in just 2 years, 1849 and 1850, yielding 450,530 barrels of oil and 3,067 tons of whalebone. Taiji whalers relied on the migration from the North to the warmer waters around Japan. The effect of this unsustainable foreign whaling was devastating for traditional whaling towns like Taiji.

Melville's prophetic call

During the 1840's, Herman Melville was undertaking research for his book *Moby Dick* by moving from ship to ship, travelling with whale hunters. Completed in 1849, Moby Dick was his sixth book; it was published in October 1851 in London, in three volumes titled *The Whale* (a month later in the US the now better-known title of Moby Dick). The message from Melville, in some editions situated in the fly leaf, is revealing:

> The moot point is, whether Leviathan can long endure so wide a chase, and so remorseless a havoc; whether he must not at last be exterminated from the waters, and the last whale, like the last man, smoke his last pipe, and then himself evaporate in the final puff.
>
> (Melville, 1950: 457)

Melville concluded that the human-induced extinction of the buffalo in the expanding North America was evidence enough to believe that the whale 'cannot escape speedy extinction'.

He acknowledged an opposing view of the time (around 1850) that the ocean is more than double the landmass of the globe and that the whales have places, such as under the Arctic ice, to seek refuge or hide. Similarly, he referred to the hunting of elephants over centuries gone by and as they were still surviving, so it must follow for whales.

However, Melville (1950: 458) observes that in the period he is writing (i.e., after the depletion of whale populations of the Atlantic and North West Pacific), 'some philosophers of the forecastle have concluded that this positive havoc has already very seriously diminished their battalions'. Melville comes down hard on humans and is optimistic about whales in his final paragraph of chapter 105:

> Wherefore, for all these things, we account the whale immortal in his species, however perishable in his individuality. He swam the seas before the continents broke water; he once swam over the site of the Tuilleries and Windsor Castle, and the Kremlin. In Noah's flood he despised Noah's Ark; and if ever the world is to be again flooded like the Netherlands, to kill off its rats, then the eternal whale will still survive, and rearing upon the topmost crest of the equatorial flood, spout his frothed defiance to the skies.
>
> (Melville, 1950: 459)

Melville could not have specifically known of the impending climate change; however, his suggestion that man will bring about his own end is prophetic and insightful. Melville's sensitive observations of the impact of hunting on whaling grounds he visited over a 20-year period stands as a unique and very early call by an environmentally sensitive inquirer. It is also worth remembering that Alfred Russell Wallace and Charles Darwin's 'theory of evolution' had been proposed less than 20 years earlier.

Meanwhile, South Pacific whaling relied on resupply in the vulnerable island nations of the area or friendly Australian ports of Hobart, Melbourne and Sydney. Hobart, Tasmania, had traditionally supported sealers and the American Whaling Industry by supplying fresh provisions. It was an active port throughout the 1850's. A painting by Henry Gritten (who lived in Hobart in 1856) hangs in the Australian Maritime Museum (author's observation)

Gold vs. whales

This section reveals, for the first time, a connection between gold being discovered in Ballarat, Australia and Perry's Expedition, nicknamed 'The Coming of the Black Ships', in Japan because of the colour of Perry's vessels. The California Gold Rush occurred in 1849 during a dynamic historical period. Plotting historic detail from 1850 to 1853 reveals a plausible explanation of the next historic event to play out in Japan. Perry acknowledges in his journal that Californian Gold and the Australian Gold Rush in 1851 made these two places significant in the world of wealth creation and, hence, the demand for whale-based products. However, the Australian and, in particular, Ballarat rush of that year was significant for a different reason.

Blainey (1987, first published 1966) describes the situation of whale ship crews abandoning ship during resupply in Australia. In a later text (Blainey,1969), he details the events of 1851and the predicament facing the owners of whaling ships visiting Australian Ports as their crews made the more lucrative and, to them, less risky decision to speculate on gold rather than the life threatening prospect of hunting whales with dwindling supply in increasingly dangerous whaling grounds. Many crews absconded at this time, including a ship's captain. Observers commented on 'a forest of masts' in Port Phillip, the Melbourne port close to the Ballarat Gold Fields (Patterson, 2013).

The whaling company owners needed to get their fleets out of the South Pacific. The sense of determination by the US to 'open' Japan had never been greater. Perry refers to the whaling industry power brokers in the introduction to his journal when he mentions 'the unanimous and urgent appeals of our commercial men' to prosecute a successful opening of Japan (cited in Hawks, 1856). This was front page news and the dominant print media of the time was excited.

In Gleason's Pictorial Drawing-Room Companion of 12 February 1853 (a newspaper published in Boston), a strong Christian message resonates in the words of an author (unknown) from an article about the Perry Expedition and its important 'intercourse' with Japan:

> The whole force under the command of Commodore M.C. Perry, will form a very powerful fleet, able to cope with any resistance that may be offered... we know Yankee grit and Yankee spirit too well to doubt the result.

A listed description followed with the words:

> an effective force of 4000 men, and above 300 guns, mostly heavy ordinance. The steamers are each to mount a couple of Paixham shell-guns, of the largest calibre, and placed on revolving trucks, so as to sweep the horizon.

On reflection, these final words are chilling and consistent with the notion of the 'Whale War'. The treaty of Kanagawa followed in 1854 and although the US was focused on gaining access and support for its whaling activities, Japan became a willing participant in a process which lead to industrial, economic and social change after an arrested beginning to Industrial revolution already underway in other parts of the world. This was a key moment in a relationship based on economic issues and alliances which the US has since dominated (Morikawa, 2009).

From 1854, Japan is open to the US, agreement with Russia follows (Hawks,1856) and the whale hunt continues at the same unsustainable rate that decimated stocks the Atlantic in the preceding century. Consequently, the Japanese traditional coastal whaling industry collapsed; the declining number of whales in the 40 years between 1835 and 1875 brought to an end over 270 years of sustainable industrial whaling. The closing scene for Taiji is a risky and calamitous final hunt in 1878. A desperate decision is made by a group run by Taiji Kakuemon, against the better judgement of the other Taiji group of Wada Kinemon (Hashiura, 1969) to chase a female Northern Right Whale (Semi Kujira) and its baby. A typhoon hits the fleet and destroys all but a few boats and kills over 100 men with only a handful of survivors making it to the safety, picked up by a passing boat. Another boat landed at Atawa (Sakurai, 2011) and 1 remaining boat with many dead and just 11 still alive hit the rocks on Kozashima. Eight survivors made it to the beach at Kozashima (author's research, Kozashima – Temple records). This event is known as Seminagare (the accident) and its tragedy resonates to this day.

Whaling heritage and tourism in Japan: creating a symbiotic relationship

A number of key themes have emerged from discussion thus far in this chapter. First, Japan is confronted with a difficult present (and what some might erroneously consider to be a difficult past) with regards to its whaling activities. Specifically, the country continues to receive a negative press for its continuing engagement in whaling, although it is not the only country to do so. Both Norway and Iceland practice whaling, the latter exporting much of its whale meat to Japan (Hogenboom, 2015), as do other communities, such as the Faroe Islanders. However, these negative attitudes often reflect misunderstanding on the part of critics. Second, Japan has a long history of whaling which was closely regulated and an important cultural practice; it was, in every respect, sustainable. Third, the sustainable whaling industry in Japan was, in effect, destroyed by the large-scale whaling undertaken by the US and Britain in the 19th century which initially decimated whale stocks around Japan and subsequently those in more northerly seas that would migrate south to Japan. It is these countries, amongst others, that are the forefront of criticism of contemporary Japanese whaling. Finally, international efforts to resolve the whaling 'problem' have reached an impasse.

Clearly, then, there is a need for a 'circuit breaker' to move the debate forward. As indicated in the introduction to this chapter, the author has for a number of years been developing a project that seeks to establish a more mutually beneficial relationship between Japan's unique whaling heritage and tourism. That is, it seeks to focus on whaling as a tourism asset that promotes appreciation of whaling culture. This is inevitably a controversial proposal given the conflicts and debates surrounding whaling, yet one that aims to present a creative and contemporary interpretation of the country's whaling history and heritage while developing awareness to neutralize contemporary arguments and debates. As such, it aims to provide context and critique without ignoring the challenges, the overall purpose being to confront the negative perceptions of whaling amongst international tourists to Japan whilst potentially influencing those observers of Japan who may currently choose to stay away in protest. At the same time, it may well offer Japan the opportunity to set the record straight with visitors by providing a proactive tourism-based approach to a subject which continues to be considered by government agencies as particularly negative in the tourism context. Such a project is timely, not least given Japan's drive to increase significantly the number of international tourists it attracts and, with its hosting of major events, The Summer Olympics and Expo 2025, when the eyes of the world will be on the country.

Developing the project

The project to create a tourism product around Japan's whaling heritage is not a formal, funded project, and nor has it followed a more traditional route typically followed in academic research, although not only has much research been undertaken into the culture and history of Japanese whaling within broader context of the county's international relations since the mid-19th century but also it is contributing to a PhD that I, the researcher, am undertaking. My professional background is in cinematography and documentary production and, as a cinematographer/researcher, most people do not believe I am not making a film; indeed, after more than 10 years of stating this as the project has progressed, many people still expect me to drop my disguise and make a film.

The real asset as a researcher in the context of this 'story' is that I am a media professional with broad experience, insights and knowledge of the industry. It is a fact that the subject and the setting offers a great opportunity for film making, and I would be delighted with that as one outcome in the long-term, but the goal is to assemble the heritage first. Only then is it possible for this research to underpin the approach of such a project. As the project has developed, my involvement in documentary and television programs and print media has been as interviewee, advisor and commentator. This process has been useful in building a network to enhance research, sharing information, creating a public profile for the work, reporting developments and contributing to knowledge (as with this chapter) and ultimately reciprocating by communicating findings and ultimately reporting the outcome when research is completed.

In addition to extensive research into the historical and related background of whaling in Japan, a key element of the project has been, and continues to be, the active process of locating, assembling and/or restoring tangible and intangible elements of whaling heritage that are currently missing or dilapidated and seeking ways to demonstrate how this can communicate across cultural divides. The *sekobune*, discussed above, is a key object of translation with, on the one hand, its role in representing history and ceremony and, on the other, its contemporary charismatic role in attracting us to a sustainable way of thinking and living (Figure 11.6).

The graphic design of each boat is deeply connected to story but also works on a different level as a graphic brand. Highlighting this early in the research prompted an initiative by the Taiji History Archive curatorial team to research and record 40 of these designs in the 'Whale Boats: Styles and Designs' exhibition in December 2011. Net whaling embodies the notion of net work, and this can be creatively appropriated to consider the ambiguity of net work and network, not just in its traditional context but in contemporary ways to stimulate imagination, communication and translation. I have produced five photographic exhibitions, including the 'Tradition Taiji'

Figure 11.6 Single sekobune, profile.
Photo: S. Wearne.

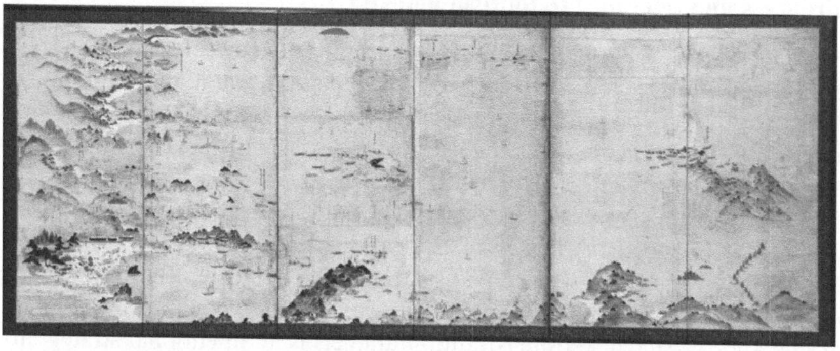

Figure 11.7 Six panels of Kishū Kumano-ura Hogei-zu (Whaling in the Kumano Sea).
Photo: S. Wearne, with permission from The Wakayama Prefectural Museum.

exhibition (2011) mentioned earlier in this chapter and three major public art projects, which have reconnected a 'network' of creative and craft based collaborators.

As a film maker, I am constantly looking for powerful visual story-telling assets and the research site has yielded many useful and engaging objects. One such example is the Byobu; rather like a 'film' from the Edo era, the six panels of Kishū Kumano-ura Hogei-zu and several other works from the same era, reveal evocative images consistent with written accounts of the whaling of Edo Japan (Figure 11.7.)

The remnants of a scroll in the Kakimoto Shrine in Kainan, Wakayama is another significant work. Both were digitized as an initiative of this project, and these images combine to create some of the strongest impressions in the resurgent narrative while eliciting an emotional response which is positively charged with respect for Japan's whaling rather than the derision of the dominant existing media discourse.

Overall, the aim of the project is to increase awareness and understanding amongst domestic and international visitors through an exhibition of both my own acquisitions and also significant existing materials combined for an improved interpretive display, including restored heritage and a living museum. Indeed, adapting my professional practice of cinematography and documentary making, I propose to conceptualize interpretive display and present a combined narrative that for the first time brings together significant but previously isolated works from Hawks (1856), Hawley (1958), Nicol (1988) and Nakazawa (2010), as well as from further research of Blainey (1987, 2003), Boyce (2008), Belich (2013), Kalland (1993) and Takemura (2010) and the significant whaling history collections of the Hashimoto family (the original Taiji whaling family) and the significant Klaus Barthelmess collection.

Progress and relevance to tourism industry

In this project, as in any cinematographic process, story is the essence of success. People need a story to engage with while, adjacent to the main subject narrative, the generic, perhaps drier topics of triple bottom line sustainability, habitat protection, climate change strategies, environmental, economic and social management and development strategies are discussed.

In June 2015, I was invited to speak at an annual conference organized by UNESCO and the Ministry of Education, Culture, Sports, Science and Technology which was held in Wakayama City. I presented the work under my working title, Tradition Taiji. This name and its accompanying graphic design and registered website (traditiontaiji.net) is an intentional strategy to test and develop an associated brand. Encouragingly, significant progress is being made. For example, now widely reported in general media and included in my Tedx Himi talk 'Tradition Taiji',[2] the proposal for UNESCO Industrial Heritage recognition for Taiji's whaling is gaining support. There is interest and enthusiasm founded on a grass roots base, from a range of secondary and tertiary education institutions, cultural entities, such as museums and libraries, as well as representatives of local politics and in senior levels of government and national institutions.

In addition, partnerships are forming to share the research to date with the community resulting in unprecedented assessment of 'the missing' (Hashiura, 1969) Taiji family text and artefacts. This prompted The Wakayama Prefectural Museum to assess these artefacts for the first time in November 2015 and resulted in an exhibit by government directive, 'Exhibition on Traditional Whaling of Taiji' (10 November to 6 December). The Prefectural Fisheries Agency and even the Police have requested access to research outcomes and seek to share ideas in an effort to broaden their own perspective.

However, the process is not always straightforward and can be one of two steps forward and one step back. For example, Wakayama Prefecture, through its tourism agency, listed Wakayama's whaling heritage as National Heritage in April 2016. Yet this did not prove to be adequate protection; just 2 months later, in June 2016, unique industrial heritage that I had fully documented was destroyed in the town of Taiji. Efforts to recover this loss continue.

Nevertheless, progress continues to be made towards achieving the project's long term objectives. One proposal, for *sekobune racing* (whale chaser propelled by eight people using the sculling oar – *Ro*), an event highlighting the spectacle, the skill, the spiritual element and superior technology of the *sekobune* and its role in net whaling and drive hunting, aims to raise awareness and educate both domestic and global audiences positively; at numerous public speaking opportunities the audience response to this proposal has been overwhelmingly positive. Indeed, a similar proposal I put forward for a Japanese theme and presence at a major bi-annual event in Tasmania in 2011 (The Australian Wooden Boat Festival) was accepted. Consequently, a Japanese *hachibune* (eight *Ro* wooden boat), along with an interpretive display, living museum, food, festivals, ceremony and restored heritage stimulated respect for the inherent sustainability ethic and wisdom on display, raising the self-esteem of those who identify with the culture presented.

More generally, the process of discussing and obtaining feedback on the project began in 2008 and led to a concerted effort, intrinsically linked to the research, to secure items of heritage. Hundreds of items, artefacts and (to date) 11 wooden boats connected to net fishing and whaling, as well as items from the fishing villages seasonal food production, have been acquired and are in storage. These objects of everyday life of coastal communities form a tangible connection to culture and inform intangible heritage and a new narrative of an old story swept aside in the period of industrial revolution of the 19th century. This is a story with strong indigenous knowledge themes linked to the first human settlement of Japan by the Jomon 'sea people' and requires a fresh (in contemporary history) yet ancient approach. A consultative and inclusive dialogic encounter (Black, 2009) can bring change. What form this will take depends on the will of stakeholders to tackle the 'messy business' of whaling. Taiaiake (2015) puts a convincing case in his 'resurgence' thinking, which he outlines in the following model:

Rooted:	'Rootedness is the land'
Accountable:	'Accountability is Community'
Transformative:	'Transformation is Culture'

In this resurgence structure, you are required to 'willingly put yourself in that messy situation' (Taiaieke, 2015).

Conclusion

Probably, the most obvious characteristic of this work, apart from its unique consideration of the Japanese position, is the critical depiction of the US. I strive for an objective view as an Australian feeling empathy with Japan, as both countries are junior partners in alliance with the US. Compared to Western whaling, Japanese whaling history, combined with an omnipresent indigenous knowledge, provides a reasonable and more appealing sustainability narrative. There exists a convincing argument, developed in this project, to contest the accepted conventional wisdom in the West that Japan's whaling is bad, cruel and barbaric. The media-driven narrative repeats this and positions Europe, the US and Australia as advanced, sensitive, sophisticated and civilized societies. Yet, Western opinion is vulnerable as it reveals intolerance, an apparent ignorance of facts and fails to respect 'others'. Japan could move to counter this in a subtle, proactive effort to bring balance and mitigate damage from sustained negative media coverage which damages investment prospects for tourism business (Baral, Baral & Morgan, 2004). Japan does not deny the past, as is evident, for example, at Wakayama History Museum, but it has so far failed to find a way to communicate the story, in any language, to its advantage (Personal communication at the Ministry of Foreign Affairs, 1 October 2016).

What this project has revealed thus far is how Japan stands against globally dominant views on whaling amidst a polarized debate which can be addressed through communicating a sustainable development narrative encouraging thought and action aligned with the core ethic and principles contained in Japanese cultural heritage of the Edo era whaling. Specifically, the birthplace of Japan's traditional whaling, Taiji in Wakayama Prefecture, provides rich content, including both tangible heritage in the form of physical places, spaces and objects and graphic image branding (discussed above) and intangible heritage in the form of traditional practice and knowledge, memory and storytelling through songs, dances and ceremonial processes. Although progress is being made in combining these into a narrative, the tangible features are in a declining state, vulnerable to natural disaster and requiring urgent action. Similarly, intangible assets also languish.

Nevertheless, the current 'cultural deficit' can be reversed if Japan can emerge with a solid ethic, based on resilience, to reclaim an enduring industrial heritage story and demonstrate it as an agent for change in the form of a tourism asset. In other countries, such as Australia (Albany, Tangalooma) and the US (New Bedford), the transition from whaling to whale tourism has been a success base on a historical narrative. Certainly, Japan has its own remarkable places (Mukojima-Taiji), objects (*sekobune)* and practice (net whaling and drive hunting), whilst Japan's food culture, including whale meat, (formally recognized by UNESCO as Intangible Cultural Heritage) celebrates traditional cuisine and respectfully offers to share it with the world. Taiji's whaling was designated as National Heritage in 2016. Building on this base and given the likely need to reinvigorate tourism post-Covid-19, the opportunity exists for Japan to present a tourism product based on a sustainable past practice which will educate and contributing to the development of culturally based sustainable tourism in the region while offering a more nuanced, respectful and knowledge-based perspective on the whaling 'problem'.

Notes

1 Personal communication with Michael Stoddart, former chief scientist of the Australian Antarctic Division, led the Census of Antarctic Marine Life in 2007 and launched the Biogeographic Atlas of the Southern Ocean (26 August 2014).
2 Available at: www.youtube.com/watch?v=qmWc_Wvykzk&feature=youtu.be

References

AFD (2019) Action for dolphins. Available at: www.afd.org.au (Accessed 1 January 2020).

Australian Antarctic Magazine (2014) Australia-Japan meeting to build Antarctic science collaboration. Available at: www.antarctica.gov.au/magazine/2011–2015/issue-26-june-2014/in-brief/australia-japan-meeting-to-build-antarctic-science-collaboration (Accessed 24 April 2020).

Baral, A., Baral, S. and Morgan, N. (2004) Marketing Nepal in an uncertain climate: Confronting perceptions of risk and insecurity. *Journal of Vacation Marketing*, 10(2), 186–192.

BBC (2019) Japan resumes commercial whaling after 30 years. BBC News, 1 July, Available at: www.bbc.co.uk/news/world-asia-48821797 (Accessed 23 April 2020).

Belich, J. (2013) *Replenishing the Earth*. Oxford: Oxford University Press.

Black, C. (2009) *The Land Is the Source of the Law*. London: Routledge.

Blainey, G. (1969) *The Rush that Never Ended.* Carlton: Melbourne University Press.

Blainey, G. (1987) *The Tyranny of Distance*. (First published 1966) South Melbourne: Sun Books.

Blainey, G. (2003) *Black Kettle and Full Moon*. Camberwell: Viking.

Blumberg, R (1985) *Commodore Perry in the Land of the Shogun.* New York: Harper Collins.

Boyce, J. (2008). *Van Diemen's Land.* Collingwood: Black Inc Publishing.

Denyer, S. (2019a) *In Japan's 'City of Whales,' An Uncertain Future After Last Antarctic Hunt*. Washington Post, 5 April. Available at: https://www.washingtonpost.com/news/world/wp/2019/04/05/feature/japan-whaling-research-hunts-end-and-defiant-commercial-era-begins/ (Accessed 27 April 2020).

Denyer, S. (2019b) *Japan Still Hunts Whales. But Now It's More Profitable to Watch Them.* Washington Post, 20 December. Available at: www.washingtonpost.com/world/2019/12/20/japan-still-hunts-whales-now-its-more-profitable-watch-them/?arc404=true (Accessed 24 April 2020).

Ellis, R. (1999) *Men and Whales.* Guilford: Lyons Press.

Hogenboom, M. (2015) Why do some countries still hunt whale? BBC, Earth Blog, 3 December. Available at: www.bbc.co.uk/earth/story/20151203-why-do-some-countries-still-hunt-whales (Accessed 27 April 2020).

Hashiura, Y. (1969) *Whaling at Taijura: A Series of Scrolls.* Tokyo: Heibonsha Ltd.

Hawks, F. L. (1856) *The Narrative of the Expedition of an American Squadron to the China Seas and Japan, Performed in the Years 1852, 1853, and 1854.* Washington: The Congress of the United States.

Hawley, F. (1948) *Japanese Whaling in Antarctic- Expedition's Return.* 4 April 1948 Report to the Times of London.

Hawley, F. (1958) *Miscellanea Japonica II, Volume 1, Whales and Whaling in Japan.* Kyoto: Kawakita Printing Co.

iAtlas (n.d.) *The Biogeographic Atlas of the Southern Ocean.* Biodiversity. Available at: http://atlas.biodiversity.aq/index.html (Accessed 26 April, 2020).

ICR (n.d.) Institute of Cetacean Research. Available at: www.icrwhale.org/eng-index.html (Accessed 25 April 2020).

Japan Whaling Association (n.d.) *History of Whaling.* Available at: www.whaling.jp/english/history.html (Accessed 25 April 2020).

Jones, N. (2008) *North to Matsumae: Australian Whalers to Japan.* Crawley: University Western Australia Press.

Kalland, A. (1993). *Japanese Whaling? End of an Era.* Surre: Curzon Press.

Kato, K. (2015) Australia's whaling discourse: Global norm, green consciousness and identity. *Journal of Australian Studies*, 39(4), 477–493.

Keith, J.A. (2011) *Civilization, Race and the Japan Expedition's Cultural Diplomacy, 1853-1854.* Vol. 35, No. 2, (April, 2011) The Journal for the Society for Historians of American Foreign Relations. Massachusetts: Wiley Periodicals, Inc.

Kozaki, D. (2017) Sea Shepherd permanently abandons Antarctic whale wars with Japanese boats. *ABC News*, 29 August. Available at: www.abc.net.au/news/2017-08-29/sea-shepherd-abandons-antarctic-whale-wars/8851890 (Accessed 1 January 2020).

Maury, M.F. (1853) *A Chart Showing the Favourite Resort of the Sperm and Right Whale Series F A Part of Maury's 'Wind and Current Charts, Series A-F.* Library of Congress. Available at: www.loc.gov/maps/?fa=contributor%3Amaury%2C+matthew+fontaine&all=true&dates=1852

McKinley, M. (2017) Cruise of the Great White Fleet. Available at: www.history.navy.mil/research/library/online-reading-room/title-list-alphabetically/c/cruise-great-white-fleet-mckinley.html (Accessed 27 April 2020).

Melville, H. (1950 (1851)) *Moby Dick or The Whale.* Toronto: Random House.

Morikawa, J. (2009). *Japan's Whaling. Power, Politics and Diplomacy.* London: Hurst.

Nakazawa, S. (2010) *Don Pur de la Nature.* Tokyo: Kodansha.

Nicol, C.W. (1988) *Harpoon.* New York: Signet.

Patterson, G. (2013) *Coastal Guide to Nature and History Port Phillip Bay.* Briar Hill: Coastal Guide Books.

Sakurai, H. (2011) *The Last Harpooner.* Taiji: Taiji Whale Museum.

SSCS (2020) Sea Shepherd Conservation Society. Available at: https://seashepherd.org (Accessed 1 January 2020).

Taiaiake, A. (2015) A Talk by Alfred Taiaiake: Research as Indigenous Resurgence. November 12, 2015. Carleton University's Centre for Aboriginal Culture and Education.

Takemura, M. (2010). *Whaling Culture.* Paper presented at Sapporo City University. (Translation H. Clerk from the Japanese original: Hogei mondai-no rekishi shakaigaku) Melbourne: Trans Pacific Press.

Wakayama Prefecture (2016) Living with Whales. Association for the Preservation of the Kumano Sea Whaling Culture, Wakayama Prefecture Tourism Promotion Division. Available at: https://kumanonada-nihonisan.jp/en/ (Accessed 25 April 2020).

Washington Post, 5 April. Available at: www.washingtonpost.com/news/world/wp/2019/04/05/feature/japan-whaling-research-hunts-end-and-defiant-commercial-era-begins/ (Accessed 27April 2020_

Watson, P. (2016) Sea Shepherd returns to Taiji, Japan; New challenges, new strategies. Huffpost, 31 August. Available at: www.huffpost.com/entry/post_13026_b_11783624 (Accessed 27 April 2020).

Watson, P. (2017) Sea Shepherd must adapt to more effectively defend dolphins. Sea Shepherd Commentary. Available at: www.seashepherd.org.uk/news-and-commentary/commentary/sea-shepherd-must-adapt-to-more-effectively-defend-dolphins.html (Accessed 24 April 2020).

12 Debating sustainability in tourism development

Resilience, traditional knowledge and community: a post-disaster perspective[1]

Kumi Kato

Introduction

In the aftermath of the East Japan Earthquake and Tsunami that literally shook the nation on 11 March 2011, the pace of recovery in the affected areas has varied. Japan is an earthquake-prone country and small tremors are a daily occurrence, but this was of an unprecedented scale. Along the 500-km coastline and its hinterland, 22,010 people were confirmed dead or missing (FDMA, 2016) and, even 6 years after the event, almost 120,000 people were still living away from their homelands (Reconstruction Agency, 2017), including those affected by the nuclear reactor meltdown at the now defunct Fukushima Daiichi Nuclear power plants. The extent and consequences of the damage to the ocean, air and land, and to human and ecological health is still unknown.

The Tohoku region covers nearly 20 percent of the nation's land, 68 percent of which is forested. Although its population is approximately 10 percent of the entire nation, primary industry production ranks fairly high on the national scale, including rice (37.7 percent), grains and vegetables (18.9 percent), forestry (15 percent), cattle farming (18.9 percent) and fisheries (17.9 percent) (MLIT, 2011). The region's coastal areas therefore are also mostly rural, and many of those affected are primary producers.

Vision policies and statements put forward by the central and regional governments all emphasize the importance of building a 'disaster resilient society', where people can live in a safe environment and that the scope of this effect should go beyond the reconstruction phase (e.g., Miyagi Prefecture, 2011). However, the nature of 'resilience' is not clearly articulated, nor how and by whom it might be built. Inevitably, the focus of reconstruction is given to hard construction, including tsunami walls, business centres and infrastructure. For example, Rikuzen-takata city, where 10 percent of the population (2,000) was lost in the tsunami, is recreating its central business district (CBD) on the original site which has been raised 15 m. However, the concept of 'shopping centre and CBD', according to a community elder, 'did not exist in rural living, where life evolved within the small communities scattered in various locations, in the mountains, on the hillsides or by the sea'.[2] The reconstruction policy does not necessarily meet local residents' views, and sometimes rather goes against their wants and needs. Ageing and population decline already occurring in the rural regions further exacerbates this gap.

In the initial recovery phase, various kinds of tourism activities were initiated, including those utilizing traditional local knowledge and historic stories related to disasters. This chapter argues that the community's resilience is located in such traditional knowledge, and is revealed in the recovery phase, which should be capitalized on to further develop the region. After defining resilience-based tourism as a fundamental form of sustainable tourism relevant to rural communities, the chapter proposes roles for tourism in post-disaster situations, and the much wider context of community development, including as a means of facilitating more diverse lifestyles for younger generations.

Resilience and sustainable tourism

Sustainability as part of systems theory is understood as an equilibrium of economic, social and environmental dimensions, and is now recognized as the fundamental principle and premise for all human activities. Among the diverse qualities of sustainable tourism, its capacity to facilitate a process of transformation socially and personally is noted (Butcher, 2008; Hunter, 1997; Reisinger, 2015). An increasing focus is given to the inter-relatedness of social and environmental sustainability, and ecological, social and community welfare (Ceballos-Lascurain in Higgins-Desbiolles, 2008; Cleverdon & Kalisch, 2000; Goodwin, 1998; Higgins-Desbiolles, 2010, 2008; Lanfant & Graburn, 1992; Minnaert, Maitland & Miller, 2007, 2009). Concepts such as *peace through tourism* (Blanchard & Higgins-Desbiolles, 2013; D'Amore, 1988) and *justice tourism* (Kassis, 2006) promote ideas of social, cultural and ecological justice, and attributes such as building 'solidarity between visitors and those visited; promotes mutual understanding and relationships based on equity, sharing and respect; supports self-sufficiency and self-determination of local communities, and maximizes local economic, cultural and social benefits' (Scheyvens, 2002: 104). The idea of 'co-transformative learning and syncretic growth' is also contained in the concept of *hopeful tourism* (Pritchard, Morgan and Ateljevic, 2011: 957–958). These qualities relate to resilience as discussed below.

Resilience as a co-transformative process

The process of transformation to a sustainable community has been recognized as a process of complex systems change, requiring community will, resources and institutional support (Carlsen, 1999; McLennan, 2011; McLennan et al, 2012; Pavlovich, 2003, 2014). It is also identified that the evolution and transformation of communities in rural areas, whose industries are predominantly natural resource-based, can be supported by tourism (Biggs, Hall & Stoeckl, 2012; McLennan et al, 2012; Miller, 2007; Popesku & Hall, 2004). For such communities' transformational process, valuing innate community resilience is the key.

Resilience, originally a dynamic systems model in mathematical science adopted to ecology, refers to 'the ability to maintain biological relationships through environmental stress' (Lew, 2014: 15). It is 'the capacity of a system to absorb disturbance and reorganize while undergoing change so as to still retain essentially the same function, structure, identity, and feedbacks' (Walker et al., 2004), and the systems here refer to the social–ecological systems (Adger et al., 2005; Folke, Colding and Berkes, 2003; Holling, 1973, 2001; Nabhan, 1997; Pricope, 2010; Walker & Salt, 2006). It is the system's capacity (an individual, a forest, a city or an economy) not only to deal with changes and continue to develop, but to use shocks and disturbances to trigger renewal and innovative thinking. Resilience thinking 'embraces learning, diversity and the belief that humans and nature are strongly coupled to the point that they should be conceived as one social- ecological system' (Stockholm Resilience Centre, 2014). The system here is conceptualized as a community, rather than an individual.

Community resilience in rural communities: social capital, reciprocity with people and land

In the specific context of disaster recovery, Cox and Perry (2011) provide a useful definition of community resilience. Community resilience is:

> the capability of a community facing a threat, survive and bounce back or, perhaps more accurately, bounce forward into a normalcy newly defined by the disaster losses and changes… . [It is] a reflection of people's shared and unique capacities to manage and adaptively respond to the extraordinary demands on resources and the losses associated with disasters… . Adaptive capacities should include the capacity to develop and maintain social capital as it is expressed through a sense of belonging, a sense of community, place attachment and participation in civil society.
>
> (Cox & Perry, 2011: 395–396)

For rural and natural resource-based communities, life evolves around the land and seas and their social capital is also place-based. Social capital is strengthened by the sense of reciprocity as a mutual obligation, responsibility and affinity in surviving the hardship. Delaney (2016) argues that social capital and sense of reciprocity are critical ingredients for social sustainability. It is also the case that reciprocity is acknowledged and expressed not only between people and community, but also between people and their place or land. Daskon (2010) demonstrates that culture and traditional values strengthen livelihood resilience, and adaptation comes from within through dynamics which are specific to values of the people. It is further argued that a multipurpose adaptive system 'not only encourages individual economic achievements in a community, but also safeguards the embedded values of their societies' (Daskon, 2010: 1097). This leads to 'an evolving complex system that co-adapts to the specifics of the particular place, and especially to the aspirations and values of local people' (Farrell & Twining-Ward, 2005: 110). Here, sense of connection with land and the community is the core of community resilience, which may be expressed in various forms, including local spiritual beliefs and religious worship, in return defining a place-based identity. As Diaz and Dayal (2008, p. 1174) suggest, 'the most catastrophic impact of natural disasters is an individual feeling of loss of place' – it is as Cox and Perry (2011: 396) define, 'a collective dislocation of community and belongings'.

The importance of 'place' as a vital linkage of social capital, resilience and identity has, however, been addressed by only few. The land-based community, its identity and its sense of place are the strengths that support their resilience, which relates to traditional faith and belief systems, as referred to in some of the post-disaster scenarios (Béné et al., 2016; Masten & Narayan, 2012; O'Grady et al., 2016; Walsh, 2007). For the Japanese, the traditional rural agricultural landscape has a strong spiritual component, both in specific beliefs (land spirits), and in the unity and balance among its components, both tangible and intangible. This land-based spirituality, supported by social capital and sense of reciprocity, is an important part of resilience, which is also closely related to traditional ecological knowledge (TEK).

Resilience and traditional ecological knowledge

TEK encompasses the knowledge, practices and beliefs developed across generations by particular traditional cultures (Berkes, 1993, 1999; Johannes, 1989), associated with particular individuals or groups and their land. It is the range of expert knowledge, wisdom and skills gained from living with a particular land and seas over a long period of time, which may be expressed in folk beliefs, rituals, festivities, monuments, oral histories and folktales, various cultural expressions, artefacts, literature and art (Kato, 2007a, b). The concept is useful, giving value to the human–nature connection of those whose livelihood is intricately connected to the natural world regardless of their ethnic or cultural background. In Japan, TEK may be referred to as '*minzoku-chi* (folklore knowledge)', '*densho-chi* (inherited knowledge)' or '*Jinbun-chi* (humanities knowledge)' and there have been studies in folklore and religious studies (Machida, 2003; Miyata, 1996; Nishigai, 2011; Nomoto, 2010; 2013a, b; Nomoto & Akasaka, 2013; Sasaki, 2006; Shinohara, 1995). They relate to concepts studied extensively in conservation management, *satoyama* and s*atoumi* (Duraipappah et al., 2012; Koike, 2014).

A critical point is that there is an underlying ethical foundation to TEK and its various expressions, which may be in the form of gratefulness, awe or fear towards the natural world (Akimichi, 2012, 2013; Hirakawa, 2012), and since 3.11, an increasing number of studies have noted the importance of human knowledge, senses and other intangible qualities (Akimichi, 2013; Comfort & Okada, 2014). In other disaster contexts, Fukao (2015) found the reconstruction of a Hindu temple being an essential part of the reconstruction of the Uyyalikuppam community in India after the 2004 Indian Ocean Tsunami, and Ishii (2016) identifies the vital role of a 'rain ritual' in reconstructing community resilience in a Northern Chinese village following a series of major social disturbances. TEK forms a 'collective social memory', which allows reflection on the past and redirection of the future, evaluating the essential values of life, which is a fundamental quality of resilience. It also plays an important role expressing local distinctiveness through a community's initiatives, creativity and leadership, which in return creates interactive and participatory opportunities in tourism.

Tourism for sustainable community development: a framework for development

Resilience is applied to tourism by Farrell and Twining-Ward (2004, 2005), who state that sustainable development must 'be viewed as an evolving complex system that co-adapts to the specifics of the particular place, and especially to the aspirations and values of local people' (Farrell and Twining-Ward, 2004: 110). For a sustainable community development through tourism valuing local initiative and benefit, community-based tourism (CBT) provides a broad-embracing framework (Ruiz-Ballesteros, 2011; Sharpley, 2000, 2009; Telfer & Sharpley, 2008). CBT is defined as tourism 'developed in local communities in innovative ways by various individuals and groups, small business owners, entrepreneurs, local association and governments' (Sharpley, 2009: 124). It aims for a community's 'empowerment and ownership, conservation of natural and cultural resources, social and economic development, and quality visitor experiences' (Hiwasaki, 2006: 677), ensuring improvement of socioeconomic conditions of those socially vulnerable (Hatton, 1999). CBT has a strong imperative of social justice, respect for local culture, heritage and traditions, and community empowerment, which clearly relates to the resilience and sustainability discussed above. This situates tourism as a development that is holistic, sustainable, locally beneficial and creative based on social–ecological systems.

Such tourism development with community resilience as a foundation can be part of resilience planning as suggested by Lew (2014: 14), who argued it to be 'a potentially more effective approach to community planning and development than the sustainability paradigm'. Lew points out that resilience has the capacity to adapt to change by attempting to build capacity to return to a desired state following both anticipated and unanticipated disruptions, whereas sustainability mitigates or prevents change by maintaining resources above a normative safe level. Further, resilience is an ongoing process of development, rather than a fixed quality or a state held by an individual or a group (Luther, 2003). Community-based sustainable tourism thus can be conceptualized as part of resilience planning.

Traditional knowledge for sustainable tourism development: cases from Iwate, Japan

With the above concept in mind, cases of community tourism development in the early recovery phase that utilizes traditional knowledge, and thus considered as opportunities for resilience development, are now discussed. They are found in the communities of Iwate in the north Tohoku region (Sanriku coast area) of Japan.

History of earthquakes and tsunami in the region

Sanriku is a stretch of coast covering Miyagi prefecture in the south and Aomori prefecture in the north. The word Sanriku originated from the former three provinces which existed in the region until the Meiji land reform in late 19th century. The region has been overwhelmed by a number of earthquakes and tsunami. Existing records date back to the *Teikan Earthquake and Tsunami*[3] (13 July 869), *Keicho-Sanriku*[4] (2 December 1611), *An'ei* (3 May 1774), *Meiji-Sanriku* (15 and 21 June 1896), *Showa-Sanriku* (3 March 1933) and the tsunami caused by the Chile earthquake (23 May 1960). More recently, the *Miyagi-oki* (12 June 1978) and *Iwate-Miyagi* (14 June 2008) earthquakes shook the region, although they did not cause tsunami. Road signs along the coast mark the 'Estimated Tsunami Inundation Area' based on accumulated data. This time, however, tsunami levels exceeded many of these marks.

Tsunami-stones and Sanriku Fukko National Park

The mid-north coast of the region is scattered with so-called tsunami stones (*Tsunami-ishi*), which are the large stones brought ashore by tsunami. They are believed to be the marker specifically warning 'not to build houses below the marked point'. One such example in Tanohata, Iwate, at 25 m above sea level and 360 m from the shore, is believed to have been brought up in the 1896 Meiji-Sanriku Tsunami. In 2011, the water stopped just below the stone, but most of the houses below that point were destroyed. Villagers say they knew about the stone as a warning sign but dismissed it as an old tale. One fisherman, who is the eighth son in his family, explained how all siblings inherited ancestor's land from their parents, who lived at the higher part of the village. 'By the time my turn came (to inherit the land), only the lower part of the land was left. Yes, we knew about the stone, but no one took it seriously, thinking we now have all sorts of technology'.[5]

In the initial recovery phase, this fisherman, along with six other 'boat captains' in the Zappa Boat Adventure Cruise using traditional fishing boats (zappa), were taking visitors to areas inaccessible from the land and telling the stories of the day the wave came. During the cruise, they explain the geographical features of the place, their tsunami experiences, including the story about the tsunami stone. The cruise not only helped fishermen to stay active and connected with their place in recovery, but also allowed them to utilize their skills and knowledge in other forms of economic activities. For visitors, it allowed them to understand how people live with the place, seeing the areas inaccessible from the land and hearing the stories of the day of the tsunami.

In the following years, the area was nominated as *Sanriku Fukko Kokuritsu-koen* (Sanriku Reconstruction National Park[6]) in May 2013, with the aim of promoting the area's geographical and natural features while commemorating the disaster and lessons learned. Part of the area was also nominated as the Sanriku Geopark[7] in the same year, including the legacy of the earthquake and tsunami. Three of the tsunami stones were included alongside the 29 disaster-related sites, such as the single pine tree that survived out of 7,000 trees,[8] and a 'Memorial Park' in Nakanohama, Miyako, which was overwhelmed by the 15 m-high tsunami. An entrance gate and part of the camping facility was preserved, and a 13 m- high observatory was built with materials from destroyed structures, so that visitors can understand the scale of the tsunami that reached the region. The park has now become the central feature of the Green Reconstruction Vision (Ministry of the Environment), including the development of a 700 km coastal trail, *Michinoku Shiokaze Trail* (Sea breeze trail).

Ban-ya at Tsukuehama (Traditional Fishing Sheds) (Tanohata, Iwate Prefecture)

Ban-ya is a wooden shed built close to the shore for fishermen to store their fishing equipment and use as a temporary shelter. Originally started as a base for managing fishing grounds during winter seasons, these sheds later became a temporary shelter where fishermen might stay overnight when necessary, while they had their houses elsewhere away from the shore or on higher ground. Ban-ya are seen in many parts of the fishing villages in Tohoku and Hokkaido regions. On the eastern coastline of Hokkaido, Nishin (herring) Ban-ya was a work place with living quarters for fishermen built during the herring fishing peak from late 1800 to early 1900. Some of them were called Nishin Goten (herring mansion) due to their large-scale and decorative features. Currently, 13 such compounds are recognized as important cultural properties, 10 of which are open for the public as museum displays.

In Tanohata, Iwate, a group of 25 ban-ya existed on the Tsukue-hama beach,[9] built after the 1933 Tsunami. After the improvement of road infrastructure (1978), fishermen stayed there less but the group of ban-ya defined the traditional landscape that was also functional. In 2006, the Ban-ya group was nominated as one of '100 historic and cultural properties to be preserved for future' by the Fisheries Agency and the National Association of Fisheries Infrastructur.[10] With the 3.11 Tsunami, however, all 25 ban-ya were destroyed and washed away. *NPO Tannohata Network*, which also facilitates the Zappa Adventure Cruise and many other ecotourism activities, is now promoting the Ban-ya Reconstruction Project, aiming to restore ban-ya not only to restore their original functions but also as a tourism facility, including display, café and restaurants with traditional-style cuisines, art gallery and activities such as charcoal and salt making. The Tanohata Network was formed by a group of young local residents wanting to promote local traditions, which they refer to as a culture of '*Han-mo-do*', literally 'people who care for the shore'.[11] They facilitate a number of outdoor activities, ranging from nature trekking, sea-shell arts, local cooking and also guided tours of the tsunami-affected areas led by local guides who have local knowledge of the area and experienced the day.

Local Carpentry and Sumita Model *(Sumita, Iwate Prefecture)*

The Sanriku coastline also has a tradition of local carpentry (*kesen-daiku*) using native cedar trees. The town of Sumita, Iwate, some 20 km inland, did not sustain serious damage from the earthquake, but was the first to complete temporary housings; 95 of them were ready to move into in less than 2 months. These houses are of exceptional quality built in the local tradition, now referred to as 'Sumita model'. It is explained that the particular technique using thick timber is energy efficient, comfortable and quiet. Importantly, Sumita was preparing for the government to build these houses in response to the South Asian Tsunami. Instead, ironically and fortunately, the houses were displayed locally.[12] An example of kesen carpentry is found in a rotunda build in a park overlooking the Sanriku coast. The small structure, built by a kesen carpenter, Sakamoto Fumio, includes a small statue of the Sea God dedicated the 'Sea Folks in the region'. At the Kesen Sakan-Sakan Densho-kan (Kesen Traditional Carpentry and Wall-building Museum), visitors can see the technique in detail and the guided by an experienced carpenter who explains details of the techniques used, the way particular timber is used for certain part of the house, and the methods to harvest particular kinds of timber to suit the purpose. On 10 December 2011, an 'eternal fire' was brought to the museum from Kobe Earthquake Memorial Site, which was built on 16 January 2000 to commemorate the Kobe Earthquake (17 January 1995).

Bahan, Horse Logging *(Ryori, Ofunato, Iwate Prefecture)*

Another wisdom found in the region is *bahan* or horse-logging. One fisherman was about to rebuild his houses and sheds lost in the Tsunami. He explained that his timber must be harvested using *bahan*, a traditional technique to retrieve felled trees using horses.

> This way, by not using the vehicle and heavy machines, you will not scar the mountain. Once the mountain is scarred, landslides and floods may happen, which will not only destroy houses and roads, but the mountain itself. Damage in the mountain will also damage the ocean.[13]

There are six *bahan* horse-handlers in the region and one of them, in his early 30s, recently moved back from Tokyo to live in his home region. According to him, *bahan* is far more efficient and productive compared to machine logging, enabling harvesting, particularly by selective logging, from steep hills in the region. In October 2011, he participated in a training seminar and competition held in England, and won the first prize, proving his skill even in the foreign environment with an unfamiliar horse. In the following year, he participated in the European Horse-logging championship, where he achieved seventh place. He says 'language is not an issue between a horse and hander – we communicate with movements and rein'. He has also started a group *Umahito* (horse people[14]) to promote sustainable ways of living – wooden products (furniture, household goods), organic vegetable (use of horse manure), therapy and healing (in pristine environment, activities with horse). The group also conducts seminars and training for people interested in a career related to horses. Educational programs for children and school groups teach about sustainable ways of living and caring for the environment, as well as providing support for young children and families affected by the disaster.

Conclusion: tourism for social and environmental sustainability

This brief review shows that in the days following the disaster, various forms of tourism activities were initiated, later developing into more systematic forms of tourism. This kind of tourism building on traditional ecological knowledge supported the affected communities in several ways. First, it allowed the communities to maintain their connection with their places; it allowed the people with various place connections to use their resources to initiate activities, which provided a vital driving force in early recovery phase. Second, such people–place connections were 'told' in various forms – personal stories, monuments, folktales, literature or arts, many of them related to the history of the place. Third, opportunities to be involved in tourism were available to people without business experience or resources. Story-telling and guided tours of the affected sites became valuable tourism experiences. It was also the case that the interactions between local communities and visitors, many of whom were also relief volunteers, were valuable not only for communities, but also for visitors. Many of them experienced a major awareness shift in life values.[15] Expressions of understanding, appreciation, empathy and compassion were important elements of recovery that supported the building and restoration of resilience, providing a foundation for further reconstruction and development. It is argued that these tourism activities provide an important foundation for the area's sustainable reconstruction and development, as it recognized identities, local history, wisdom, skills and knowledge that have allowed living with a particular kind of environment which has evolved over a long period of time.

Tourism development as a process of community transformation and empowerment may be envisioned as including the following three closely related areas: first, identifying and developing locally distinct features, particularly the traditional wisdom of coping with natural disasters, which can be an important opportunity for social, cultural and economic development in a sustainable manner; second, opportunities to enhance local residents' own appreciation of and connection with the place as the basis for younger generations to settle in regional areas, providing one solution for social, cultural and economic decline of rural areas. It is often the younger generations who take initiatives in conceptualizing new tourism destinations, activities and products and promoting them strategically, as well as bringing new community structures to traditional communities who may have resisted change. Tourism could be a new business for younger generations moving to rural areas in pursuit of a better environment to live, and raise a family. Such a move may be supported by MAFF (Ministry of Agriculture, Forestry and Fisheries) in its scheme to facilitate rural resettlement. And third, recognition of traditional wisdom can promote a transition to a less energy-intense and less materially dependent society, promoting more sustainable use of resources. Most importantly, traditional wisdom should be acknowledged as an expression of ethics towards the natural world that allows us to reflect on the way we live. Wisdom including a sense of awe, respect, fear and gratitude forms a valuable foundation for living more sustainably and ethically, providing sustainable alternatives in lifestyle.

Tourism can provide a positive turn for the communities socially, culturally and spiritually. For this, a 'more ethical, philosophic and reflexive approach to tourism knowledge is needed' (Tribe, 2009). Pritchard, Morgan and Ateljevic (2011: 942) suggest that such knowledge is 'directly related to the challenge of creating a more just and sustainable world'. Sustainability, as Holling (2001: 395) states, is 'the capacity to create, test and maintain adaptive capacity. Development is the process of creating, testing and maintaining opportunity', thus sustainable development refers to the 'goal of fostering adaptive capabilities and create opportunities' (Holling, 2001: 395). Tourism's holistic and creative capacity will be able to foster this, which also asserts tourism's social contribution and responsibility today.

Ethical engagement with a place, contained within traditional knowledge, will be one important foundation for communities in disaster recovery, and in much wider social contexts today as they aim for sustainability. In terms of tourism planning and development, communities need to be inspired and supported to envision a future that is more sustainable and fulfilling for the communities themselves, using powerful new stories as an intrinsic part of the place.

Notes

1 A previous version of this chapter appears in *Tourism Planning & Development*, *15*(1), 55–67.
2 Personal communication (20 December 2013).
3 Japan's earliest record of earthquake is the one on 22 August 416 in Nara region as described in Japan's first Chronology Nishonshoki (c.720), which also refers to an earthquake in northern region (National Astrological Centre, Chronological Scientific Table, 2011) www.rikanenpyo.jp/
4 Sanriku (lit. three coastlands) today refers to the coastlines of Miyagi, Iwate and Aomori. A major toll was recorded in each sanriku earthquake: Keicho-sanriku (1,783), Meiji-Sanriku (22,000) and Showa-Sanriku (3,064).
5 Personal communication, 24 August 2011.
6 An existing national park, prefectural park and two other protected areas were restructured to this national park. Another park was included in March 2015, and further extension plan is in place.
7 As of 2014, 36 locations in the Japan Geopark Network, seven of which are also listed in the Global Network of National Geoparks. Geoparks feature locally distinct culture and industry associated with the geophysical characteristics of the area and grassroots initiatives such as local guides with story-telling and interpretative skills.
8 Named as 'Miraculous Pine Tree'. It however deteriorated significantly, and cut down and replaced with a concrete replica in 2013. The act was controversial.
9 The Tsukue-hama originates from an indigenous Ainu word Tsuku-etsu, meaning a cape of small hill. In Northern Tohoku, like Hokkaido, many place names originate in Ainu language.
10 Available at www.gyokou.or.jp/100sen/index.html.
11 Personal communication, 25 August 2011.
12 Personal communication, 25 August 2011. Sumita Emission ZERO, 2000 (Suminta Town Office, 2000).
13 Personal communication, 25 August 2011.
14 Umahito (horse people) HP http://umahito.com/#.
15 Personal communication, 24 August 2011.

References

Adger, W., Hughes, T., Carl F., Carpenter, S. and Rockström, J. (2005) Social-ecological resilience to coastal disasters. *Science*, New Series, *309*, No. 5737 (Aug. 12, 2005), 1036–1039.

Akimichi, T. (2012) *Foundation of Japanese Environmental Thoughts – Enquiring from Folklore Knowledge (Nihon no kankyo shiso-no Kiso- Jinbunchi karano toi)* Tokyo: Iwanami.

Akimichi, T. (Ed.) (2013) *Environmental History of Japan, 5: Nature Usage and Destruction* (*Kankyo-no Nihonshi, 5 Shizenriyo to Hakai*) Tokyo: Yoshikawa-kobunkan.

Béné, C., Al-Hassan, R. M., Amarasinghe, O., Fong, P., Ocran, J., Onumah, E. and Mills, D. J. (2016) Is resilience socially constructed? Empirical evidence from Fiji, Ghana, Sri Lanka, and Vietnam. *Global Environmental Change,* 38, 153–170.

Berkes, F. (1993) Traditional ecological knowledge in perspective. In Inglis, J. (Ed.), *Traditional Ecological Knowledge: Concepts and Cases.*Ottawa: International Program on Traditional Ecological Knowledge and International Development Research Centre, pp. 1–9.

Berkes, F. (1999) *Sacred Ecology: Traditional Ecological Knowledge and Resource Management.* New York: Routledge.

Biggs, D., Hall, M. and Stoeckl, N. (2012) The resilience of formal and informal enterprises to disaster: Reef tourism in Phuket, Thailand. *Journal of Sustainable Tourism*, 20(5), 645–665.

Blanchard, L. and Higgins-Desbiolles, F. (Eds) (2013) *Peace Through Tourism. Promoting Human Security Through International Citizenship*. London: Routledge.

Butcher, J. (2008) Ecotourism as life-politics, *Journal of Sustainable Tourism*, 16(3), 315–326.

Carlsen, J. (1999) A systems approach to island tourism destination management. *Systems Research and Behavioral Science*, 16(4), 321–327.

Cleverdon, R. and Kalisch, A. (2000) Fair trade in tourism. *International Journal of Tourism Research*, 2, 171–187.

Comfort, L. and Okada, A. (2014) Coping with catastrophe: Black Swan in Northereastern Japan, March 11, 2011. In N. Kapucu, C. Hawkins and F. Rivera (Eds.), *Disaster Resilience. Interdisciplinary Perspective*. New York: Routledge.

Cox, R. and Perry, K. (2011) Like a fish out of water: Reconsidering disaster recovery and the role of place and social capital in community disaster resilience. *Journal of American Psychology*, 48, 395–411.

D'Amore, L.J. (1988) *Tourism: The World's Peace industry*. Paper presented at Tourism: A Vital Force for Peace, 1st Global Conference, 23–27 Oct, Montreal, Canada.

Daskon, C. (2010) Cultural resilience – the role of cultural traditions in sustaining rural livelihood: A case from rural Kandyan villages in Central Sri Lanka. *Sustainability*, 2, 1080–1100.

Delaney, A. (2016) Social sustainability in post-3.11 coastal Japan: The significance of social capital. In S. Assmann (Ed.), *Sustainability in Contemporary Rural Japan. Challenges and Opportunities*. New York: Routledge, pp. 3–18.

Diaz, J. and Dayal, A. (2008) Sense of place: A model for community based psychological support programs. *The Australian Journal of Disaster and Trauma Studies*, 1, 1174–4707.

Duraipappah, K., Takeuchi, K., Watanabe, M., Nishi, M. and Nakamura, K. (Eds.) (2012) *Satoyama-Satoumi Ecosystems and Human Well-Being: Socio-Ecological Production Landscapes of Japan*. New York: United Nations University Press.

Farrell, B. and Twining-Ward, L. (2004) Reconceptualising tourism. *Annals of Tourism Research*, 31(2), 274–295.

Farrell, B. and Twining-Ward, L. (2005) Seven steps towards sustainability: Tourism in the context of new knowledge. *Journal of Sustainable Tourism*, 13(2), 109–122.

FDMA (2016) *2011 East Japan Earthquake*. Fire and Disaster Management Agency No. 153. Available at www.fdma.go.jp/bn/153.pdf (in Japanese). (Accessed 7 July 2017).

Folke, C., Colding, J. and Berkes, F. (2003) Synthesis: Building resilience and adaptive capacity in social–ecological systems. In F. Berkes, J. Colding, and C. Folke (Eds), *Navigating Social–Ecological Systems: Building Resilience for Complexity and Change*. Cambridge: Cammbridge University Press, pp. 352–387.

Fukao, J. (2015) Community relocation and recreation of cultural environment. A case of Southern Indian region affected by the Indian Ocean Tsunami. In I. Hayashi (Ed.), *Disaster Reconstruction in the Asia Pacific Nations. Humanitarian Support, Community Relocation, Disaster Prevention and Culture*. Tokyo: Akashi Publishing.

Goodwin, H. (1998) *Sustainable Tourism and Poverty Elimination*. www.haroldgoodwin.info/resources/dfidpaper.pdf (Accessed 25 July 2014)

Hatton, M. (1999) *Community-Based Tourism in the Asia-Pacific*. Toronto: Canadian Tourism Commission.

Higgins-Desbiolles, F. (2008) Justice tourism and alternative globalisation. *Journal of Sustainable Tourism,* 16(3), 345–364.

Higgins-Desbiolles, F. (2010) Justifying tourism. In S. Cole and N. Morgan (Eds), *Tourism and Inequality*. Wallingford: CABI, pp. 194–211.

Hirakawa, M. (Ed.) (2012) *Environmental History of Japan 1: Japanese History and Environment: People and the Nature*. Tokyo: Yoshikawa-Kobunkan.

Hiwasaki, L. (2006) Community-based tourism: A pathway to sustainability for Japan's protected areas. *Society & Natural Resources,* 19(8), 675–692.

Holling, C. (1973) Resilience and stability of ecological systems. *Annual Review of Ecological Systems*, 4, 1–23.

Holling, C. (2001) Understanding the complexity of economic, ecological, and social systems. *Ecosystems*, 4(5), 390–405.

Hunter, C. (1997) Sustainable tourism as an adaptive paradigm. *Annual Review of Ecological Systems*, 24(4), 850–867.

Ishii, Y. (2016) Resilience seen in a village, Northern China. Revitalisation of rain ritual. In A. Kawakita and Y. Nishi (Eds.), *Resilience as History. War, Independence, Disaster* (Rekishi-to shiteno rejiriansu, in Japanese). Kyoto: Kyoto University Press, pp. 97–127.

Johannes, R. (1989) *Traditional Ecological Knowledge: A Collection of Essays*. Gland, Switzerland: IUCN.

Kassis, R. (2006) The Palestine and justice tourism. *Alternative Tourism Group,* www.atg.ps/print.php?page=palestinianandjusticetourism. (Accessed 24 August, 2014).

Kato, K. (2007a) Prayer for the whales. Spirituality and ethics for former whaling community. Intangible cultural heritage for sustainability. *International Journal of Cultural Property*, 14, 283–313.

Kato, K. (2007b) Community, connection and conservation. Intangible cultural values in natural heritage. The case of Shirakami-sanchi World Heritage Area. *International Journal of Heritage Studies*, 12(5), 458–473.

Koike, O. (2014) Rural landscape conservation in Japan. Lessons from Satoyama conservation program in Kanagawa Prefecture. In N. Kaneko, S. Yoshiura and M. Kobayashi (Eds), *Sustainable Living with Environmental Risks*. Tokyo: Sprinter, 253–263.

Lanfant, M.F. and Graburn, N. (1992) International tourism reconsidered: The principle of the alternative. In V. Smith and W. Eadington (Eds), *Tourism Alternatives*. Philadelphia: University of Pennsylvania Press, pp. 88–112.

Lew, A. (2014) Scale, change and resilience in community tourism planning. *Tourism Geographies: An International Journal of Tourism Space, Place and Environment*, 16(1), 14–22.

Luther, S. (2003) *Resilience and Vulnerability*. Cambridge: Cambridge University Press.

Machida, S. (2003) *Spiritual Power of the Mountain. What Japanese See There (yamano reiryoku. Hihonjin-wa sokoni nanio mitaka)*. Tokyo: Kodansha.

Masten, A. and Narayan, A. (2012) Child development in the context of disaster, war and terrorism: Pathways of risk and resilience. *Annual Review of Psychology*, 63, 227–257.

McLennan, C. J. (2011) *The Tourism Transformation Process: An Enquiry into the Three Main Process Phases*. Unpublished doctoral thesis. St Lucia: The University of Queensland.

McLennan, C. J., Ruhanen, L., Ritchie, B. and Pham, T. (2012) Dynamics of destination development: Investigating the application of transformation theory. *Journal of Hospitality & Tourism Research*, 36(2), 164–190.

Miller, D. S. (2007) Disaster tourism and disaster landscape attractions after Hurricane Katrina. *International Journal of Culture, Tourism and Hospitality Research*, 2(2), 115–131.

Minnaert, L., Maitland, R., and Miller , G. (2007) Social tourism and ethical foundations. *Tourism Culture & Communication,* 7, 7–17.

Minnaert, L., Maitland, R. and Miller, G. (2009) Tourism and social policy. The value of social tourism. *Annals of Tourism Research*, 36(2), 316–334.

Miyagi Prefecture (2011) Miyagi Prefecture Disaster Recovery Plan. Available at: www.pref.miyagi.jp/site/ej-earthquake/fukkou-keikaku.html

Miyata, N. (1996) *Ethno-Shinto: Dynamism of the Folklore Beliefs*. Tokyo: Shunju-sha.

MLIT (2011) *Overview of Tohoku Region* (Data). Ministry of Land, Infrastructure, Transport and Tourism. Available at: www.mlit.go.jp/singikai/kokudosin/sankou2.pdf (Accessed 10 March 2015)

Nabhan, G. P. (1997) *Culture of Habitat. On Nature, Culture and Story.* Berkeley, CA: Counterpoint Press.

Nishigai, K. (2011). *Mountain Worship in Eastern Japan and Worshiper Groups (ko).* Tokyo: Iwata Shoten.

Nomoto, K. (2010) *Regaining Land Spirituality: Exploring Folklore Connecting with Nature*. Tokyo: Iwanami.

Nomoto, K. (2013a) *Landscape of Gods and Nature: Reading Worshiping Environment.* Tokyo: Kodansha.

Nomoto, K. (2013b) *Natural Disaster and Folklor*e. Tokyo: Shinwa-sha.

Nomoto, K. and Akasaka, N. (Eds) (2013) *Exploring Traditional Knowledge in Daily Living.* Tokyo: Tamagawa University Publishing.

O'Grady, K. A., Orton, J. D., White, K. and Snyder, N. (2016) A way forward for spirituality, resilience, and international social science. *Journal of Psychology & Theology*, 44(2), 166–172.

Pavlovich, K. (2003) The evolution and transformation of a tourism destination network: The Waitomo caves, New Zealand. *Tourism Management*, 24(2), 203–216.

Pavlovich, K. (2014) A rhizomic approach to tourism destination evolution and transformation. *Tourism Management*, 41, 1–8.

Popesku, J. and Hall, D. (2004) Sustainability as the basis for future tourism development in Serbia. In D. Hall, (Ed.), *Tourism and Transition. Governance, Transformation and Development*. Wallingford: CABI.

Pricope, N. G. (2010) Resilience. In B. Warf (Ed.), *The Encyclopedia of Geography.* Thousand Oaks: Sage Publishing, pp. 2437–2441

Pritchard, A., Morgan, N. and Ateljevic, I. (2011) Hopeful tourism. A new transformative perspective. *Annals of Tourism Research,* 38(3), 941–963.

Reconstruction Agency (2017) Number of people living away from home. Available at: www.reconstruction.go.jp/topics/main-cat2/sub-cat2-1/20170328_hinansha.pdf

Reisinger, Y. (Ed.) (2015) *Transformational Tourism. Host perspectives.* Wallingford: CABI.

Ruiz-Ballesteros, E. (2011) Social-ecological resilience and community-based tourism: An approach from Agua Blanca, Ecuador. *Tourism Management*, 32, 655–666.

Sasaki, T. (2006) *Mountain Gods and Japanese: Exploring Foundation of Japanese Culture from Mountain Worshiping* (*Yamano-kami-to Nihonjin*). Tokyo: Yosen-sha.

Scheyvens, R. (2002) *Tourism for Development: Empowering Communities.* Harlow: Prentice Hall.

Sharpley, R. (2000) Tourism and sustainable development: Exploring the theoretical divide. *Journal of Sustainable Tourism*, 8(1), 1–9.

Sharpley, R. (2009) *Tourism, Development and the Environment: Beyond Sustainability?* London: Earthscan.

Shinohara, T. (1995) *Ethnological Records of Sea and Mountain* (*Umi-to Yama-no Minzokushi*). Tokyo: Yoshikawa-kobunkan.

Stockholm Resilience Centre. (2014) *What Is Resilience?* Available at: www.stockholmresilience.org. (accessed 15 August, 2015)

Telfer, D. and Sharpley, R. (2008) *Tourism and Development in the Developing World.* London: Routledge.

Tribe, J. (2009) Philosophical issues in tourism. In J. Tribe (Ed.), *Philosophical Issues in Tourism.* Bristol: Channel View Publications, pp. 3–22.

Walker, B. and Salt, D. (2006) *Resilience Thinking: Sustaining Ecosystems and People in a Changing World.* Washington: Island Press.

Walker, B., Holling, C. S., Carpenter, S. R. and Kinzig, A. (2004) Resilience, adaptability and transformability in social–ecological systems, *Ecology and Society*, 9(2), 5. Available at: www.ecologyandsociety.org/vol9/iss2/art5. (Accessed 15 August 2015).

Walsh, F. (2007) Traumatic loss and major disasters: Strengthening family and community resilience. *Family Process*, 46(2), 20–27.

13 International tourists in Japan

Their increasing numbers and vulnerability to natural hazards

Hayato Nagai, Kaede Sano, Brent W. Ritchie and Takashi Yoshino

Introduction

The world has experienced many natural disasters in recent years, including earthquakes, floods, hurricanes and tsunamis, some of which have affected the tourism industry owing to its vulnerability to these major events (Pforr, 2009; Ritchie & Campiranon, 2015). Indeed, a wide range of literature has acknowledged that natural disasters, particularly large-scale incidents, significantly impact the flow of tourism owing to the immense damage they cause to facilities and infrastructure and their negative effect on tourists' perceptions of destinations, which is particularly strong if they occur during a tourist's visit (Beirman, 2003; Cooper & Erfurt, 2007; Lehto, Douglas & Park, 2008). Additionally, tourists, especially international tourists, are regarded as vulnerable to natural disasters owing to their lack of familiarity with the destination and its possible hazards (Ritchie, 2008).

This chapter focuses on Japan as an international tourism destination. The country received more than 28 million inbound tourists in 2017, the highest number in its history (JNTO, 2019), and the tourism sector is now seen as an important driver of economic growth. In particular, tourism is expected to contribute to the Japan's regional revitalization efforts to combat the decrease in the size of its population as well the challenges posed by its rapidly ageing society. Hence, both the central and local governments have made an effort to attract tourists to not only major urban destinations but also regional destinations and, as a result, many international tourists, especially repeat visitors, now travel to the regional areas of Japan to engage in a variety of cultural- and nature-based activities (JTA, 2018b). Given that the government aims to attract 40 million inbound tourists annually by 2020 and 60 million by 2030 (Prime Minister of Japan and His Cabinet, 2016), the level of such regional visitation is likely to increase in future years.

Although Japan has successfully attracted inbound tourists in recent years and the country is generally perceived as a safe destination by international tourists (Handler, 2016; Nagai, Ritchie, Sano & Yoshino, 2020), they are actually at high risk of exposure to a range of natural hazards. In fact, Japan is ranked fourth (after Vanuatu, Tonga and the Philippines) in the World Risk Report's list of natural hazard-exposed countries (UNU-EHS, 2014). Several studies have investigated the impact of natural disasters on tourism in Japan (e.g., Chew & Jahari, 2014; Handler, 2016; Henderson, 2015; Muskat, Nakanishi & Blackman, 2015), typically focusing on large-scale events, such as the Great East Japan earthquake and tsunami, which heavily impacted tourism (JNTO, 2019). However, there is a constant risk of other (often smaller-scale) types of natural disasters in various locations throughout the country (Cabinet Office, 2016). It is, therefore, possible that international tourists will experience natural disasters while visiting Japan, and the likelihood of this will increase if Japan continues to attract international tourists in ever greater numbers. Hence, the purpose of this chapter is to provide a broad overview of Japan's natural hazards and discuss future challenges that the country will face as inbound tourism rapidly increases. It first outlines the current situation regarding Japan's inbound tourism sector and then presents an overview of the various types of natural disasters that occur in Japan. Finally, the vulnerability of international tourists is explored in depth using a recent natural disaster in Osaka as a case study.

Inbound tourism in Japan

Especially since the launch of the Visit Japan Campaign in 2003, the Japanese government has actively promoted inbound tourism and expended considerable effort to attract international tourists through a range of strategies, such as the enhancement of national-level promotional activities. For example, the Japan National Tourism Organization (JNTO) has developed multilingual websites in more than ten languages and actively provides up-to-date information through social media, such as Facebook, Instagram and Weibo. The websites and social media channels have been tailored to each key target market to attract tourists (JNTO, n.d.). In addition, the government has actively promoted Japan as a tourism destination by organizing various events overseas. For example, active promotional activities were held at the Japanese Pavilion at Expo Milano 2015, and JNTO hosted a Visit Japan tourism seminar for media and industry professionals in New York in September 2015, which the Prime Minister of Japan attended. Furthermore, the government supported promotional activities conducted by local governments, including events organized in cooperation with Japanese diplomatic facilities overseas (JTA, 2016a).

In addition to such active promotional activities, the government has also worked to respond to the increase in tourists from various cultural backgrounds. For example, as an increasing number of Muslim tourists, especially those from Malaysia and Indonesia, visit Japan, it has become essential to provide a Muslim-friendly travel environment. Thus, the Action Plan for Muslim Tourists in Japan was formulated in 2018 to promote Muslim-friendly destinations across the nation by training local practitioners to provide culturally appropriate services and providing information on Halal restaurants or places of worship to tourists (JTA, 2018b; Yamagata, 2019). In addition, in 2015, the Japan Tourism Agency (JTA) launched a program to support and establish local and regional level destination management organizations to respond to the rapidly changing tourism environment (JTA, 2016b). Other national-level strategies, such as attracting low-cost carriers and cruise ships, relaxing visa requirements for visitors from countries like China and Thailand, encouraging international business events and increasing the number of tax-free shopping environments, have also been implemented (JTA, 2018b). As a result, the share of low-cost carriers providing international flights to Japan has grown from 1.5 percent in 2009 to 21.7 percent in 2017, which has significantly contributed to the number of tourists that visit Japan, especially those from Asian countries (Ministry of Land, Infrastructure, Transport and Tourism, 2018).

A series of government strategies, many of which involve cooperation between local governments and private sectors, have also contributed to the steady increase in the number of international tourists visiting the country. The nation received more than 28 million international tourists in 2017, compared to only about 5 million in 2003. Thus, the inbound tourism market has strongly expanded in the last 15 years. In addition, in 2017, international visitors spent about 4.4162 trillion Japanese yen (approximately 40 US billion dollars), a 17.8 percent increase from the previous year (JTA, 2018b). Figure 13.1 indicates the trend in inbound tourist numbers to Japan since 2003.

According to the JTA (2018b), not only did the country receive its highest number of international tourists in 2017, but it also had the highest record of

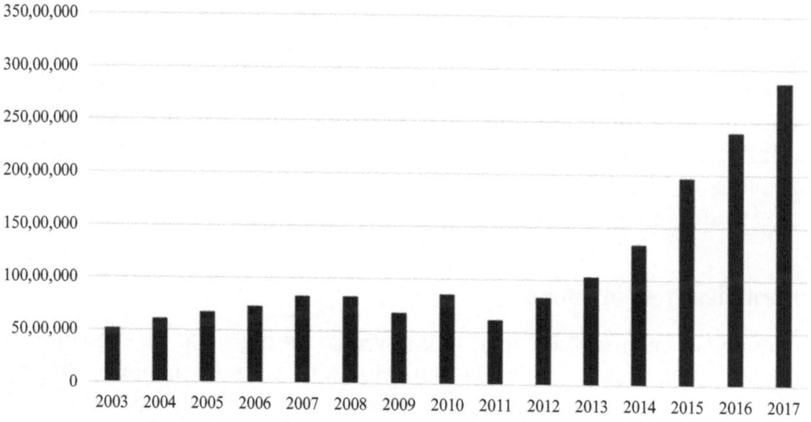

Figure 13.1 Number of international tourists visiting Japan, 2003–2017.
Source: Adapted from JNTO (2019).

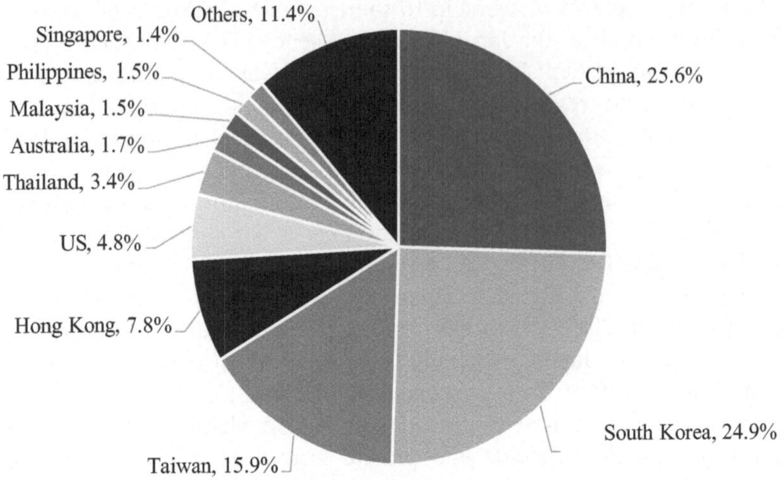

Figure 13.2 Top ten source markets for Japanese inbound tourism, 2017.
Source: Adapted from JNTO (2018).

international tourists among the 20 major inbound markets, including China, South Korea, Taiwan, Hong Kong, the United States, Thailand, Australia, Malaysia, the Philippines, Singapore, Indonesia, the United Kingdom, Vietnam, Canada, France, Germany, India, Italy, Spain and Russia. Approximately 24 million tourists were from Asian countries, an increase of 21.1 percent from the previous year and 84.8 percent of the total number of inbound tourists. The growth rate of the East-Asian market (China, South Korea, Hong Kong and Taiwan) has been the most significant – the market recorded a 21.9 percent growth in 2017, with over 20 million tourists – but the number of tourists from ASEAN countries, including Thailand, Singapore, Malaysia, Indonesia, the Philippines and Vietnam, has also continued to increase. Among Western countries, the highest number of tourists came from the United States (4.8 percent), followed by Australia (1.7 percent). Figure 13.2 shows Japan's top ten inbound markets in terms of visitor numbers and their market shares in 2017.

Natural disasters in Japan

As previously mentioned, Japan is known for its high risk of exposure to a range of natural hazards (UNU-EHS, 2014); approximately 20 percent of all earthquakes with a magnitude greater than 6.0 have occurred in Japan (Cabinet Office, 2014). In this section, available data from government departments, including the Cabinet Office and the Fire and Disaster

Management Agency, will be used to provide an overview of natural disasters in Japan. Secondary sources, such as newspaper articles, will also be consulted. This review focuses on natural disasters that have occurred in the past 25 years (1993–2017), especially catastrophic events that caused fatalities and missing persons.

According to the Cabinet Office (2018), between 1993 and 2017, a total of 32,307 individuals died or went missing in Japan owing to natural disasters. The majority of these losses were the result of two devastating natural disasters, namely, the Great Hanshin-Awaji earthquake of 1995 and the Great East Japan earthquake and tsunami of 2011. However, other types of smaller-scale natural disasters also occurred in this period. For example, water-related disasters, such as storms, floods and typhoons, resulted in 1,605 people losing their lives or going missing. When Typhoon Talas hit the Kii Peninsula in September 2011, 98 individuals died or went missing (Ministry of Land, Infrastructure, Transport and Tourism, Kinki Regional Development Bureau, 2013). Snow-related disasters, such as snowstorms and avalanches, have also had an impact, resulting in 1,262 people dying or going missing. Although the number of fatalities is smaller than in the aforementioned disasters and the frequency of occurrence is lower, 68 individuals also died or went missing owing to volcanic eruptions. Table 13.1 shows the breakdown of the number of fatalities and missing persons by natural disaster during the selected period.

As Table 13.1 reveals, while earthquakes and tsunamis do claim a significant number of lives, unlike water- and snow-related disasters this does not occur every year. Moreover, these calamities are geographically dispersed. Figure 13.3 shows the number of fatalities and missing persons caused by natural disasters between 2013 and 2017 in each prefecture. Kumamoto Prefecture on Kyushu island, which is located southwest of the main island, had the highest number of fatalities and missing persons during the selected period, largely due to the Kumamoto Earthquake that struck in April 2016 in which 267 persons lost their lives (Cabinet Office, 2018). In total, 14 of the 47 prefectures experienced more than 10 fatalities or missing persons during the selected period, while only three prefectures – Nara Prefecture, located in the Kansai region of the main island, and the Saga and Nagasaki Prefectures on Kyushu island – did not have any fatalities or missing persons. This highlights that the entire country is at high risk of exposure to a range of natural hazards.

Although there are no data regarding the specific number of tourists involved in each disaster, it is known that tourists were involved in several events. For example, more than 50 hikers were killed by Mount Ontake's eruption in 2014 (O'Connor, 2014), and two foreign skiers were killed in an avalanche in Nagano Prefecture in 2015 (The Japan Times, 2015). International media widely reported these events and implied that tourists could be the victim of natural disasters or small-scale tragedies at any destination throughout Japan.

Table 13.1 Number of fatalities and missing persons caused by natural disasters in Japan, 1993–2017

Year	*Water-related*	*Earthquake/Tsunami*	*Volcano*	*Snow-related*	*Other*	*Total*
1993	183	234	1	9	11	438
1994	8	3	0	21	7	39
1995	19	6,437	4	14	8	6,482
1996	21	0	0	28	35	84
1997	51	0	0	16	4	71
1998	80	0	0	28	1	109
1999	109	0	0	29	3	141
2000	19	1	0	52	6	78
2001	27	2	0	59	2	90
2002	20	0	0	26	2	48
2003	48	2	0	12	0	62
2004	240	68	0	16	3	327
2005	43	1	0	98	6	148
2006	87	0	0	88	2	177
2007	14	16	0	5	4	39
2008	22	24	0	48	7	101
2009	76	1	0	35	3	115
2010	31	0	0	57	1	89
2011	136	22,203	0	125	2	22,466
2012	52	0	0	138	0	190
2013	75	0	0	92	6	173
2014	112	0	63	108	0	283
2015	28	0	0	49	0	77
2016	45	267	0	32	0	344
2017	59	0	0	77	0	136
Total	1,605	29,259	68	1,262	113	32,307

Source: Adapted from Cabinet Office (2018).

International tourists and natural disasters in Japan

The tourism literature has long acknowledged the vulnerability of tourists, especially international tourists, to natural disasters owing to their often limited knowledge of the local environment and hazards as well as because of potential communication challenges which may hinder them from receiving accurate, real-time information at the destination (Mair, Ritchie & Walters, 2016; Pforr, 2009; Ritchie, 2008). Using the 2018 Osaka earthquake as a case study, this section highlights issues that international tourists may face when a natural disaster happens in Japan and discusses future challenges that the country – including at the local level – faces in the rapidly changing tourism environment.

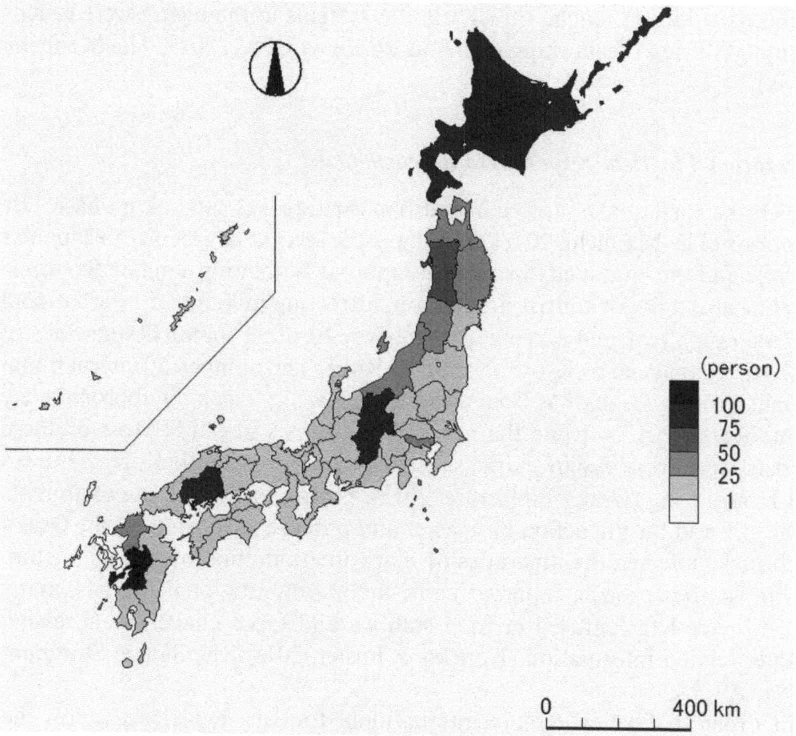

Figure 13.3 Number of deaths and missing persons caused by natural disasters in Japan organized by prefecture, 2013–2017.
(Note: The figure was created by the authors using the GIS software MANDARA).
Source: Adapted from Fire and Disaster Management Agency (2018).

The case of the 2018 Osaka earthquake: overview

On 18 June 2018, at 7:58 AM Japan Standard Time, a magnitude-6.1 earthquake struck in northern Osaka Prefecture, which is located in the Kansai region of the main island of Japan. Although no tsunami was generated by the earthquake, intense shaking (above the lower 5 on the shindo intensity scale, which measures the degree of shaking at a particular place on the Earth's surface) was felt in both the Osaka Prefecture and nearby cities in the Kyoto, Shiga, Hyogo and Nara Prefectures (Cabinet Office, 2019). By 5 July 2018, 4 deaths, 434 injured persons, 9 completely destroyed houses and 87 partially destroyed houses had been reported (Headquarters for Earthquake Research Promotion, 2018). Although power was recovered on the same day, up to around 170,000 buildings lost power following the earthquake. Other essential services, such as water and gas supplies, were also affected across the region. In addition, public transportation systems in the region were heavily disrupted during the morning rush hour (Cabinet Office, 2019; The Mainichi, 2018a).

International tourists' responses to the earthquake

The Osaka earthquake showed how urban earthquakes can disrupt basic city functions (The Mainichi, 2018a). Osaka is the second largest city in Japan as a whole and the largest city in western Japan. It is not only a major economic hub but also a major tourist destination, attracting millions of international tourists every year and serving as a gateway to other major destinations in western Japan, such as Kyoto, Nara and Kobe. The number of international tourists visiting Osaka has been steadily increasing, reaching approximately 11 million in 2017 – triple the number of tourists in 2014. Most of these tourists were from Asian markets, especially China, South Korea, Taiwan and Hong Kong (Osaka Prefecture, 2018). Owing to the volume of tourists in the city and the impact on businesses and public transportation, the Osaka earthquake affected the itineraries of many international tourists (Johnston, 2018). The news media reported panic among international tourists, many of who were left confused at train stations and faced challenges accessing disaster-related information (Kaneko & Foster, 2018; The Sankei Shimbun, 2018a, 2018b).

In order to further explore international tourists' experiences with the Osaka earthquake, this study employed netnography, a qualitative method for understanding online communities and communications (Kozinets, 2002). As they comprise the highest share of the inbound market, Chinese tourists were specifically considered in this exploration. Following the criteria suggested by Kozinets (2002, 2015), Mafengwo, one of the most popular and active travel-related online communities in China, was selected as the online platform for this study. To identify blog posts related to the Osaka earthquake, 'Japan', 'Osaka' and 'earthquake' were used as keywords. In total, 43 blog posts about experiences of the Osaka earthquake were identified on Mafengwo.

The earthquake occurred during the Dragon Boat Festival, a traditional Chinese festival. Because of this three-day holiday (16–18 June), many Chinese tourists were visiting Osaka when the earthquake struck. Analysis of the blog posts written by Chinese tourists clearly showed that many were frightened and confused during the event, especially if it was their first time experiencing an earthquake. The following blog entries (originally written in Chinese but translated for this study) were posted by Chinese tourists on Mafengwo:

> June 18, 2018, I woke up due to shaking. At the beginning, I thought someone had knocked on my door, but it was actually an earthquake. Because we hadn't experienced an earthquake before, we didn't know what had happened. When I accessed WeChat to check my colleagues' greetings

> for the Dragon Boat Festival, the hotel bed began to shake strongly. I fell into a panic. Then, almost at the same time, the building's alarms alerted us. We heard a Japanese announcement, but we could not understand it. We quickly changed our clothes and ran down to the ground floor without stopping, even though we had stayed on the 26th floor!
>
> [Female tourist, blog name 'Wangxiaoman', posted on 27 June 2018]

> What happened?! I have never experienced this situation. So scared! Unfortunately, we stayed on the highest floor of the hotel. I was woken up by strong shaking. Even now, I still feel dizzy. I just want to be back home! I think my mom will never travel to Japan with me again!
>
> [Female tourist, blog name 'bubble', posted on 29 June 2018]

> This was the most 'impressive' day for me! I felt strong shaking when I was sleeping. At the same time, the alarms from our mobile phones alerted us. What? It's an earthquake? Because we hadn't experienced an earthquake before, we didn't know what had happened or what we should do. Suddenly, one of my friends cried, 'An earthquake has occurred!' At this moment, we heard the announcement from the hotel; however, it was in Japanese, so we could not understand it. Did they ask us to go outside?! We quickly put our baggage away and ran down to the hotel lobby. So scared!
>
> [Female tourist, blog name 'lily', posted on 28 June 2018]

Many Chinese tourists highlighted that the announcement made by the hotel after the earthquake began was only in Japanese, which meant they were not able to understand it and did not know what to do. In addition, a lack of previous experience with earthquakes seemed to enhance their panic. For this reason, the Chinese tourists tended to rely on Chinese online communities, such as Weibo, to acquire information and guidance, but some of the information posted on social media was not accurate, resulting in increased panic among some tourists:

> We encountered the earthquake in Kyoto. The shaking was very strong. At the same time, I received a Japanese message on my mobile phone. Then, I checked Weibo for more information and found some media analysis that said there would be a stronger earthquake the following week: a situation similar to the Kumamoto earthquakes that occurred in 2016. I also found someone who wrote on Weibo that a lot of Nara's deer ran to Osaka before the earthquake. Because of the different negative information on Weibo regarding the earthquake, I was very frightened. Now I know that most of the information was rumour, so I suggest that other Chinese tourists use Weibo as a reference, but they should not trust it completely.
>
> [Male tourist, blog name 'Hezi', posted on 28 June 2018]

> This was the second day after we had arrived in Osaka. After the earthquake, I found several rumours on the Internet. Someone said that they could not find a direct flight to China and that they had to pay a high price for tickets, but they just wanted to leave Osaka, Japan, for safety. I immediately checked flight tickets online and found that the flight tickets for that day were already sold out. Moreover, the next day's one-way ticket prices had risen to 10,000 RMB. I have to say that the rumours made us more panicked than the earthquake!
>
> [Female tourist, blog name 'kelly', posted on 5 December 2018]

Because the earthquake occurred on 18 June, the last day of the Dragon Boat Festival, many Chinese tourists had planned to travel back home. However, some stated that they did not know how to travel to the airport, as the public transport system was disrupted after the earthquake. The following blog posts describe two female tourists' experiences traveling to the airport that day:

> June 18 was the day we were due to return to China. We didn't expect that even the shuttle bus would be out of service. How could we get to the airport? Fortunately, some routes of the subway were back to normal. However, we were not familiar with the public transport system. When we got to the airport, we had only two hours left before our departure.
>
> [Female tourist, blog name 'pearl', posted on 27 July 2018]

> Although our flight was scheduled for the afternoon, we decided to arrive at Kansai International Airport earlier. However, we then found out that the ticket-selling machine was out of service due to the earthquake. Even the shuttle bus had stopped! Because we didn't know when the shuttle bus would resume, we waited for it for two hours at the bus terminal. All of the public transport systems were down. We wanted to call a taxi; however, we found that the waiting line was so long. My baggage was 20 kilograms! You can imagine how terrible it was to climb the stairs without any operating elevators due to the earthquake!
>
> [Female tourist, blog name 'lily', posted on 28 July 2018]

Challenges in an era of rapid growth in international tourism

The blog entries posted by tourists who experienced the Osaka earthquake revealed several challenges faced by Japan, a country at high risk of exposure to natural hazards. As widely acknowledged in the literature (Hoogenraad, Eden & King, 2004; Pforr, 2009; Ritchie, 2008), the Osaka case reaffirmed international tourists' vulnerability to natural disasters, especially the challenges they face in accessing accurate and updated information in a timely manner compared to tourists in a group with a tour guide. As mentioned in the blog entries above, Chinese tourists did not understand the Japanese announcements made at their hotels after the earthquake, which increased their uncertainty and made them seek information from online communities in their native language. A Japanese newspaper reported that many accommodation facilities in Kyoto, a major international tourism destination in Japan, provided information in English after the earthquake and major panic among international tourists was not observed. However, it was also noted that this situation might have been related to the limited earthquake damage in Kyoto (The Mainichi, 2018b). The same article reported the story of a tourist from Argentina; although staff at her accommodations were able to communicate in English, they did not have up-to-date information, so she had to seek information online (The Mainichi, 2018b). This story highlights the importance of language as well as the provision of accurate and timely information during a disaster. Given the large percentage of tourists from non-English-speaking countries, hazard-related information provided in their native languages in a timely manner is essential.

The official websites of destination management organizations and local governments have been considered useful tools for disseminating information to tourists (Ritchie, 2009). In the case of Osaka, government agencies responded after the earthquake, and efforts were made to provide accurate and up-to-date information. Specifically, the Osaka Tourism and Convention Bureau provided disaster-related information, including public transport information, in five different languages (Japanese, English, Chinese, Korean and Thai) through their social media and websites, which were accessed about 10,000 times on both the 18th and 19th of June (The Sankei Shimbun, 2018a). However, a limited number of phone calls regarding earthquake and disaster information were received from tourists at the Bureau's multilingual call centre, which offers services in eight different languages. The Osaka Prefecture also established 24-h telephone support services in the early evening on 18 June. Although it welcomed phone calls from tourists, the majority of phone calls were from foreign nationals living in Osaka who sought information about, for example, gas services and evacuation warnings (The Nikkei, 2018). In addition, the use of mobile safety apps appeared to be limited. The Japan Tourism Agency has been providing smartphone apps that are targeted for international tourists visiting Japan, called 'Safety tips', since 2014. As shown in Figure 13.4, the app was promoted at tourism information centres, airports, terminal stations and other major sites through which tourists pass (JTA, 2016b). However, dissemination of the apps to international tourists has been an issue (The Sankei Shimbun, 2018b). As Nagai et al. (2020) report, based on interviews with departing tourists at the Kansai International Airport in April 2018, hazard information search behaviour among tourists was low, and few knew about the safety app. It is thus necessary to conduct further investigations to better understand tourists' disaster information-seeking behaviours, a research field that is still lacking a comprehensive understanding (Cahyanto et al., 2016).

The analysis also indicated that rumours were disseminated on social media following the earthquake, which caused anxiety among international

Figure 13.4 Promotion of the mobile safety app 'Safety tips' at JR Namba Station in Osaka.
Photo: H. Nagai.

tourists. Social media plays an important role in spreading information during a disaster, and it has a great potential to help communities collectively cope with such disasters (Oh, Agrawal & Rao, 2013). However, the rapid spread of rumours and false information on social media is a critical issue in crisis communication (Wang & Zhuang, 2018). This is not unique to the tourist context; it can also affect local residents, even though they are generally regarded as more knowledgeable about local information. For example, the spread of false information on Twitter was observed in Japan during the Great East Japan earthquake and tsunami (Murayama, Mukai, Nishioka & Saito, 2013). In addition to the Chinese tourists mentioned above, several local newspapers reported that international tourists frequently used online sources in their home countries to seek information during a disaster (The Mainichi, 2018b; The Nikkei, 2018). However, this meant that international tourists faced difficulties accessing accurate information in a timely manner, and some were unable to understand what was happening in the city in which they were staying.

Thus, it is essential to provide onsite multilingual support during a disaster to minimize disruption among international tourists and help avoid the excessive spread of false information in online communities. In addition to accommodations, multilingual information should be provided at public transportation venues, such as train stations and airports, as a large number of international tourists were stuck at these sites without accurate information during the Osaka earthquake (The Sankei Shimbun, 2018b). Several initiatives have aimed to address this issue, such as that at the Kansai International Airport in Osaka. The airport was not physically damaged by the Osaka earthquake, but it was struck by Typhoon Jebi in September 2018. The airport was shut down due to flooding, and approximately 8,000 individuals, including airport staff, were stranded at the airport (Kyodo News, 2018). It became clear after the event that the airport's preparation for multilingual risk communication was limited. To improve the situation and provide effective risk communication, the airport announced in December 2018 that several measures would be implemented, including multilingual alert information-providing systems (Kansai Airports, 2018). The national government has also announced strategies to promote multilingual information services during disasters through multiple channels, including social media, mobile apps and call centres (JTA, 2018a).

Conclusion

This chapter has attempted to provide an overview of natural disasters in Japan and has explored the impacts (including the potential future impacts) of these events on tourism. The available secondary sources indicated that the major earthquakes and tsunamis have been the worst natural disasters (based on total deaths) in the past 25 years, but other types of natural disasters also consistently claimed lives. However, smaller-scale disasters, including smaller earthquakes, have not been thoroughly explored in the tourism literature to date. The 2018 Osaka earthquake, a recent case, highlighted common challenges, especially communication challenges, faced by international tourists during a disaster and confirmed the widely accepted belief that international tourists are vulnerable to natural disasters. Although the Osaka case did not claim the lives of any international tourists, Japan is at high risk of exposure to a range of natural hazards. As outlined in the beginning of this chapter, Japan has actively promoted tourism over the past few decades, and individuals from a variety of cultural backgrounds are expected to visit the nation, including regional destinations, in the coming years. Thus, it is likely that international tourists will be involved in future natural disasters, especially if tourism continues to grow as the government expects.

The literature has acknowledged that the development and implementation of a disaster strategy requires relevant agencies to adopt a coordinated team-based approach (Faulkner, 2001; Ritchie, 2009). The same is true for developing and providing effective communication during disasters, as it requires much effort to coordinate effective communication between authorities and industry stakeholders (Ashcroft, 1997; Beirman, 2002). Efforts to develop multilingual and effective risk communication environments have been rapidly expanding across Japan, and they are expected to increase destinations' disaster preparedness (JTA, 2018b). This chapter reaffirms the need for continuous development of both national- and regional-level measures, especially those targeted to international tourists, in order to respond to the rapidly changing environment of Japan.

Acknowledgements

This work was supported by JSPS KAKENHI Grant Number JP17H02250.

References

Ashcroft, L. (1997) Crisis management: Public relations. *Journal of Managerial Psychology,* 12(5), 325–332.

Beirman, D. (2002) Marketing of tourism destinations during a prolonged crisis: Israel and the Middle East. *Journal of Vacation Marketing,* 8(2), 167–176.

Beirman, D. (2003) *Restoring Tourism Destinations in Crisis: A Strategic Marketing Approach*. Wallingford: CABI.

bubble (2018) Quanjiazongdongyuan Daban 6.1ji dizhen chutiyan [The first time experiencing the magnitude 6.1 earthquake at Osaka with my family] [Blog post, 29 June]. Available at: www.mafengwo.cn/i/9660085.html (Accessed 12 December 2019).

Cabinet Office (2014) *Heisei 26 nen ban bosai hakusho [White Paper: Disaster Management in Japan, 2014]*. Available at: www.bousai.go.jp/kaigirep/hakusho/h26/ (Accessed 12 December 2019).

Cabinet Office (2016) *Heisei 28 nen ban bosai hakusho [White Paper: Disaster Management in Japan, 2016]*. Available at: www.bousai.go.jp/kaigirep/hakusho/h28/ (Accessed 12 December 2019).

Cabinet Office (2018) *White Paper: Disaster Management in Japan, 2018*. Available at: www.bousai.go.jp/kaigirep/hakusho/pdf/H30_hakusho_english.pdf (Accessed 12 December 2019).

Cabinet Office (2019) *Reiwa gan nen ban bosai hakusho [White Paper on Disaster Management in Japan, 2019]*. Available at: www.bousai.go.jp/kaigirep/hakusho/r1.html (Accessed 12 December 2019).

Cahyanto, I., Pennington-Gray, L., Thapa, B., Srinivasan, S., Villegas, J., Matyas, C. and Kiousis, S. (2016) Predicting information seeking regarding hurricane evacuation in the destination. *Tourism Management,* 52, 264–275.

Chew, E. Y. T. and Jahari, S. A. (2014) Destination image as a mediator between perceived risks and revisit intention: A case of post-disaster Japan. *Tourism Management,* 40, 382–393.

Cooper, M. and Erfurt, P. (2007) Tsunamis, earthquakes,volcanism and other problems: Disasters, responses and Japanese tourism. In E. Laws, B. Prideaux and K. Chon (Eds), *Crisis Management in Tourism*. Wallingford: CABI, pp. 234–251

Faulkner, B. (2001) Towards a framework for tourism disaster management. *Tourism Management,* 22(2), 135–147.

Fire and Disaster Management Agency (2018) *Chiho bosai gyosei no genkyo [Status of Regional Disaster Management Administration]*. Available at: www.fdma.go.jp/publication/bousai/#046832 (Accessed 12 December 2019).

Handler, I. (2016) The impact of the Fukushima disaster on Japan's travel image: An exploratory study on Taiwanese travellers. *Journal of Hospitality and Tourism Management,* 27, 12–17.

Headquarters for Earthquake Research Promotion (2018) *2018 nen 6 gatsu 18 nichi Osaka fu hokubu no jishin no hyoka [The Evaluation of the Northern Osaka Earthquake on 18 June 2018]*. Available at: www.static.jishin.go.jp/resource/monthly/2018/20180618_osaka_2.pdf (Accessed 12 December 2019).

Henderson, J. C. (2015) The Great East Japan Earthquake and tourism: A preliminary case study. *Tourism Recreation Research, 38*(1), 93–98.

Hezi, Z. (2018) Dabandizhenzhong de guanxi linshi zhuanguandong zhilv [I changed my travel plan from Kansai to Kanto due to the Osaka earthquake] [Blog post, 28 June]. Available at: www.mafengwo.cn/i/9644945.html (Accessed 12 December 2019).

Hoogenraad, W., Eden, R. and King, D. (2004) Cyclone awareness amongst backpackers in Northern Australia. *Australian Journal of Emergency Management,* 19(2), 25–29.

JNTO (n.d.). *Uebusaito ya sns o tsujita honichi kanko joho no teikyo [Providing Tourism Information Through Websites and SNS]*. Japan National Tourism Organization. Available at: www.jnto.go.jp/jpn/inbound_market/webpr/index.html (Accessed 12 December 2019).

JNTO (2019) *Bijitto Japan jigyo kaishi iko no honichi kyaku su no suii [Number of International Tourists After the Launch of the Visit Japan Campaign]*. Japan National Tourism Organization. Available at: www.jnto.go.jp/jpn/statistics/marketingdata_tourists_after_vj.pdf (Accessed 12 December 2019).

Johnston, E. (2018). At least four killed, more than 300 injured after strong earthquake rattles Osaka. *Japan Times*, 18 June. Available at: www.japantimes.co.jp/news/2018/06/18/national/strong-m5-9-temblor-reaching-lower-6-japan-scale-rocks-northern-osaka/#.XfDTUWy7Lid (Accessed 11 December 2019).

JTA (2016a) *White Paper on Tourism in Japan, 2016*. Tokyo, Japan: Japan Tourism Agency.

JTA (2016b) *Asu no nihon o sasaeru kanko bijon [Tourism Vision to Support the Future of Japan]*. Available at: www.mlit.go.jp/common/001126601.pdf (Accessed 12 December 2019).

JTA (2018a) *For More Safe and Secure Travel in Japan*. Available at: www.mlit.go.jp/kankocho/news08_000255.html (Accessed 12 December 2019).

JTA (2018b) *White Paper on Tourism in Japan, 2018*. Tokyo, Japan: Japan Tourism Agency.

Kaneko, K. and Foster, M. (2018) Magnitude 6.1 quake in Japan's Osaka area kills four, halts factories. *Reuters*, 18 June. Available At: https://uk.reuters.com/article/uk-japan-quake/magnitude-6-1-quake-in-japans-osaka-area-kills-four-halts-factories-idUKKBN1JD13H (Accessed 11 December 2019).

Kansai Airports (2018) *Saigai taisaku ni tuite [Disaster Prevention]*. Available at: www.kansai-airports.co.jp/news/2018/2663/J181213_Disaster_prevention_.pdf (Accessed 12 December 2019).

kelly (2018) Daishang mushangdarenmen yiqiqu lvyou – Liuyue daban dizhen zhilv [Travel with mom: Experienced the Osaka earthquake in June] [Blog post, 5 December]. Available at: www.mafengwo.cn/i/11328670.html (Accessed 12 December 2019).

Kozinets, R. V. (2002) The field behind the screen: Using netnography for marketing research in online communities. *Journal of Marketing Research,* 39(1), 61–72.

Kozinets, R. V. (2015) *Netnography: Redefined,* 2nd Edn. London: Sage Publications.

Kyodo News. (2018) Kansai airport closure likely prolonged, typhoon death toll rises to 11. *Kyodo News,* 5 September. Available at: https://english.kyodonews.net/news/2018/09/66d903e758df-5000-typhoon-stranded-people-being-transported-out-of-kansai-airport.html (Accessed 11 December 2019).

Lehto, X., Douglas, A. C. and Park, J. (2008) Mediating the effects of natural disasters on travel intention. *Journal of Travel & Tourism Marketing,* 23(2–4), 29–43.

lily (2018) Diyici kuangchi kuangchi de dabanyou jiujinglile dizhen [I experienced the Osaka earthquake the first time I traveled to Osaka] [Blog post, 28 June]. Available at: www.mafengwo.cn/i/9648869.html (Accessed 12 December 2019).

Mair, J., Ritchie, B. W. and Walters, G. (2016) Towards a research agenda for post-disaster and post-crisis recovery strategies for tourist destinations: A narrative review. *Current Issues in Tourism,* 19(1), 1–26.

Ministry of Land, Infrastructure, Transport and Tourism (2018) *Waga kuni ni okeru LCC no sannyu sokushin [Promotion of Entry of LCC in Japan].* Available at: www.mlit.go.jp/koku/koku_tk1_000025.html (Accessed 12 December 2019).

Ministry of Land, Infrastructure, Transport and Tourism, Kinki Regional Development Bureau. (2013). *2011 nen Kii Hanto daisuigai [The Kii Peninsula Great Flood in 2011].* Available at: www.kkr.mlit.go.jp/bousai/qgl8vl0000008ajd-att/kiihantou-kirokushi.pdf (Accessed 12 December 2019).

Murayama, Y., Mukai, M., Nishioka, D. and Saito, Y. (2013) A model for decision making in retweet which caused spreading of false rumor in emergencies. *Proceedings of Multimedia, Distributed, Cooperative, and Mobile Symposium 2013.* Tokyo: Information Processing Society of Japan, pp. 873–879. Available at: https://ipsj.ixsq.nii.ac.jp/ej/?action=pages_view_main&active_action=repository_view_main_item_detail&item_id=97256&item_no=1&page_id=13&block_id=8

Muskat, B., Nakanishi, H. and Blackman, D. (2015) Integrating tourism into disaster recovery management: The case of the Great East Japan Earthquake and Tsunami 2011. In B. W. Ritchie and K. Campiranon (Eds), *Tourism Crisis and Disaster Management in the Asia-Pacific.* Wallingford: CABI, pp. 97–115.

Nagai, H., Ritchie, B. W., Sano, K. and Yoshino, T. (2020) International tourists' knowledge of natural hazards. *Annals of Tourism Research.* 80, Article 102690

O'Connor, R. (2014) Mount Ontake eruption in Japan: Final moments of hikers killed by volcano captured in poignant photos. *The Independent,* 5 October. Availaboe at: www.independent.co.uk/news/world/asia/mount-ontake-eruption-in-japan-final-moments-of-hikers-killed-by-volcano-captured-in-poignant-photos-9775110.html (Accessed 11 Dcember 2019).

Oh, O., Agrawal, M. and Rao, H. R. (2013) Community intelligence and social media services: A rumor theoretic analysis of Tweets during social crises. *MIS Quarterly,* 37(2), 407–426.

Osaka Prefecture (2018) *Osaka no kanko doko ni tsuite [Tourism Trend in Osaka].* Available at: www.pref.osaka.lg.jp/attach/33874/00294944/02shiryou1.pdf (Accessed 12 December 2019).

pearl (2018) Diyici jingli ribendaban liujidizhen de lvxing [The travel that I experienced magnitude 6 in Osaka] [Blog post, 27 July]. Available at: www.mafengwo.cn/i/9971626.html (Accessed 12 December 2019).

Pforr, C. (2009) Crisis management in tourism: A review of the emergent literature. In C. Pforr and P. Hosie (Eds), *Crisis Management in the Tourism Industry: Beating the Odds?* Farnham: Ashgate, pp. 37–52.

Prime Minister of Japan and His Cabinet (2016) *The Prime Minister in Action: Meeting of the Council for the Development of a Tourism Vision to Support the Future of Japan*. Available at: http://japan.kantei.go.jp/97_abe/actions/201603/30article1.html (Accessed 11 December 2019).

Ritchie, B. W. (2008) Tourism disaster planning and management: From response and recovery to reduction and readiness. *Current Issues in Tourism,* 11(4), 315–348.

Ritchie, B. W. (2009) *Crisis and Disaster Management for Tourism*. Bristol: Channel View Publications.

Ritchie, B. W. and Campiranon, K. (2015) Major themes and perspectives. In B. W. Ritchie and K. Campiranon (Eds), *Tourism Crisis and Disaster Management in the Asia-Pacific*. Wallingford: CABI, pp. 3–11.

The Japan Times (2015) Avalanches claim four on slopes of Nagano, Niigata. *The Japan Times*, 19 January. Available at: www.japantimes.co.jp/news/2015/01/19/national/avalanches-claim-four-slopes-nagano-niigata-trio-may-missing/#.XfDUB2y7Lic (Accessed 11 December 2019).

The Mainichi. (2018a). Editorial: Osaka quake stark reminder of chaos caused by major urban temblors. *The Mainichi,* 19 June. Available at: https://mainichi.jp/english/articles/20180619/p2a/00m/0na/026000c (Accessed 11 December 2019).

The Mainichi (2018b) Jishin: Osaka shindo 6 jaku funai no fushosha 18 nin ni. Jutaku higai kakudai 213 ken. Yoshin ha kei 24 kai. [Earthquake: Osaka magnitude lower 6, 18 people injured, 213 houses damaged in Osaka Prefecture, 24 aftershocks] *The Mainichi*, 20 June Available at: https://mainichi.jp/articles/20180620/ddl/k26/040/386000c (Accessed 12 December 2019)

The Nikkei (2018) Saigaiji no gaikokujin kanko kyaku kea fujubun Osaka jishin de kadai [Limited support of international tourists during a disaster] *The Nikkei,* 20 June. Available at: www.nikkei.com/article/DGXMZO31994260Q8A620C1AC1000/ (Accessed 12 December 2019).

The Sankei Shimbun. (2018a) Hirogaru fuan aitsugu kyanseru honichi kyaku heno joho hasshin kadai ni [Spreading uncertainty, increasing cancellation: Providing information to international tourists is an issue]. *The Sankei Shimbun*, 25 June. Available at: www.sankei.com/west/news/180625/wst1806250120-n1.html (Accessed 12 December 2019).

The Sankei Shimbun (2018b) Osaka hokubu jishin 3 kagetsu sutsu kesu kakae tomadou gaikoku jin kanko kyaku 'joho nammin' do sukuu jichitai ya kigyo de taisaku susumu [Three months after the Osaka earthquake: International tourists confused with their suitcases. How to protect them with limited information access. Measures provided by local governments and private sectors]. *The Sankei Shimbun*, 18 September. Available at: www.sankei.com/west/news/180918/wst1809180014-n1.html (Accessed 12 December 2019).

UNU-EHS. (2014) *World Risk Report 2014*. Available at: https://i.unu.edu/media/ehs.unu.edu/news/4070/11895.pdf (Accessed 11 December 2019).

Wangxiaoman, X. M. (2018) Guimiyou yushang 618 daban dizhen [Encountered the 618 Osaka Earthquake in girls' travel] [Blog post, 27 June]. Available at: www.mafengwo.cn/i/9640101.html (Accessed 12 December 2019).

Wang, B. and Zhuang, J. (2018) Rumor response, debunking response, and decision makings of misinformed Twitter users during disasters. *Natural Hazards,* 93(3), 1145–1162.

Yamagata, A. (2019) Perceptions of Islam and Muslims in contemporary Japan. *New Voices in Japanese Studies,* 11, 1–25.

14 The expansion of peer-to-peer accommodation rentals in Japan

Issues and challenges

Takahiro Ikeji and Hayato Nagai

Introduction

The supply of peer-to-peer (P2P) accommodation is rapidly expanding on a global basis. One of the most well-known P2P accommodation platforms is Airbnb which, founded in 2008, has since played a major role in popularizing the concept of P2P accommodation all over the world (Airbnb, 2018a). However, Airbnb, as well as other platforms, are also becoming increasingly affected by the tightening of regulations, not least because most P2P accommodation operates in general residences and many cities have struggled with managing this expansion. As a consequence, many popular tourism destinations, such as New York and Paris, have enacted regulations to manage the development of P2P accommodation which, in turn, have in some cases led to lawsuits; Paris, for example, has sued Airbnb for what are considered to be illegal advertisements (BBC, 2019) whilst the company has taken legal action against New York's requirement that it provides details of its hosts in the city (Bloomberg, 2018). Moreover, the basis of regulations for P2P accommodation are not all the same; as indicated by Nieuwland and van Melik (2020), some regulations aim to protect affordable housing, as in the case in Berlin and London, whereas other cities, such as Amsterdam and Barcelona, seek to preserve the quality of life of their residents. It should be noted that yet other cities, such as San Francisco, have implemented regulations to meet both objectives. In short, P2P accommodation is increasingly becoming regulated according to local challenges and objectives.

P2P accommodation has been growing in popularity in Japan along with the rapid increase in the number of inbound tourists. Such accommodation facilities are often called *mimpaku* (民泊) in Japanese, which translates as 'residential houses where tourists can stay'. In the past few years, the number of inbound tourists to Japan has more than tripled, from 10 million in 2013 to 31 million in 2018 (Japan National Tourism Organization, 2019) – see also Chapter 1. P2P accommodation has played an important role in this growth, as evidenced by the fact that the number of international tourists to Japan using Airbnb reached almost 6 million in 2017 (Airbnb, 2018b). Since it had been forecast that there would be a significant shortage of traditional accommodation in Japan (Mizuho Research Institute, 2017), the development of P2P accommodation has allowed the country to keep pace with the rapid increase in inbound tourism

However, the rapid expansion of the P2P accommodation sector in Japan has given rise to a number of issues being highlighted in the media, such as excessive noise and safety concerns (The Nikkei, 2017a). These problems are consistent with what has been reported in previous studies (e.g., Gurran & Phibbs, 2017; Jordan & Moore, 2018). Other issues, such as an increase in housing prices, seem to be less serious; this most likely reflects the abundance of unused properties in Japan (something arguably unique to that country). Indeed, as a result of a declining population and younger people moving to cities, the proportion of empty houses, known as *akiya*, increased from 9.8 percent in 1993 to 13.5 percent in 2013 (Statistics Japan, 2013) and, according to some reports, that proportion will continue to grow, particularly in rural areas (Jozuka, 2019). The core issue related to P2P accommodation in Japan, therefore, seems to be determining how to prevent the deterioration of residents' quality of life.

In an attempt to control the expansion of P2P accommodation, Japan enacted a new regulation called the Private Housing Accommodation Law (P2P Law) in June 2018. This law limited the maximum number of operating days to 180 per year and required owners to register their P2P accommodation properties with the local government. Interestingly, just before the enforcement of the P2P law, nearly 80 percent of all properties in Japan listed on Airbnb were deleted, resulting in the cancellation of a significant number of fully booked reservations (Negishi, 2018). Airbnb released an announcement that this decision was made based on a requirement from the Japan Travel Agency (Airbnb, 2018c); according to Airbnb, this requirement from the government was issued without any advance notice.

Since then, one of the government's objectives in regulating P2P accommodation in Japan has been to prevent the (further) deterioration of local residents' quality of life. This has resulted in the enactment of new laws designed to regulate P2P accommodation as well as the deregulation of the existing law. In other words, the government has been seeking to regulate the development of P2P accommodation to limit the impact on residents' quality of life whilst, at the same time, easing the restrictions within the existing law in order to meet the accommodation needs of growing numbers of inbound tourists.

Following a review of the development of P2P accommodation in general, this chapter outlines the current situation regarding the sector in Japan in particular. It goes on to consider the evolution of regulations related to P2P accommodation in the country as well as potential future challenges, and then a case study of Kyoto is presented as the city has enacted one of the strictest sets of city-level regulations for P2P accommodation. It should also be noted that this chapter offers a unique Japanese perspective on the development of and challenges associated with P2P accommodation as most previous studies into the phenomenon have been conducted in a Western context (e.g., Adamiak, 2018; Blal, Singal & Templin, 2018; Tussyadiah & Park, 2018), with few focusing on P2P accommodations in Asia more generally.

The global expansion of P2P accommodation

It has been claimed that Airbnb posts more rooms on its website than the top five hotel brands combined (Hartmans, 2017) whilst, according to Airbnb itself, on average, more than 2 million people are staying at an Airbnb accommodation every night (Airbnb, 2018a). However, the popularity of P2P accommodation is not limited to Airbnb in particular; the supply of P2P accommodation in general has expanded rapidly around the world with online travel agents, such as Booking.com, also including inventories of P2P accommodation on their websites (Vynck, 2018). Based on the size of the contemporary market for P2P accommodation, therefore, it can be said that it is becoming a general accommodation choice rather than simply a unique accommodation type.

The main difference between traditional and P2P accommodation is that the former operates in commercial facilities whereas the latter usually operates in general residences – though 'bed and breakfast' facilities fall somewhere between the two. In many countries, accommodation facilities are required to be registered with the government (Nieuwland & van Melik, 2020); moreover, owing to land use restrictions, areas where formal accommodation facilities can operate are frequently restricted. P2P accommodation, however, most commonly operates within general residences which are, of course, designed for residential purposes. And because general residences are, by definition, typically located in residential areas, it follows that P2P accommodation is usually to be found in areas other than where conventional accommodation facilities are located, such as in suburban neighbourhoods. For example, Heo, Blal and Choi (2019) compared hotel and Airbnb locations in Paris, revealing that hotels and P2P accommodation are to be found in quite different areas in the city. Similarly, Gutiérrez, García-Palomares, Romanillos and Salas-Olmedo (2017) compared Airbnb accommodation and hotels in Barcelona and found that P2P accommodation is primarily located in areas where there were no accommodation facilities previously.

These differences are sometimes viewed as positive aspects of P2P accommodation (Prayag & Ozanne, 2018). Hosts offering P2P accommodation have the opportunity to earn additional income by utilizing space that is not otherwise used. It is also argued that the economic benefits of tourism are spread more widely than they otherwise would be because P2P accommodation is, as noted above, typically located in areas where there are no hotels. In addition, the unique aspects of staying at P2P accommodation, such as authentic interactions between guests and hosts – an attribute which is often advertised by Airbnb (Airbnb, 2017) – are considered by some to be the motivations for using this form of accommodation (Tussyadiah & Pesonen, 2016). In other words, some argue that it offers a new experience that cannot be offered by traditional accommodation facilities.

In addition to the benefits it brings to individuals (both guests and hosts), P2P accommodation can also contribute to destination competitiveness. More specifically, one of the characteristics of the hotel industry is that it is difficult to hold inventory. In other words, accommodation providers cannot 'store' their unoccupied rooms for the future use when there is high demand, reflecting what is widely considered to be a principal characteristic of service products in general, namely, perishability (Kandampully & Solnet, 2015). Thus, it is challenging for destinations to meet or respond to sudden increases or frequent fluctuations in demand whilst, in addition, most destinations experience high seasonality owing to constraints inherent on both the demand (e.g., most Japanese consumers take a New Year's holiday) and supply sides (e.g., tourists usually visit skiing destinations during the winter). Because the P2P accommodation sector does not require the construction of new facilities, it is relatively easy to achieve flexibility in supply in accordance with fluctuations in demand. In this context, Fairley and Dolnicar (2018) suggest, for example, that P2P accommodation is suited to addressing the lack of capacity when holding big events. Alternatively, P2P accommodation might also support a destination following a disaster; Hajibaba, Karlsson and Dolnicar (2017) found that, if natural disasters result in a shortage of accommodation, there is a segment of residents willing to offer rooms to tourists. In short, because of the ease with which its supply shifts in line with fluctuations in demand, P2P accommodation enables destinations to meet the demand more flexibly.

Conversely, P2P accommodation attracts criticism for a number of reasons (see, e.g., Lee, 2016; Nieuwland & van Melik, 2020). For instance, it is considered that providers of P2P accommodation avoid or are not covered by existing laws and regulations. In other words, because the P2P accommodation sector differs from the traditional accommodation concept, which is often the case for innovative products and services, administrations have failed to update relevant regulations in line with its development and expansion; hence, P2P accommodation providers are criticized by some for paying less tax than traditional accommodations (Heo et al., 2019), which has resulted in a significant loss of tax revenue in many areas. In addition, because regulations concerning existing accommodation facilities, such as the requirement to install a sprinkler system, do not apply, there have been concerns regarding safety. However, such concerns with regards to safety are inherent in P2P accommodation transactions and, thus, suppliers (including the platform) have an incentive to address them, not least because for an organization such as Airbnb or, indeed, for individual hosts, a perceived lack of safety regulations may result in a decline in reputation. Consequently, it might be expected that, even if the government does not intervene, the sector will self-regulate with regards to health and safety issues.

The same cannot be said, however, in relation to the arguably more significant problem of the negative externalities that the development of P2P accommodation imposes on many cities. That is, neither P2P platforms nor hosts have any incentive to address widely acknowledged problems, such as increases in noise, gentrification or the excessive growth in the number of tourists. In this case, intervention from the government is required.

The negative externalities of P2P accommodation can be categorized broadly into, on the one hand, housing issues and, on the other hand, neighbourhood issues (Nieuwland & van Melik, 2020). Housing issues occur because increasing demand from tourists for accommodation provides a strong incentive for owners to transform residential properties into P2P facilities. Housing data indicate that an increase in the supply of P2P accommodation raises rental prices more generally, sometimes resulting in local residents having to move from cities as they can no longer afford their rents (Lee, 2016). As discussed by Telfer (2019: 220), this problem has been particularly acute in Iceland where, in addition to finding difficulty in finding affordable housing, some 'local people have been evicted from rental units to make way for higher-paying tourists'. Neighbourhood issues, in contrast, are more related to the quality of local residents' living environment, with P2P accommodation being criticized for leading to a reduction in the quality of life of local communities

It is, therefore, essential to analyse the nature of the problems resulting from the development of P2P accommodation that particular cities experience, as these inevitably determine the type of regulations introduced to address them. In this context, Nieuwland and van Melik (2020) compared the regulation of the P2P accommodation sector in 11 Western cities, finding that their regulations vary widely from region to region. For example, regarding their 'intensity', London imposes relatively moderate regulations whereas Anaheim in California imposes strong regulations which completely prohibit P2P accommodation. Furthermore, as previously indicated, regulations in some destinations are implemented in order to maintain the quality of the living environment for residents, whereas in others, the objective of regulation is to maintain affordable housing prices. For example, zoning control is essential if the aim is to preserve the residents' quality of life. Inevitably, the design and aim of regulations will influence the supply of P2P accommodation in different ways (Yang & Mao, 2019) although, interestingly, it has been found that laws which impose a tax or require registration do not influence its supply (Yang & Mao, 2019).

It should be noted that regulations can be difficult to enforce owing to some elements of the inherent nature of P2P accommodation (Nieuwland & van Melik, 2020). For example, because it operates in residential housing, it can be difficult for governments to identify where P2P accommodation is located. At the same time, given that P2P accommodation is usually, though not always, operated by private citizens as opposed to businesses, it is assumed that they display a low legal compliance rate, such as the in case of individual income tax compliance, which is often documented (Kirchler, Hoelzl & Wahl, 2008). When designing regulations, therefore, a cost–benefit analysis is necessary because the effectiveness of regulations may be called into question (Yang & Mayo, 2019).

Given the rapid growth in the supply of P2P accommodation, it is not surprising that it has attracted academic attention (Prayag & Ozanne, 2018). Many studies have focused on consumers, such as their motivations (Mahadevan, 2018), travel patterns (Tussyadiah & Pesonen, 2016) and satisfaction levels (Liang, Choi, & Joppe, 2018). In contrast, residents' attitudes have, to date, been less frequently investigated (Garau-Vadell, Gutiérrez-Taño & Díaz-Armas, 2019; Jordan & Moore, 2018; Yeager, Boley, Woosnam & Green, 2020). Nevertheless, as indicated by Nunkoo and Smith (2013), the government requires a certain amount of support from residents to ensure the effectiveness of their policies. Hence, given that an increasing number of cities are expected to impose regulations regarding P2P accommodation, it is important to investigate residents' attitudes.

In the following sections, the manner in which P2P accommodation is perceived in Japan is explored. Japan is an interesting context, not least because the number of international tourists has been rapidly increasing in recent years which has led to intergroup conflicts between tourists and residents as well as a lack of accommodation capacity. In addition, transformations in Japanese regulations related to the accommodation sector following the expansion of P2P accommodation are also discussed.

P2P accommodations in Japan

Recent tourism development in Japan

Over the past 5 years, there has been a significant increase in the number of international tourists in Japan (Japan National Tourism Organization, 2019), which has caused some disruption in the accommodation market. As pointed out in Chapter 1, before 1990 the number of international tourists was less than 3 million; the figure slowly increased to 8 million by 2007 and, by 2012, was still fluctuating around that number. At the same time, the number of domestic tourists had not changed significantly owing to the maturity of the national economy. Hence, the demand for accommodation had been rather stable for some time. However, because the number of international tourists, especially from Asian countries, has increased significantly over recent years, from 10 million in 2013 to 31 million in 2018 (Japan National Tourism Organization, 2019), this situation has changed. Figure 14.1 shows the growth in the inbound tourism market in the past decade. Importantly, not only has the volume of tourists changed but so too has the customer base of the accommodation sector. That is, many facilities that used to host predominantly Japanese tourists must now meet the demands of international tourists as well. The government has acknowledged this growth and announced that tourism is one of the most important industries for the nation's continuous economic development (Prime Minister of Japan and His Cabinet, 2016).

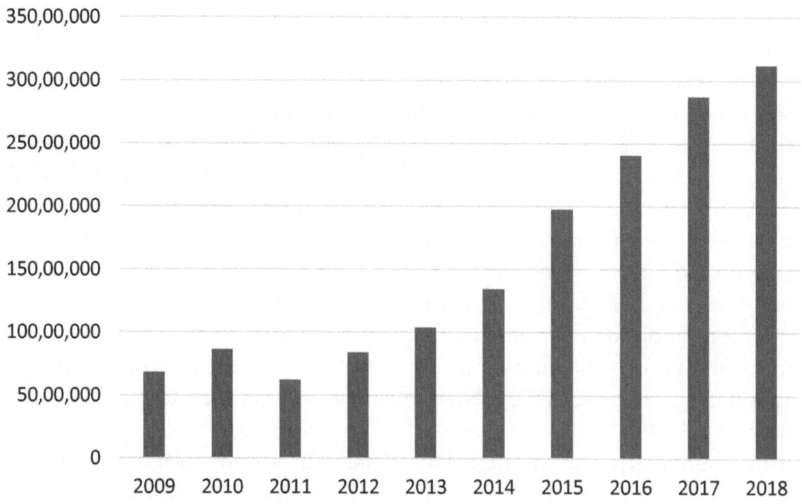

Figure 14.1 Growth in inbound tourism to Japan.
Source: Adapted from Japan National Tourism Organization (2019).

One of the challenges pointed out by both the tourism industry and the government is the lack of accommodation capacity (Japan Tourism Agency, 2017; Mizuho Research Institute, 2017). Because the number of international tourists is growing rapidly, constructing new accommodation to meet the new demand is difficult. In addition, Japan will be hosting a number of large-scale international events, such as the Olympics and Paralympics in 2020 (postponed to 2021) and the World Expo in 2025, which are expected to bring a significant number of attendees, further exacerbating the lack of accommodation capacity. Hence, the development of P2P accommodation could be useful in addressing this accommodation shortfall; as pointed out by Fairley and Dolnicar (2018), P2P accommodation offers an efficient means of dealing with a sudden increase in demand when the construction of new facilities is not required. Indeed, seeking to construct new accommodation to meet a sudden increase in demand would be not be an appropriate policy because there is a high probability that such an increase in demand is temporary.

The challenge of utilizing P2P accommodation, however, is that it could be viewed negatively by residents. As is the case with most tourism-related development, residents' opinions are important (Nunkoo & Smith, 2013) and if P2P accommodation is unpopular amongst local people, it is difficult for either central or local government to set policies for its development. As already noted, previous studies have identified negative impacts, such as the lack of a sense of community and excessive increases in noise, resulting from P2P development (Jordan & Moore, 2018).

Indeed, the negative consequences of P2P accommodation have been reported in the Japanese media (The Nikkei, 2017a). More specifically, as the number of international tourists has increased rapidly, some destinations have been seen to suffer from so-called 'over-tourism', often referred to in Japan as *kanko-kogai* (観光公害), which translates as 'tourism pollution'. The development and spread of P2P accommodation is often cited as one of the main causes of *kanko-kogai*, as it blurs the boundary between tourists and residents. The terms *iho-mimpaku* (違法民泊) and *yami-mimpaku* (ヤミ民泊) are also used, both translated as 'illegal P2P accommodation'. Furthermore, P2P accommodations is often associated in the Japanese media with serious crimes; The Nikkei (2017b), for example, employed the sensational title 'P2P Accommodations Become a Breeding Place of Crime' to describe the perceived danger associated with its development. It argued that P2P accommodation is more prone to illegal activities as most facilities do not check the identity of guests.

More specifically, one of the reasons that P2P accommodation in Japan was initially controversial was that it could not be regulated prior to the enactment of the P2P accommodation law. That law is discussed in the next section.

The change in accommodation laws and the enactment of the P2P accommodation law

Accommodation facilities in Japan had previously been regulated solely by the Japanese Hotels and Inns Act. Under this law, accommodation facilities were classified into two categories: fully equipped accommodation, including both Western-style and Japanese-style hotels (*ryokan*) and simplified accommodation, such as hostels. In the case of fully equipped accommodation, it was mandatory to provide the necessary facilities, such as a toilet and shower, in each room. The law also imposed restrictions, such as the minimum number of rooms required (more than 10 rooms for hotels and more than five rooms for *ryokan*), the floor space per room, and the installation of a front desk.

In contrast, simplified accommodation facilities which, in Japan take a number of forms, including pensions, guest houses and capsule hotels, benefited from more relaxed requirements. There were no regulations regarding the minimum number of rooms, whilst the requirements regarding the installation of a front desk and provision of bathing facilities were less rigid compared to those for fully equipped accommodation. However, the total floor area was required to be more than 33 m^2 and, in addition, as with fully equipped accommodation, simplified accommodation facilities are also subject to urban planning laws which stipulate the zoning area.

P2P accommodation was, therefore, controversial because it occupied a grey area in the existing law, as is the case in many countries. On the one hand, the government initially argued that P2P accommodation providers should be bound by the Japanese Hotels and Inns Act because they earned profits by conducting the same business as traditional form of accommodations

(Ministry of Health, Labour and Welfare, 2018). On the other hand, others argued that it was less certain or logical that the P2P accommodation sector should be regulated by the Japanese Hotels and Inns Act. For example, Fukui (2016) argued that because the concept of P2P accommodation is based upon people renting their vacant spaces to others, it is clearly different from the concept of traditional accommodations. Thus, there was no judicial precedent, whilst opinions among scholars was also divided.

In June 2018, following the enactment of the P2P law (the Private Housing Accommodation Law), the legal position of the P2P accommodation sector became clear. P2P accommodation became legally distinct from existing accommodation facilities and is regulated under the new law, which requires those offering P2P accommodation services to register their properties as such. Table 14.1 summarizes the key elements of this new law. Significantly, although it is a national law, local governments are permitted to impose further restrictions as long as they are necessary to prevent negative impacts on local residents' quality of life. Furthermore, not only are individual hosts but also P2P platforms, such as Airbnb, required to register under the new law. As a consequence, the position of Airbnb has been from simply an online platform to a key stakeholder; the company must monitor whether properties on its website have legal rights. Having said that, the enforcement of regulation is challenging, as P2P transactions are, of course, conducted on the Internet. Hence, Japan Tourism Agency (2019) reports that almost 16 percent of P2P accommodation facilities are illegal or cannot be confirmed as not illegal, whilst several local governments have established regulatory departments specifically to crack down on illegal P2P accommodation (The Nikkei, 2018a).

The enactment of the new law related to P2P accommodation was widely publicized in the media, but the changes to regulations were not limited to that; traditional accommodation facilities benefited from a degree of deregulation. For example, a number of restrictions in the existing law, such as the

Table 14.1 Comparison of Japanese Hotels and Inns Act and Private Housing Accommodation Law

	Japanese Hotels and Inns Act	*Private Housing Accommodation Law*
Competent authority	Ministry of Health, Labour and Welfare	Ministry of Land, Infrastructure, Transport and Tourism Ministry of Health, Labour and Welfare Japan Tourism Agency
Type of licensing	Permission	Notification
Operation in a residential zone	Banned	Allowed
Number of maximum business days	None	180 per year

minimum number of rooms and the minimum floor area, were relaxed. This deregulation has had a significant impact on the P2P accommodation sector as well, in as much as previously, the Japanese Hotels and Inns Act made it impossible for P2P accommodation providers to operate if they adhered to this law. As mentioned above, there was a minimum total floor space requirement of 33 m^2 which, unsurprisingly, was not compatible with most P2P accommodation because usually only one or two rooms are offered. However, following the relaxation of this provision within the Japanese Hotels and Inns Act, it became a viable option to operate under this law. In other words, strict regulations were enacted for P2P accommodation operators but, at the same time, several provisions in the existing law were eased, enabling the P2P accommodation sector to operate under the Japanese Hotels and Inns Act.

The case of Kyoto

Kyoto has always been one of the most popular destinations in Japan for domestic tourists and also for foreign tourists; the city is included in the itineraries of many tours for international visitors. Kyoto was an ancient capital of Japan (Tokyo became the county's capital in 1868) and because it was not bombed during Second World War, tourists can enjoy viewing traditional Japanese-style houses as well as its well-known temples (most notably, Kinkaku-ji, or the Golden Pavilion) while exploring the city. In 2018, Kyoto prefecture announced that the annual number of tourists visiting the city had reached 85 million and the economic impact of tourism-related consumption was more than 1.3 trillion yen, indicating the importance of the tourism industry in Kyoto (Kyoto Prefecture, 2019). In the past few years, the popularity of Kyoto has been further driven by foreign tourists. Figure 14.2 summarizes the increase of the inbound market in Kyoto. In 2018, 4.6 million foreign tourists were staying in Kyoto Prefecture, more than three times that of 2013 (1.1 million nights). The biggest market is mainland China, followed by Taiwan, the United States of America and South Korea.

However, this excessive volume of tourists can cause problems. The Japanese media have reported that, similar to other cities such as Barcelona and Amsterdam, over-tourism is occurring in Kyoto (The Nikkei, 2018b). Not unexpectedly, the term *kanko-kogai* is used to describe the current situation in the city, the term being used to indicate that Kyoto is receiving far more tourists than it is able to accommodate. The principal consequence is that the day-to-day life and environment of local residents is suffering significant negative impacts, with specific problems reported by local people being issues with transportation, excessive noised and poor air quality. Because of the sudden increase in the numbers of tourists visiting and staying in Kyoto, the public transportation system has become overcrowded with foreign tourists, preventing residents from going about their daily business (Kyoto Shimbun, 2019). Furthermore, because of the high demand for accommodation, a number of new hotels have been constructed, which has resulted in noise and

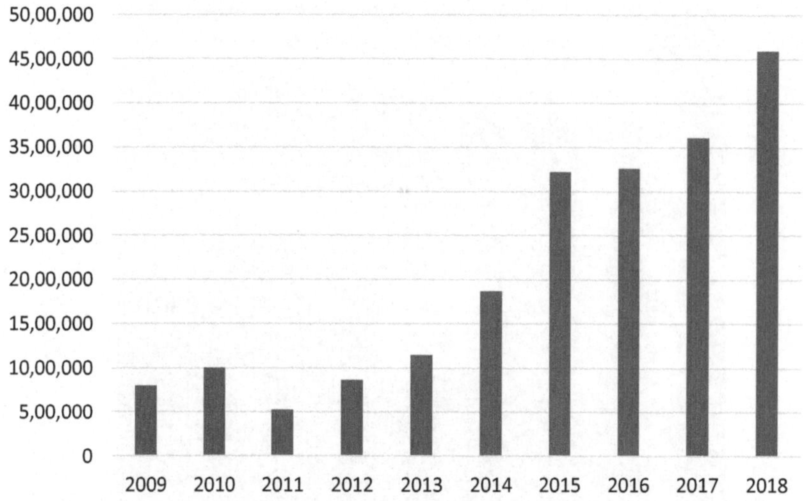

Figure 14.2 Growth in number of international visitors staying in Kyoto Prefecture.
Source: Adapted from Kyoto Prefecture (2019).

air pollution (Kyoto Shimbun, 2019). The new hotels include international luxury hotel brands such as Park Hyatt and Aman.

Of course, a variety of factors have brought about the surge of tourists in Kyoto and the adverse effects accompanying it, but the development of P2P accommodation in particular is often blamed for creating the problems (Kyoto Shimbun, 2018). The biggest issue is that, in the past, formal accommodation facilities and the general residential areas were located in relatively different places and, hence, interactions between tourists and residents were limited. However, with the growth in the supply of P2P accommodation in primarily residential areas, tourists now share those spaces with residents. Hence, an increasing number of residents have begun to feel that their daily lives are being disturbed. This is not limited to obvious problems, such as noise; the mere existence of strangers in their neighbourhoods makes residents feel uncomfortable (Jordan & Moore, 2018). Figures 14.3 and 14.4 offer examples which illustrate the negative opinions of the community with regards to P2P accommodation; similar posters can be found in abundance around Kyoto, pointing to widespread negative attitudes. The signs translate as: 'Against mimpaku (P2P accommodation)' and 'Mimpaku (P2P accommodation) is not acceptable'.

In addition, it is highly likely that the manner in which the media covers debates surrounding the growth in P2P accommodation in Kyoto has an influence on the perceptions of local residents. For example, the term 'illegal P2P accommodation' is often used (Kyoto Shimbun, 2018), which might create unconscious bias amongst readers. Interestingly, Mody, Suess and Dogru

Figure 14.3 Protest sign in a residential area of Kyoto.
Photo: Takahiro Ikeji.

(2018) found that residents' perceptions of P2P accommodation in the United States are not as negative as those reported in the media; in Japan, however, the situation may be different, although further research is necessary. Nevertheless, the majority of P2P accommodation customers in Japan are international tourists (Japan Tourism Agency, 2018), indicating that the proportion of Japanese citizens who have rented such accommodations remains low. If people have not used P2P accommodation, it is perhaps not surprising that their perceptions are influenced by the media which often adopts a negative perspective on the P2P phenomenon.

Under such circumstances, strict measures have been taken against the development of P2P accommodation in Kyoto. As previously mentioned, Japanese municipalities are able to impose additional regulations on P2P

Figure 14.4 Protest sign in a residential area of Kyoto.
Photo: Takahiro Ikeji.

accommodation as long as they are deemed necessary to protect the living environment of residents. Consequently, in residential areas in Kyoto, P2P accommodation can only operate for up to 60 days, from 15 January to 16 March. Moreover, in order to prevent the use of illegal P2P accommodation, the city has made significant efforts to control the sector. For example, a hotel tax was introduced in October 2018, and it was announced that a proportion of the revenue raised would be allocated to measures of preventing the illegal operation of P2P accommodation (Kyoto City, 2018). Specifically, 150 million Japanese yen will be allocated to strengthen the system for reporting and monitoring the illegal and inappropriate use of P2P accommodations.

Consequently, the number of P2P accommodation facilities registered in Kyoto has decreased significantly. Although the total number of such facilities prior to the introduction of regulations was uncertain, it is reported that around 2,000 facilities were mandated to close their businesses between 2016 and 2018 (Airstair, 2019). At the same time, however, the number of simplified

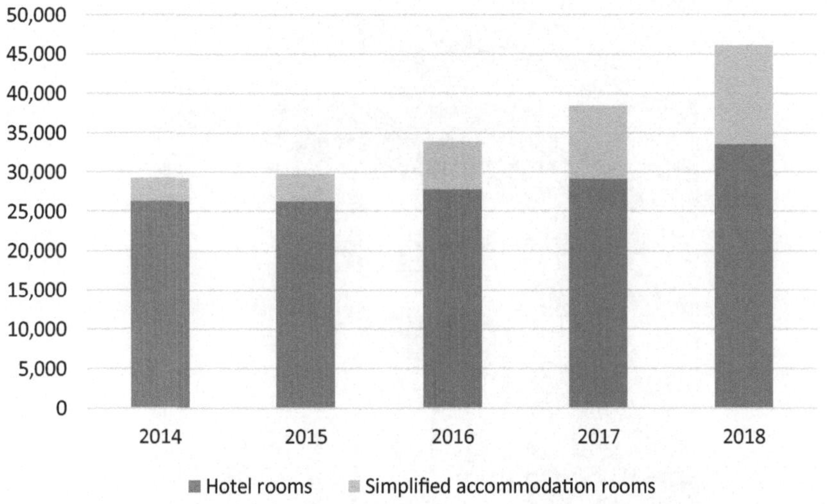

Figure 14.5 Growth in hotel and simplified accommodation rooms in Kyoto.
Source: Adapted from Kyoto City (2019).

accommodation rooms increased from 2,929 in 2014 to 12,539 in 2018 (Kyoto City, 2019). In other words, it would appear that former P2P accommodation facilities are being converted into simplified accommodations, which are registered under the Japanese Hotels and Inns Act (see Figure 14.5). As discussed, the key difference is that P2P accommodation can only operate between 15 January and 16 March without any locational restrictions, whereas simplified accommodations can operate without any time restrictions but outside of residential areas.

This change is intended to relieve some problems that residents experience, as one of the main problems related to P2P accommodation facilities is considered to be their location. In other words, if the P2P accommodation is not located in residential areas, it becomes more similar to traditional accommodation. However, it is still unclear whether residents' negative attitude towards P2P accommodation will be relieved, or whether the overall volume of tourists in Kyoto will continue to be controversial. It should also be noted that although Kyoto is presented in this chapter as a representative case, similar situations are occurring in other cities elsewhere in Japan (The Nikkei, 2018b).

Conclusion

This chapter has discussed the current P2P accommodation rental market in Japan. A particular theme to emerge is that the emergence of P2P accommodation has the potential to affect the regulation of the accommodation industry as a whole. In Japan, the change in regulations can be understood from two perspectives. The first aspect is straightforward: regulating P2P accommodation. Some elements of the P2P law are very strict, such as restricting the maximum number of operating days to 180, which makes it difficult for P2P accommodation operators to earn a profit. However, this point should not be argued on its own but rather examined alongside the second perspective: the deregulation of the Japanese Hotels and Inns Act. As explained in this chapter, the existing law on accommodation was relaxed, enabling P2P accommodation to operate under the existing law. Hence, the P2P law itself might be too strict but, combined with the relaxed law, a large number of P2P accommodation facilities can still function throughout the year as long as they obtain permission under the regular law.

In short, it is evident that the regulations for the accommodation sector in Japan should be reconsidered and redesigned in light of the current accommodation market. Specifically, the supply of P2P accommodation is beneficial to Japan in as much as it mitigates the shortage of accommodation without the need to construct new facilities. Yet, the development of P2P accommodation has given rise to a number of issues, not least conflicts between tourists and residents. By combining the two changes in regulations, the government has attempted to balance these two dilemmas but, as this chapter has demonstrated, this cannot be done by simply enacting a new law for P2P accommodation without considering the entire regulation system.

It may therefore be appropriate to rethink the regulation of accommodations as a whole; without any amendments, regulations can become outdated. The emergence of P2P accommodation has caused such a situation because the development of this form of accommodation has been a 'disruptive innovation' enabled by the development of new technology. As explained in this chapter, regulation related to P2P accommodation is often examined from one perspective: how can the government regulate such a new disruption? Similarly, proponents of P2P accommodation advocate the benefits of this new concept without considering the existing broader context of the accommodation industry as a whole. Examining the case of Japan in particular could provide a balanced approach to the emergence of P2P accommodation more generally.

In addition, the change in regulations for accommodations in Japan might pose a question regarding the current distinction between traditional forms of accommodation and the P2P accommodation sector. In other words, that distinction might become subtle. For example, there are some cases in which a single private house is rented as an accommodation facility under the Japanese Hotels and Inns Act, which became possible owing to the deregulation of the previous law. There are many studies that have regarded all the properties listed on Airbnb as P2P accommodation (e.g., Martin-Fuentes, Fernandez, Mateu & Marine-Roig, 2018; So, Oh & Min, 2018). However, many commercial properties use Airbnb in a similar way as online travel agents (Crommelin, Martin & Troy, 2018). Furthermore, online travel agents, such as Booking.com, have begun listing P2P accommodation on their websites (Vynck, 2018). These highlight the rapid change in the industry and suggest that continuous, detailed monitoring of the market is required to understand the up-to-date situations and issues surrounding P2P accommodation.

References

Adamiak, C. (2018) Mapping Airbnb supply in European cities. *Annals of Tourism Research*, 71, 67–71.

Airbnb (2017) Airbnb encourages travelers to 'Live Like Locals' in Goa. *Airbnb.* Available at: https://press.airbnb.com/airbnb-encourages-travelers-to-live-like-locals-in-goa/ (Accessed 8 November 2018).

Airbnb (2018a) Fast facts. *Airbnb*. Available at: https://press.airbnb.com/fast-facts/ (Accessed 11 January 2019).

Airbnb (2018b) Hosuto no minasama heno sapoto to nihon no Airbnb komyunitei hatten ni mukete [Towards support for hosts and development of Japan's Airbnb community]. *Airbnb*. Available at: http://tsite.jp/r/cpn/airbnb/news/20180614-3.html (Accessed 3 January 2019).

Airbnb (2018c) Nihon he no goryoko o yotei sareteiru gesuto no minasama he no sapoto ni tsuite [About the support of guests traveling in Japan]. *Airbnb.* Available at: https://press.airbnb.com/ja/supporting-travelers-injapan/ (Accessed 20 September 2018).

Airstair (2019) Kyoto shi, mutodoke mimpaku nado yaku 2 senken o haigyo ni oikomu sono kage de, ryokangyoho no shukuhakushisetsu ga kyuzo [Kyoto City drove approximately 2,000 cases of unregistered P2P accommodations to out of business, but the number of accommodations under the Japanese Hotels and Inns Act has rapidly increased], *Airstair,* 5 January. Available at: https://airstair.jp/kyoto-minpaku-2000/ (Accessed 18 November 2019).

BBC (2019) Paris to sue Airbnb over 'illegal ads'. *BBC News*, 11 February. Available at: www.bbc.com/news/technology-47198746 (Accessed 9 December 2019).

Blal, I., Singal, M. and Templin, J. (2018) Airbnb's effect on hotel sales growth. *International Journal of Hospitality Management*, 73, 85–92.

Bloomberg. (2018) Airbnb sues New York city to block user-data bill over privacy. *Bloomberg*, 24 August. Available at: www.bloomberg.com/news/articles/2018-08-24/airbnb-sues-new-york-city-to-block-user-data-bill-over-privacy (Accessed 9 December 2019).

Crommelin, L., Martin, C. and Troy, L. (2018) Airbnb regulation needs to distinguish between sharing and plain old commercial letting. *The Conversation,* 5 June. Available at: https://theconversation.com/airbnb-regulation-needs-to-distinguish-between-sharing-and-plain-old-commercial-letting-97613 (Accessed 10 December 2019).

Fairley, S. and Dolnicar, S. (2018) Supporting events. In S. Dolnicar (Ed.), *Peer-to-Peer Accommodation Networks*. Oxford: Goodfellow Publishers, pp. 109–119.

Fukui, H. (2016) Legal issues and policy on renting private homes and apartments. *The Japanese Journal of Real Estate Sciences*, 30(2), 37–44.

Garau-Vadell, J. B., Gutiérrez-Taño, D. and Díaz-Armas, R. (2019) Residents' support for P2P accommodation in mass tourism destinations. *Journal of Travel Research,* 58(4), 549–565

Gurran, N. and Phibbs, P. (2017) When tourists move in: How should urban planners respond to Airbnb? *Journal of the American Planning Association*, 83(1), 80–92.

Gutiérrez, J., García-Palomares, J. C., Romanillos, G. and Salas-Olmedo, M. H. (2017) The eruption of Airbnb in tourist cities: Comparing spatial patterns of hotels and peer-to-peer accommodation in Barcelona. *Tourism Management,* 62, 278–291.

Hajibaba, H., Karlsson, L. and Dolnicar, S. (2017) Residents open their homes to tourists when disaster strikes. *Journal of Travel Research*, 56(8), 1065–1078.

Hartmans, A. (2017) Airbnb now has more listings worldwide than the top five hotel brands combined. *Business Insider*, 8 August. Available at: www.businessinsider.com/airbnb-total-worldwide-listings-2017–8 (Accessed 29 October 2018).

Heo, C. Y., Blal, I. and Choi, M. (2019) What is happening in Paris? Airbnb, hotels, and the Parisian market: A case study. *Tourism Management*, 70, 78–88.

Japan National Tourism Organization (2019) Honichi gaikyaku su [Number of foreign tourists to Japan]. *Japan National Tourism Organization*. Available at: www.jnto.go.jp/jpn/statistics/data_info_listing/pdf/190116_monthly.pdf (Accessed 18 November 2019).

Japan Tourism Agency (2017) Heisei 29 nendo shukuhaku shisetsu imbaundo taio shien jigyo no keikaku o nintei [The authorization of 'Accommodation support project corresponding to the increase in inbound tourists' in 2017]. Available at: www.mlit.go.jp/kankocho/topics06_000111.html (Accessed 14 March 2019).

Japan Tourism Agency (2018) Jutaku shukuhaku jigyo no shukuhaku jisseki ni tsuite [The performance of housing accommodation business]. Available at: www.mlit.go.jp/kankocho/topics06_000170.html (Accessed 30 November 2018).

Japan Tourism Agency (2019) 9 gatsu matsu jiten ni okeru mimpaku bukken no tekihosei no kakunin kekka ni tsuite [Confirmation result of the legality of P2P accommodations (As of the end of September)]. Available at: www.mlit.go.jp/common/001271484.pdf (Accessed 22 March 2019).

Jordan, E. J. and Moore, J. (2018) An in-depth exploration of residents' perceived impacts of transient vacation rentals. *Journal of Travel and Tourism Marketing*, 35(1), 90–101.

Jozuka, E. (2019) Japan has so many vacant homes it's giving them away. *CNN*, 15 January. Available at: https://edition.cnn.com/2018/12/05/asia/japan-vacant-akiya-ghost-homes/index.html (Accessed 9 December 2019).

Kandampully, J. and Solnet, D. (2015) *Service Management Principles for Hospitality and Tourism,* 2nd ed. Dubuque: Kendall Hunt.

Kirchler, E., Hoelzl, E. and Wahl, I. (2008) Enforced versus voluntary tax compliance: The 'slippery slope' framework. *Journal of Economic Psychology*, 29(2), 210–225.

Kyoto City (2018) Shukuhakuzei o zaigen toshite jujitsu kyoka suru torikumi [Measures increasing revenues from accommodation tax]. Available at: www.city.kyoto.lg.jp/gyozai/cmsfiles/contents/0000227/227727/1037.pdf (Accessed 3 January 2019).

Kyoto City (2019) Kyoto shi ryokangyo kyoka kensu no suii [The change in the number of registered accommodations in Kyoto]. Available at: www.city.kyoto.lg.jp/hokenfukushi/cmsfiles/contents/0000193/193116/20191009suii.pdf (Accessed 18 November 2019).

Kyoto Prefecture (2019) Heisei 30 nen kanko irikomi kyakusu oyobi kanko shohi gaku ni tsuite [The number of visitors and tourism consumption in 2018]. Available at: www.pref.kyoto.jp/kanko/research/30report3.html (Accessed 18 November 2019).

Kyoto Shimbun (2018) Kanko kogai shimin to masatsu kyoto yamuoezu gaikokujin seigen no mise mo ['Tourism Pollution'. Problems with residents. Some stores in Kyoto limited foreign visitors], *Kyoto Shimbun*, 28 October. Available at: www.kyoto-np.co.jp/articles/-/5065 (Accessed 18 November 2019).

Kyoto Shimbun (2019) Basu ni norenai jumin kujo kyoto basu kanko kyaku noriba bunri he [Residents' complaints about not getting on the bus. Kyoto City Bus decided to separate the bus stop for residents and tourists], *Kyoto Shimbun,* 26 April. Available at: www.kyoto-np.co.jp/articles/-/6941 (Accessed 20 November 2019).

Lee, D. (2016) How Airbnb short-term rentals exacerbate Los Angeles's affordable housing crisis: Analysis and policy recommendations. *Harvard Law & Policy Review*, 10(1), 229–253.

Liang, L. J., Choi, H. C. and Joppe, M. (2018) Exploring the relationship between satisfaction, trust and switching intention, repurchase intention in the context of Airbnb. *International Journal of Hospitality Management*, 69, 41–48.

Mahadevan, R. (2018) Examination of motivations and attitudes of peer-to-peer users in the accommodation sharing economy. *Journal of Hospitality Marketing and Management*, 27(6), 679–692.

Martin-Fuentes, E., Fernandez, C., Mateu, C. and Marine-Roig, E. (2018) Modelling a grading scheme for peer-to-peer accommodation: Stars for Airbnb. *International Journal of Hospitality Management*, 69, 75–83.

Ministry of Health Labour and Welfare (2018) Mimpaku sabisu to ryokangyoho ni kansuru Q & A [Q&A on peer-to-peer accommodation services and Japanese Hotels and Inns Act]. Available at: www.mhlw.go.jp/stf/seisakunitsuite/bunya/0000111008.html (Accessed 21 September 2018).

Mizuho Research Institute (2017) Estimate of hotel room shortage in 2020. Available at: www.mizuho-ri.co.jp/publication/research/pdf/report/report17-0922.pdf (Accessed 2 April 2019).

Mody, M., Suess, C. and Dogru, T. (2018) Not in my backyard? Is the anti-Airbnb discourse truly warranted? *Annals of Tourism Research*, 74, 198–203.

Negishi, M. (2018) Japan legalizes Airbnb, and thousands of listings vanish. *Wall Street Journal*, 8 June. Available at: www.wsj.com/articles/japan-legalizes-airbnb-and-thousands-of-listings-vanish-1528457217 (Accessed 20 September 2018).

Nieuwland, S. and van Melik, R. (2020) Regulating Airbnb: How cities deal with perceived negative externalities of short-term rentals. *Current Issues in Tourism*, 27(7), 811–825.

Nunkoo, R. and Smith, S. (2013) Political economy of tourism: Trust in government actors, political support, and their determinants. *Tourism Management*, 36, 120–132.

Prayag, G. and Ozanne, L. K. (2018) A systematic review of peer-to-peer (P2P) accommodation sharing research from 2010 to 2016: Progress and prospects from the multi-level perspective. *Journal of Hospitality Marketing and Management*, 27(6), 1–30.

Prime Minister of Japan and His Cabinet (2016) Asu no nihon o sasaeru kanko bijon [Tourism vision to support the future of Japan]. Available at: www.kantei.go.jp/jp/singi/kanko_vision/pdf/honbun.pdf (Accessed 14 March 2019).

So, K. K. F., Oh, H. and Min, S. (2018) Motivations and constraints of Airbnb consumers: Findings from a mixed-methods approach. *Tourism Management*, 67, 224–236.

Statistics Japan (2013) Heisei 25nen jutaku tochi tokei chosa kekka no yoyaku [Summary of land survey in 2013]. Available at: www.stat.go.jp/data/jyutaku/2013/10_1.html (Accessed 2 April 2019).

Telfer, D. (2019) Tourism and (re)development in developed nations. In R. Sharpley and D. Harrison (Eds), *A Research Agenda for Tourism and Development*. Cheltenham: Edward Elagar, pp. 206–231.

The Nikkei (2017a) Kinrin ni meiwaku kakenu mimpaku ni [Peer-to-peer accommodation should not create problems with local residents]. *The Nikkei,* 12 July. Available at: www.nikkei.com/article/DGXKZO18751900R10C17A7EA1000/ (Accessed 11 December 2019).

The Nikkei. (2017b) Mimpaku akuyo hanzai no su ni mimotokakunin amai bukken o nerau [Peer-to-peer accommodation become a breeding place of crime], *The Nikkei,* 11 August. Available at: www.nikkei.com/article/DGKKASDG24HBL_Q7A810C1CC1000/ (Accessed 20 November 2018).

The Nikkei (2018a) Mimpaku haran no makuake ugoku jichitai nuguenu yami [Peer-topeer accommodation began with uncertainty: Municipalities are on the move]. *The Nikkei,* 11 July. Available at: www.nikkei.com/article/DGKKZO32847440Q8A710C1EA1000/ (Accessed 20 November 2018).

The Nikkei (2018b). Nantokashiteyo! Kanko kogai. Jumin, honichikyaku he keihatsu suishin [Tourism pollution needs to be solved: Promoting awareness among residents and inbound tourists], *The Nikkei,* 28 November. Available at: www.nikkei.com/article/DGKKZO37757870U8A111C1PP8000/ (Accessed 6 February 2019).

Tussyadiah, I. P. and Park, S. (2018) When guests trust hosts for their words: Host description and trust in sharing economy. *Tourism Management*, 67, 261–272.

Tussyadiah, I. P. and Pesonen, J. (2016) Impacts of peer-to-peer accommodation use on travel patterns. *Journal of Travel Research,* 55(8), 1022–1040.

Vynck, G. (2018) Booking.com has 5 million non-hotel listings, topping Airbnb. *Bloomberg,* 10 April. Available at: www.bloomberg.com/news/articles/2018-04-10/booking-com-has-5-million-non-hotel-listings-topping-airbnb (Accessed 26 September 2018).

Yang, Y. and Mao, Z. (2019) Welcome to my home! An empirical analysis of Airbnb supply in US cities. *Journal of Travel Research*, 58(8), 1274–1287.

Yeager, E. P., Boley, B. B., Woosnam, K. M. and Green, G. T. (2020) Modeling residents' attitudes toward short-term vacation rentals. *Journal of Travel Research,* 59(6), 955–974.

Postscript

Richard Sharpley and Kumi Kato

In the earlier stages of writing this book, the travel and tourism sector globally was operating normally. Certainly, increasing concerns were being expressed with regards to the environmental impact of flying or, more precisely, the contribution of aircraft emissions to global warming, whilst the challenge of 'overtourism' was not only attracting increasing academic, policymaker and media attention – at least three academic texts focusing on overtourism were published in 2019 alone (Dodds & Butler, 2019; Milano, Cheer & Novelli, 2019, Pechlaner, Innerhofer & Erschbamer, 2019) – but also eliciting a variety of planning and policy responses from destinations. At the same time, there were some notable business failures in the sector; the collapse of Thomas Cook in particular sent shock waves across the industry. Nevertheless, it was, for all intents and purposes, business as usual in tourism, with international arrivals continuing their inexorable growth. In 2018, for example, over 1.4 billion international arrivals were recorded, an increase of 5 percent on the previous year (UNWTO, 2019), and in 2019 the figure reached 1.5 billion (UNWTO, 2020).

Over the first 3 months of 2020, however, the coronavirus outbreak that originated in Wuhan, China, evolved into a global pandemic and, at the time of writing this postscript, the travel and tourism sector worldwide faces an existential crisis. Many countries have closed their borders to international visitors and, at the same time, have restricted the mobility of their own citizens; indeed, reports suggest that one-third of the world's population is living under some form of 'lockdown' (Kaplan, Frias & McFall-Johnsen, 2020). The consequences for tourism have been immediate and dramatic. With the collapse in demand, airlines have grounded most if not all of their fleets, with knock-on impacts for airports and all other businesses associated with the air travel; cruises have been cancelled, hospitality providers, from international hotel chains to local guest houses and bed-and-breakfasts, as well as restaurants, cafes and bars, have been obliged to (hopefully temporarily) cease trading, and the survival of the myriad of other businesses that directly and indirectly meet the needs of tourists is in doubt. Putting it bluntly, the coronavirus pandemic and national responses to it have, in effect, closed down travel and tourism worldwide, in so doing threatening the livelihood of the millions of people who work in the sector. Indeed, according to one report, up to 50 million jobs, or the equivalent of 14 percent of all travel and tourism employment globally, are at risk (WTTC, 2020). Moreover, the economic future of many destinations has been thrown into doubt, particularly those such as warm-water island destinations that are highly dependent on tourism; equally, the excessive dependency of many countries on particular tourist markets, such as Japan's increasing reliance on China as a source of international visitors (see Chapter 1), has been starkly revealed. In short, within the space of just a few months the challenge of overtourism has become one of no tourism.

The longer-term consequences of the coronavirus pandemic on and the responses of the travel and tourism sector – and indeed of tourists themselves – are difficult if not impossible to predict, although this has not prevented considerable speculation (and a degree of wishful thinking) within academic circles. Historically, tourism has demonstrated remarkable resilience to major crises; following the global financial crisis of 2008, for example, international arrivals and receipts declined by 4 and 6 percent, respectively in 2009 but, by the following year, the figures for both were significantly higher than pre-crisis levels. However, it goes without saying that the current crisis facing tourism is deeper, more far-reaching and potentially more transformative than any previous one. Never before in the history of modern tourism have so many people around the world been denied (albeit temporarily) freedom of movement and, consequently, never before has almost the entire travel and tourism sector been brought to a standstill. Hence, predictions on the speed and extent of recovery cannot be based on past experience. In fact, many express the hope that past experience will not be repeated; that is, that a new, more responsible, community focused, considerate form of tourism might emerge.

Inevitably, much will depend on the duration of the coronavirus pandemic and the varying policies that countries adopt in relaxing restrictions on people's mobility, re-opening businesses, schools and so on in what is most likely to be only a gradual return towards normality or, perhaps, a new normality. What is certain is that, for some considerable time, efforts will be made to maintain some degree of 'social distancing' in most areas of social life in order to prevent a resurgence in coronavirus cases. In the airline sector, for example, the UK-based low cost carrier EasyJet has suggested that once it is able to recommence operations, middle seats (in rows of three) will, at least in the shorter term, be left empty to facilitate 'distancing', in so doing encouraging more people to fly (Jordan, 2020). Much will also depend on the policies adopted by governments with regards to financial support for travel and tourism businesses, both large and small; it is unlikely that the sector will survive unscathed, with consolidation and rationalization of the airline industry in particular being predicted (Calder, 2020).

Also inevitably, destinations and travel and tourism businesses will respond with marketing campaigns, special promotions and other activities to tempt tourists back, although many people may initially choose to take domestic holidays to avoid health risks. And in the longer term, it may well be back to business as usual; such is the number of livelihoods that depend on tourism, such is the economic significance of the tourism industry, and such is the desire of people to consume tourism that is difficult to imagine otherwise. At the same time, however, there may be challenges. The cost of air travel will, in all likelihood, rise significantly in the coming years; the availability of low-cost air travel, in terms of both costs and routes, will become more limited. Equally, the overall scale of the sector might be limited as businesses rationalize operations to become more profitable and destinations implement policies for more effective tourism planning to avoid the return of overtourism. Conversely, land-based independent travel might enjoy a resurgence.

This remains, however, speculation. Nevertheless, in the context of this book, what is certain is that Japan may have to reassess its policies for international tourism, specifically its growth targets as outlined in Chapter 1. Arrivals in 2020 will be significantly lower than in 2019 and although the Summer Olympics have, at the time of writing, been rescheduled for 2021 and will stimulate an increase in arrivals, the annual growth rates witnessed in recent years are unlikely to return. Moreover, the country's dependence on the Chinese market may well require reassessment and opportunities for market diversification should be explored. More positively, the challenge of overtourism, as discussed in Chapter 14, may be resolved by a reduction in arrivals while, in general, the current crisis offers an opportunity for Japan to consider alternative, non–growth-based tourism policies to address the wider socioeconomic challenges the country faces. More particularly, it also offers the opportunity for the values of regions and communities to be reconsidered, not as another economic resource to be exploited by tourism but as a rich community life and environment that can be enjoyed through and contribute to tourism. Hence, the meaning of a nation that is 'good to live, and good visit' (*Sunde-yoshi, otozurete-yoshi*) as promoted by the Japan Tourism Agency might need to be adjusted. Whether these opportunities are grasped, however, remains to be seen.

Richard Sharpley and Kumi Kato
May 2020

References

Calder, S. (2020) The future of flying remains unclear but at least there is one. The Independent, 17 April. Available at: www.independent.co.uk/travel/news-and-advice/future-of-flying-airlines-easyjet-business-travel-flights-news-a9469291.html (Accessed 18 April 2020).

Dodds, R. and Butler, R. (Eds) 2019. *Overtourism: Issues, Realities and Solutions*. Berlin: De Gruyter Oldenbourg.

Jordan, D. (2020) Coronavirus: EasyJet to leave middle plane seats empty. BBC News, 16 April. Available at: www.bbc.co.uk/news/business-52306868 (Accessed 23 April 2020).

Kaplan, J., Frias, L. and McFall-Johnsen, M. (2020) A third of the global population is on coronavirus lockdown. Business Insider, 21 April. Available at: www.businessinsider.com/countries-on-lockdown-coronavirus-italy-2020–3?r=US&IR=T (Accessed 22 April 2020).

Milano, C., Cheer, J.M. and Novelli, M. (Eds) (2019) *Overtourism: Excesses, Discontents and Measures in Travel and Tourism*. Wallingford: CABI.

Pechlaner, H., Innerhofer, E. and Erschbamer, G. (Eds) (2019) *Overtourism: Tourism Management and Solutions*. London: Routledge.

UNWTO (2019) *International Tourism Highlights 2019 Edition.* Madrid: UN World Tourism Organisation. Available at: www.e-unwto.org/doi/pdf/10.18111/9789284421152 (Accessed 22 April 2020).

UNWTO (2020) *World Tourism Barometer No. 18, January 2020*. Madrid: UN World Tourism Organisation. Available at: www.unwto.org/world-tourism-barometer-n18-january-2020 (Accessed 22 April 2020).

WTTC (2020) Coronavirus puts up to 50 million Travel and Tourism jobs at risk says WTTC. World Travel & Tourism Council. Available at: www.wttc.org/about/media-centre/press-releases/press-releases/2020/coronavirus-puts-up-to-50-million-travel-and-tourism-jobs-at-risk-says-wttc/ (Accessed 23 March 2020).

Index

agritourism/agritourists 143, 151, 156
Airbnb 257–60, 265, 271
akiya 258
Amsterdam 257, 266
anime 33
Arima 7
atrocity 180
authentic/authenticity 32, 51, 144, 195, 196, 259

ban-ya 230–1
Barcelona 59, 257, 259, 266
Basic Tourism Law 10, 31, 168
Berlin 257
big data 92
Booking.com 259, 272
bubble economy 51–3, 71
Buddhism 185
bushido 184, 185

chado 146
chanoyu 146
Chiang Rai 142, 146, 148
China 3, 30, 110, 146, 153, 184, 241, 242, 246, 248, 266, 276, 277
Chiran Peace Museum 15, 180, 183, 184, 186, 192, 197
climate change 15, 34, 120, 131, 134, 213, 218
Comprehensive Resort Areas Development Act 22, 23, 30, 53, 70, 107
coronavirus 16, 276, 277
Covid-19 48, 221
culture, tourism 2, 5, 7, 33

dark heritage 179
dark tourism 32, 180, 181, 183, 191, 192, 196
deindustrialization 47, 52, 59
Destination Management Organizations (DMO) 12, 151, 170
disaster 2, 16, 28, 34, 69, 168, 173, 181, 220, 224, 226, 230, 232, 233, 239, 240, 242, 245, 248, 252, 260
disinheritance 182, 183
dissonant heritage 182, 183
domestic tourism 2–13, 27, 29, 31–3, 55, 56, 105, 107, 109, 111, 129, 141, 153, 162, 171, 174, 186, 191, 192, 196, 205, 218, 262, 266, 278

earthquake 1, 16, 30, 34, 168, 172, 224, 229, 231, 239, 240, 242, 244, 245, 248, 249, 252
ecological consciousness 15, 136
ecological knowledge 16, 227, 228, 232
Edo 5–8, 124, 164, 205, 207, 210, 211, 217, 220
emissions 34, 276

Facebook 204, 240
feudalism 65
forest bathing 70, 74
Forestry Agency 65, 67, 76
'forests for people' 14, 64, 68, 73–74, 77, 80
forest therapy 74, 75
Fukuoka 57, 165
Fukushima 2, 34, 224

gastronomic tourism 49
gentrification 50
Germany 1, 30, 242
globalization 49, 53, 112
global warming 276
golf 8, 53, 67, 68, 93, 95, 107

Graburn, N. 2, 4, 26, 28, 29, 32, 105
green tourism 140–3, 148, 150, 156

Hakone 8
Haneda Airport 8
Harris Treaty 8, 9, 17
Himeji Naval Air Corps 192
Hiroshima 31–3, 57, 67
honeymoon 103, 105–6, 114
Hong Kong 3, 30, 242, 246
hopeful tourism 225
horse logging 231–3
hot springs/*onsen* 5–7, 12, 56, 120, 124, 126, 129, 169

identity, national 32, 84, 165, 182, 183, 192
International Year of Mountains 120
Ise Shrine 5, 6, 163, 164, 168
I-turn 154, 156
Iwakiri, S. 104
Iwate 229–32

Japan Academic Society of Tourism 19, 22
Japan Airlines 10
Japan Air Transport Corporation 8
Japan Alps 121, 129, 133–5
Japan Foundation for International Tourism 22
Japan Ground 212
Japan Institute of Tourism Research 19, 22
Japan Sport Agency 82
Japan Sport Tourism Alliance 82
Japan Tourism Agency (JTA) 11–13, 241
Japan Tourist Bureau 9, 20
Japanese Hotels and Inns Act 264–6, 270, 271
J-turn 154, 156
justice tourism 225

Kagoshima 141, 180, 188
kamikaze 15, 180, 183, 188, 190–2, 194–6
Kamikochi valley 121, 123, 124, 127–9, 132, 134, 135
Kanagawa, Treaty of 210, 214
kankō 19, 21, 24, 35, 36, 55–7, 266
kanko-kogai 264
Kasai 50, 180, 192, 194
Kigali Genocide Memorial 182
Kii 32, 164, 166, 176, 169, 170, 243
kō 6
Kobe 7, 8, 52, 54, 59, 231, 246
Koizumi 11, 53, 54
Kōyasan 5
Kumano 5, 15, 160, 165, 168, 174, 204
Kyoto 3, 6–8, 13, 16, 48, 54–9, 67, 78, 153, 164, 169, 245–7, 249, 258, 266–70
Kyushu 6, 34, 104, 165, 186, 243

London 9, 212, 257, 261
low-cost carrier 48, 241

Manchuria 184
mass tourism 5, 13, 76, 105, 128, 164
Meiji 5, 7, 64, 65, 124, 128, 129, 229
Melville, H. 208, 212–13
Midway, Battle of 184
mimpaku 257
miyage 6
Miyazaki 15, 103–15
mobility 31, 60, 276, 277
Moby Dick 208, 212

Nagasaki 8, 32, 57, 153, 165–7, 243
Nara 145–7, 149, 234, 243, 245, 246
natural hazard 30, 239, 240, 242–4, 248, 251
New York 9, 240, 257
noodle culture 33

Okinawa 32, 106, 185, 186, 192, 195, 196
Olympic Games 4, 9, 10, 12, 31, 47, 51, 54, 82–4, 90, 102, 104, 108, 111, 114, 115, 125, 127, 129, 130, 216, 264, 279
Omine, Mount 167
onsen/hot springs 5–7, 12, 56, 120, 124, 126, 129, 169
Osaka 7, 12, 51, 55, 74, 78, 106, 113, 192, 240, 244–9, 252
outdoor fitness 109, 110, 112
overtourism 3, 48, 59, 60, 78, 179, 276–8

Pacific War 180, 183–5, 191
Paris 257
peace-building 181, 182
peace, tourism and 179–83
Pearl Harbor 9, 184
peer-to-peer/P2P accommodation 16, 257–66, 267–72
Perry, M. 208–11, 213, 214
pilgrimage 5–8, 15, 29, 32, 127, 160–75

policy, tourism 2–5, 11, 12, 20, 21, 25, 27, 32, 34, 55, 65–7, 69, 70, 72, 73, 76, 180
politics/political influence 28, 35, 59, 92, 109, 112, 114, 182, 205, 218
post-conflict site 181
post-industrial era 49, 50, 53, 113
powerspot 167, 168, 174
precinct tourism 50, 59
Private Housing Accommodation Law 258, 265

reciprocity 226, 227
reconciliation 15, 179–83, 192
resilience 16, 24, 35, 119, 134, 220, 224, 229, 233
Resort Law 10, 23, 30, 70, 71, 107, 112
revitalization 3, 12, 14, 24, 25, 28–30, 35, 53, 56, 69, 70, 94, 143, 144, 166, 168, 169, 172, 174, 239
rural economy 109, 113, 151
ryokan 6, 8, 25, 265

Samurai 5, 185
San Francisco 10, 257
Seagaia Ocean Dome 106–8
Seagaia Resort 10, 107
Sea Shepherd Conservation Society 200, 203, 209
sekobune 207, 208, 216, 217, 219, 220
sendatsu 6, 167
Shikoku 32, 163–8
Shinbashi 7
Shinto/Shintoism 168, 185, 187, 207
slow tourism 160, 173, 174
smartphone 249
snowfall 131–4
social capital 35, 135, 226, 227
South Korea 3, 30, 90, 95, 153, 242, 246, 266
Special Attack Force 180, 184
spirituality 15, 160–4, 167, 171, 174
Spirit World 161
surf capital 112
Surf City Miyazaki 104–6, 109–10, 112, 113
sustainability 13–16, 28–31, 33, 34, 93, 142, 155, 160, 172, 218, 219, 224, 234

Taiji 16, 200, 201, 203–5, 207, 211, 212, 214, 216, 218–20
Taiwan 3, 27, 29, 30, 153, 242, 246, 266
Tanabe 168–71
Tawara 142, 143, 146–51, 155, 156
tea spaces 145, 146, 149, 151
Ten Million Plan 42
Thailand 1, 142, 148–50, 155, 156, 241, 242
Thomas Cook 276
tokkō 180, 185, 187
Tokugawa 5, 164, 205, 207
Tokyo 6–9, 31, 47, 51–6, 59, 67, 78, 82–4, 90, 102, 106, 109, 111, 113, 114, 128, 200–2, 204, 209, 211, 212, 232
Torihama, T. 187
Tourism Nation Promotion Basic Plan 3, 4, 11, 23, 168, 169
tsunami 1, 16, 30, 34, 168, 172, 224, 228–31, 240, 243–5, 250, 251
tsurizumu 25

UK 1, 10, 65, 202
urbanization 11, 51, 54, 59, 68, 161
US/USA 8, 17, 27, 103, 203, 209, 212, 214, 242
U-turn 154, 156
Uzurano 15, 180, 186, 192–6

volcanic/volcanism 34, 120, 123, 125, 133, 243, 244

Wakayama 15, 16, 29, 106, 144, 148, 155, 160, 169, 196, 200, 201, 204, 205, 217–20
walking 70, 89, 163, 174
Weibo 240, 247
Welcome Plan 21 3, 11, 31
Weston, W. 123, 127, 128
whale war 208–9
World Masters Games 12, 82, 83

Yasukini Shrine 185
Yokohama 7, 8, 9, 52, 55, 57, 59
Yokosuka 106